The Curve of Binding Energy

" The Energy of Fission & Fusion "

Edited by Paul F. Kisak

Contents

Chapter 1

Nuclear binding energy

Nuclear binding energy is the energy that would be required to disassemble the nucleus of an atom into its component parts. These component parts are neutrons and protons, which are collectively called nucleons. The binding energy of nuclei is due to the attractive forces that hold these nucleons together and this is usually a positive number, since most nuclei would require the expenditure of energy to separate them into individual protons and neutrons. The mass of an atomic nucleus is usually less than the sum of the individual masses of the constituent protons and neutrons (according to Einstein's equation E=mc^2) and this 'missing mass' is known as the mass defect, and represents the energy that was released when the nucleus was formed.

The term nuclear binding energy may also refer to the energy balance in processes in which the nucleus splits into fragments composed of more than one nucleon. If new binding energy is available when light nuclei fuse, or when heavy nuclei split, either process can result in release of this binding energy. This energy may be made available as *nuclear energy* and can be used to produce electricity as in (nuclear power) or in a nuclear weapon. When a large nucleus splits into pieces, excess energy is emitted as photons (gamma rays) and as the kinetic energy of a number of different ejected particles (nuclear fission products).

The nuclear binding energies and forces are on the order of a million times greater than the electron binding energies of light atoms like hydrogen.[1]

The mass defect of a nucleus represents the mass of the energy of binding of the nucleus, and is the difference between the mass of a nucleus and the sum of the masses of the nucleons of which it is composed.[2]

1.1 Introduction

Nuclear binding energy is explained by the basic principles involved in nuclear physics.

1.1.1 Nuclear energy

An absorption or release of nuclear energy occurs in nuclear reactions or radioactive decay; those that absorb energy are called endothermic reactions and those that release energy are exothermic reactions. Energy is consumed or liberated because of differences in the nuclear binding energy between the incoming and outgoing products of the nuclear transmutation.[3]

The best-known classes of exothermic nuclear transmutations are fission and fusion. Nuclear energy may be liberated by atomic fission, when heavy atomic nuclei (like uranium and plutonium) are broken apart into lighter nuclei. The energy from fission is used to generate electric power in hundreds of locations worldwide. Nuclear energy is also released during atomic fusion, when light nuclei like hydrogen are combined to form heavier nuclei such as helium. The Sun and other stars use nuclear fusion to generate thermal energy which is later radiated from the surface, a type of stellar nucleosynthesis. In any exothermic nuclear process, nuclear mass might ultimately be converted to thermal energy, given off as heat.

In order to quantify the energy released or absorbed in any nuclear transmutation, one must know the nuclear binding energies of the nuclear components involved in the transmutation.

1.1.2 The nuclear force

Electrons and nuclei are kept together by electrostatic attraction (negative attracts positive). Furthermore, electrons are sometimes shared by neighboring atoms or transferred to them (by processes of quantum physics), and this link between atoms is referred to as a chemical bond, and is responsible for the formation of all chemical compounds.[4]

The force of electric attraction does not hold nuclei together, because all protons carry a positive charge and repel each other. Thus, electric forces do not hold nuclei together, because they act in the opposite direction. It has been established that binding neutrons to nuclei clearly requires a non-electrical attraction.[4]

Therefore, another force, called the *nuclear force* (or *residual strong force*) holds the nucleons of nuclei together. This force is a residuum of the strong interaction, which binds quarks into nucleons at an even smaller level of distance.

The nuclear force must be stronger than the electric repulsion at short distances, but weaker far away, or else different nuclei might tend to clump together. Therefore it has short-range characteristics. An analogy to the nuclear force is the force between two small magnets: magnets are very difficult to separate when stuck together, but once pulled a short distance apart, the force between them drops almost to zero.[4]

Unlike gravity or electrical forces, the nuclear force is effective only at very short distances. At greater distances, the electrostatic force dominates: the protons repel each other because they are positively charged, and like charges repel. For that reason, the protons forming the nuclei of ordinary hydrogen—for instance, in a balloon filled with hydrogen—do not combine to form helium (a process that also would require some protons to combine with electrons and become neutrons). They cannot get close enough for the nuclear force, which attracts them to each other, to become important. Only under conditions of extreme pressure and temperature (for example, within the core of a star), can such a process take place.[5]

1.1.3 Physics of nuclei

The nuclei of atoms are found in many different sizes. In hydrogen they contain just one proton, in deuterium or heavy hydrogen a proton and a neutron; in helium, two protons and two neutrons, and in carbon, nitrogen and oxygen - six, seven and eight of each particle, respectively. A helium nucleus weighs less than the sum of the weights of its components. The same phenomenon is found for carbon, nitrogen and oxygen. For example, the carbon nucleus is slightly lighter than three helium nuclei, which can combine to make a carbon nucleus. This illustrates the mass defect.

Mass defect

The "mass defect" can be explained using Albert Einstein's formula $E = m\,c^2$, expressing the equivalence of energy and mass. By this formula, adding energy also increases mass (both weight and inertia), whereas removing energy decreases mass.

If a combination of particles contains extra energy—for instance, in a molecule of the explosive TNT—weighing it reveals some extra mass, compared to its end products after an explosion. (The weighing must be done after the products have been stopped and cooled, however, as the extra mass must escape from the system as heat before its loss can be noticed, in theory.) On the other hand, if one must inject energy to separate a system of particles into its components, then the initial weight is less than that of the components after they are separated. In the latter case, the energy injected is "stored" as potential energy, which shows as the increased mass of the components that store it. This is an example of the fact that energy of all types is seen in systems as mass, since mass and energy are equivalent, and each is a "property" of the other.

The latter scenario is the case with nuclei such as helium: to break them up into protons and neutrons, one must inject energy. On the other hand, if a process existed going in the opposite direction, by which hydrogen atoms could be combined to form helium, then energy would be released. The energy can be computed using $E = \Delta m\,c^2$ for each nucleus, where Δm is the difference between the mass of the helium nucleus and the mass of four protons (plus two electrons, absorbed to create the neutrons of helium).

For elements heavier than oxygen, the energy that can be released by assembling them from lighter elements decreases, up to iron. For nuclei heavier than iron, one actually releases energy by breaking them up into 2 fragments. That is how energy is extracted by breaking up uranium nuclei in nuclear power reactors.

The reason the trend reverses after iron is the growing positive charge of the nuclei. The electric force may be weaker than the nuclear force, but its range is greater: in an iron nucleus, each proton repels the other 25 protons, while the nuclear force only binds close neighbors.

As nuclei grow bigger still, this disruptive effect becomes steadily more significant. By the time polonium is reached (84 protons), nuclei can no longer accommodate their large positive charge, but emit their excess protons quite rapidly in the process of alpha radioactivity—the emission of helium nuclei, each containing two protons and two neutrons. (Helium nuclei are an especially stable combination.) Because of this process, nuclei with more than 94 protons are not found naturally on Earth (see periodic table). The isotopes beyond uranium (atomic number 92) with the longest half-lives are plutonium-244 (80 million years) and curium-247 (16 million years).

Solar binding energy

The nuclear fusion process works as follows: five billion years ago, the new Sun formed when gravity pulled together a vast cloud of gas and dust, from which the Earth and other planets also arose. The gravitational pull released energy and heated the early Sun, much in the way Helmholtz proposed.

Thermal energy appears as the motion of atoms and molecules: the higher the temperature of a collection of particles, the greater is their velocity and the more violent are their collisions. When the temperature at the center of the newly formed Sun became great enough for collisions between nuclei to overcome their electric repulsion, and bring them into the short range of the attractive nuclear force, nuclei began to stick together. When this began to happen, protons combined into deuterium and then helium, with some protons changing in the process to neutrons (plus positrons, positive electrons, which combine with electrons and are destroyed). This released nuclear energy now keeps up the high temperature of the Sun's core, and the heat also keeps the gas pressure high, keeping the Sun at its present size, and stopping gravity from compressing it any more. There is now a stable balance between gravity and pressure.

Different nuclear reactions may predominate at different stages of the Sun's existence, including the proton-proton reaction and the carbon-nitrogen cycle—which involves heavier nuclei, but whose final product is still the combination of protons to form helium.

A branch of physics, the study of controlled nuclear fusion, has tried since the 1950s to derive useful power from nuclear fusion reactions that combine small nuclei into bigger ones, typically to heat boilers, whose steam could turn turbines and produce electricity. Unfortunately, no earthly laboratory can match one feature of the solar powerhouse: the great mass of the Sun, whose weight keeps the hot plasma compressed and confines the nuclear furnace to the Sun's core. Instead, physicists use strong magnetic fields to confine the plasma, and for fuel they use heavy forms of hydrogen, which burn more easily. Magnetic traps can be rather unstable, and any plasma hot enough and dense enough to undergo nuclear fusion tends to slip out of them after a short time. Even with ingenious tricks, the confinement in most cases lasts only a small fraction of a second.

Combining nuclei

Small nuclei that are larger than hydrogen can combine into bigger ones and release energy, but in combining such nuclei, the amount of energy released is much smaller compared to hydrogen fusion. The reason is that while the overall process releases energy from letting the nuclear attraction do its work, energy must first be injected to force together positively charged protons, which also repel each other with their electric charge.[5]

For elements that weigh more than iron (a nucleus with 26 protons), the fusion process no longer releases energy. In even heavier nuclei energy is consumed, not released, by combining similar sized nuclei. With such large nuclei, overcoming the electric repulsion (which affects all protons in the nucleus) requires more energy than what is released by the nuclear attraction (which is effective mainly between close neighbors). Conversely, energy could actually be released by breaking apart nuclei heavier than iron.[5]

With the nuclei of elements heavier than lead, the electric repulsion is so strong that some of them spontaneously eject positive fragments, usually nuclei of helium that form very stable combinations (alpha particles). This spontaneous break-up is one of the forms of radioactivity exhibited by some nuclei.[5]

Nuclei heavier than lead (except for bismuth, thorium, uranium, and plutonium) spontaneously break up too quickly to appear in nature as primordial elements, though they can be produced artificially or as intermediates in the decay chains of lighter elements. Generally, the heavier the nuclei are, the faster they spontaneously decay.[5]

Iron nuclei are the most stable nuclei (in particular iron-56), and the best sources of energy are therefore nuclei whose weights are as far removed from iron as possible. One can combine the lightest ones—nuclei of hydrogen (protons)—to form nuclei of helium, and that is how the Sun generates its energy. Or else one can break up the heaviest ones—nuclei of uranium or plutonium—into smaller fragments, and that is what nuclear power reactors do.[5]

1.1.4   Nuclear binding energy

An example that illustrates nuclear binding energy is the nucleus of ^{12}C (Carbon 12), which contains 6 protons and 6 neutrons. The protons are all positively charged and repel each other, but the nuclear force overcomes the repulsion and causes them to stick together. The nuclear force is a close-range force (it is very strongly inversely proportionate to distance), and virtually no effect of this force is observed outside the nucleus. The nuclear force also pulls neutrons together, or neutrons and protons.[6]

The energy of the nucleus is negative with regard to the energy of the particles pulled apart to infinite distance (just like the gravitational energy of planets of the solar system), because energy must be utilized to split a nucleus into its individual protons and neutrons. Mass spectrometers have measured the masses of nuclei, which are always less than the sum of the masses of protons and neutrons that form them, and the difference—by the formula $E = m c^2$—gives the binding energy of the nucleus.[6]

Nuclear fusion

The binding energy of helium is the energy source of the Sun and of most stars. The sun is composed of 74 percent hydrogen (measured by mass), an element whose nucleus is a single proton. Energy is released in the sun when 4 protons combine into a helium nucleus, a process in which two of them are also converted to neutrons.[6]

The conversion of protons to neutrons is the result of another nuclear force, known as the weak (nuclear) force. The weak force, like the strong force, has a short range, but is much weaker than the strong force. The weak force tries to make the number of neutrons and protons into the most energetically stable configuration. For nuclei containing less than 40 particles, these numbers are usually about equal. Protons and neutrons are closely related and are sometimes collectively known as nucleons. As the number of particles increases toward a maximum of about 209, the number of neutrons to maintain stability begins to outstrip the number of protons, until the ratio of neutrons to protons is about three to two.[6]

The protons of hydrogen combine to helium only if they have enough velocity to overcome each other's mutual repulsion sufficiently to get within range of the strong nuclear attraction. This means that fusion only occurs within a very hot gas. Hydrogen hot enough for combining to helium requires an enormous pressure to keep it confined, but suitable conditions exist in the central regions of the Sun, where such pressure is provided by the enormous weight of the layers above the core, pressed inwards by the Sun's strong gravity. The process of combining protons to form helium is an example of nuclear fusion.[6]

The earth's oceans contain a large amount of hydrogen that could theoretically be used for fusion, and helium byproduct of fusion does not harm the environment, so some consider nuclear fusion a good alternative to supply humanity's energy needs. Experiments to generate electricity from fusion have so far have only partially succeeded. Sufficiently hot hydrogen must be ionized and confined. One technique is to use very strong magnetic fields, because charged particles (like those trapped in the Earth's radiation belt) are guided by magnetic field lines. Fusion experiments also rely on heavy hydrogen, which fuses more easily, and gas densities can be moderate. But even with these techniques far more net energy is consumed by the fusion experiments than is yielded by the process.[6]

The binding energy maximum and ways to approach it by decay

In the main isotopes of light nuclei, such as carbon, nitrogen and oxygen, the most stable combination of neutrons and of protons are when the numbers are equal (this continues to element 20, calcium). However, in heavier nuclei, the disruptive energy of protons increases, since they are confined to a tiny volume and repel each other. The energy of the strong force holding the nucleus together also increases, but at a slower rate, as if inside the nucleus, only nucleons close to each other are tightly bound, not ones more widely separated.[6]

The net binding energy of a nucleus is that of the nuclear attraction, minus the disruptive energy of the electric force. As nuclei get heavier than helium, their net binding energy per nucleon (deduced from the difference in mass between the nucleus and the sum of masses of component nucleons) grows more and more slowly, reaching its peak at iron. As nucleons are added, the total nuclear binding energy always increases—but the total disruptive energy of electric forces (positive protons repelling other protons) also increases, and past iron, the second increase outweighs the first. Iron-56 (^{56}Fe) is the most efficiently bound nucleus[6] meaning that it has the least average mass per nucleon. However, nickel-62 is the most tightly bound nucleus in terms of energy of binding per nucleon. (Nickel-62's higher energy of binding does not translate to a larger mean mass loss than Fe-56, because Ni-62 has a slightly higher ratio of neutrons/protons than does iron-56, and the presence of the heavier neutrons increases nickel-62's average mass per nucleon).

To reduce the disruptive energy, the weak interaction allows the number of neutrons to exceed that of protons—for instance, the main isotope of iron has 26 protons and 30 neutrons. Isotopes also exist where the number of neutrons differs from the most stable number for that number of nucleons. If the ratio of protons to neutrons is too far from stability, nucleons may spontaneously change from proton to neutron, or neutron to proton.

The two methods for this conversion are mediated by the weak force, and involve types of beta decay. In the simplest beta decay, neutrons are converted to protons by emitting a negative electron and an antineutrino. This is always possible outside a nucleus because neutrons are more massive than protons by an equivalent of about 2.5 electrons. In the opposite process, which only happens within a nucleus, and not to free particles, a proton may become a neutron by ejecting a positron. This is permitted if enough energy is available between parent and daughter nuclides to do this (the required energy difference is equal to 1.022 MeV, which is the mass of 2 electrons). If the mass difference between parent and daughter is less than this, a proton-rich nucleus may still convert protons to neutrons by the process of electron capture, in which a proton simply electron captures one of the atom's K orbital electrons, emits a neutrino, and becomes a neutron.[6]

Among the heaviest nuclei, starting with tellurium nuclei (element 52) containing 106 or more nucleons, electric forces may be so destabilizing that entire chunks of the nucleus may be ejected, usually as alpha particles, which consist of two protons and two neutrons (alpha particles are fast helium nuclei). (Beryllium-8 also decays, very quickly, into two alpha particles.) Alpha particles are extremely stable. This type of decay becomes more and more probable as elements rise in atomic weight past 106.

The curve of binding energy is a graph that plots the binding energy per nucleon against atomic mass. This curve has its main peak at iron and nickel and then slowly decreases again, and also a narrow isolated peak at helium, which as noted is very stable. The heaviest nuclei in nature, uranium ^{238}U, are unstable, but having a lifetime of 4.5 billion years, close to the age of the Earth, they are still relatively abundant; they (and other nuclei heavier than iron) may have formed in a supernova explosion [7] preceding the formation of the solar system. The most common isotope of thorium, ^{232}Th, also undergoes α particle emission, and its half-life (time over which half a number of atoms decays) is even longer, by several times. In each of these, radioactive decay produces daughter isotopes that are also unstable, starting a chain of decays that ends in some stable isotope of lead.[6]

1.2 Determining nuclear binding energy

Calculation can be employed to determine the nuclear binding energy of nuclei. The calculation involves determining the *mass defect*, converting it into energy, and expressing the result as energy per mole of atoms, or as energy per nucleon.[2]

1.2.1   Conversion of mass defect into energy

Mass defect is defined as the difference between the mass of a nucleus, and the sum of the masses of the nucleons of which it is composed. The mass defect is determined by calculating three quantities.[2] These are: the actual mass of the nucleus, the composition of the nucleus (number of protons and of neutrons), and the masses of a proton and of a neutron. This is then followed by converting the mass defect into energy. This quantity is the nuclear binding energy, however it must be expressed as energy per mole of atoms or as energy per nucleon.[2]

1.3   Fission and fusion

Nuclear energy is released by the splitting (fission) or merging (fusion) of the nuclei of atom(s). The conversion of nuclear mass-energy to a form of energy, which can remove some mass when the energy is removed, is consistent with the mass-energy equivalence formula:

$\Delta E = \Delta m\, c^2,$

in which,

ΔE = energy release,

Δm = mass defect,

and c = the speed of light in a vacuum (a physical constant).

Nuclear energy was first discovered by French physicist Henri Becquerel in 1896, when he found that photographic plates stored in the dark near uranium were blackened like X-ray plates (X-rays had recently been discovered in 1895).[8]

Nuclear chemistry can be used as a form of alchemy to turn lead into gold or change any atom to any other atom (though this may require many intermediate steps).[7] Radionuclide (radioisotope) production often involves irradiation of another isotope (or more precisely a nuclide), with alpha particles, beta particles, or gamma rays. Nickel-62 has the highest binding energy per nucleon of any isotope. If an atom of lower average binding energy is changed into two atoms of higher average binding energy, energy is given off. Also, if two atoms of lower average binding energy fuse into an atom of higher average binding energy, energy is given off. The chart shows that fusion of hydrogen, the combination to form heavier atoms, releases energy, as does fission of uranium, the breaking up of a larger nucleus into smaller parts. Stability varies between isotopes: the isotope U-235 is much less stable than the more common U-238.

Nuclear energy is released by three *exoenergetic* (or exothermic) processes:

- Radioactive decay, where a neutron or proton in the radioactive nucleus decays spontaneously by emitting either particles, electromagnetic radiation (gamma rays), or both. Note that for radioactive decay, it is not strictly necessary for the binding energy to increase. What is strictly necessary is that the mass decrease. If a neutron turns into a proton and the energy of the decay is less than 0.782343 MeV (such as rubidium-87 decaying to strontium-87), the average binding energy per nucleon will actually decrease.

- Fusion, two atomic nuclei fuse together to form a heavier nucleus

- Fission, the breaking of a heavy nucleus into two (or more rarely three) lighter nuclei

1.4   Binding energy for atoms

The binding energy of an atom (including its electrons) is not the same as the binding energy of the atom's nucleus. The measured mass deficits of isotopes are always listed as mass deficits of the neutral atoms of that isotope, and mostly in MeV. As a consequence, the listed mass deficits are not a measure for the stability or binding energy of isolated nuclei, but for the whole atoms. This has very practical reasons, because it is very hard to totally ionize heavy elements, i.e. strip them of all of their electrons.

This practice is useful for other reasons, too: Stripping all the electrons from a heavy unstable nucleus (thus producing a bare nucleus) changes the lifetime of the nucleus, indicating that the nucleus cannot be treated independently (Experiments at the heavy ion accelerator GSI). This is also evident from phenomena like electron capture. Theoretically, in orbital models of heavy atoms, the electron orbits partially inside the nucleus (it doesn't *orbit* in a strict sense, but has a non-vanishing probability of being located inside the nucleus).

A nuclear decay happens to the nucleus, meaning that properties ascribed to the nucleus change in the event. In the field of physics the concept of "mass deficit" as a measure for "binding energy" means "mass deficit of the neutral atom" (not just the nucleus) and is a measure for stability of the whole atom.

1.5 Nuclear binding energy curve

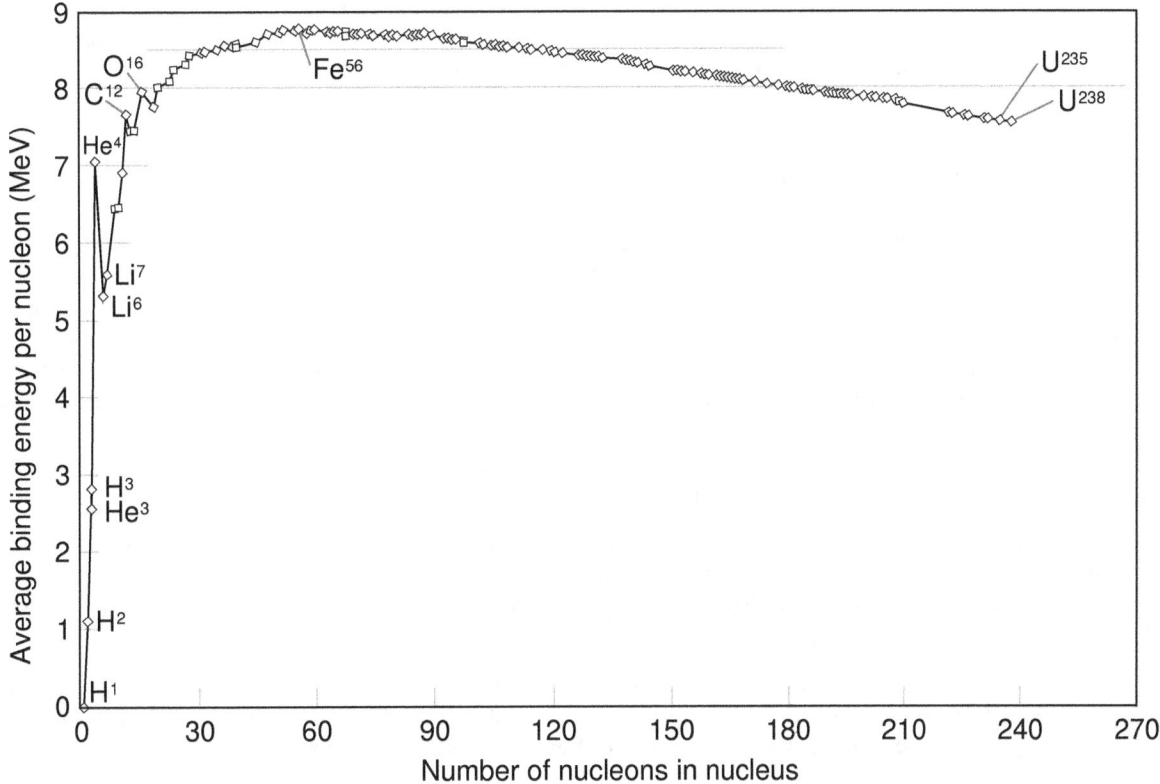

In the periodic table of elements, the series of light elements from hydrogen up to sodium is observed to exhibit generally increasing binding energy per nucleon as the atomic mass increases. This increase is generated by increasing forces per nucleon in the nucleus, as each additional nucleon is attracted by other nearby nucleons, and thus more tightly bound to the whole.

The region of increasing binding energy is followed by a region of relative stability (saturation) in the sequence from magnesium through xenon. In this region, the nucleus has become large enough that nuclear forces no longer completely extend efficiently across its width. Attractive nuclear forces in this region, as atomic mass increases, are nearly balanced by repellent electromagnetic forces between protons, as the atomic number increases.

Finally, in elements heavier than xenon, there is a decrease in binding energy per nucleon as atomic number increases. In this region of nuclear size, electromagnetic repulsive forces are beginning to overcome the strong nuclear force attraction.

At the peak of binding energy, nickel-62 is the most tightly bound nucleus (per nucleon), followed by iron-58 and iron-56.[9] This is the approximate basic reason why iron and nickel are very common metals in planetary cores, since they are produced profusely as end products in supernovae and in the final stages of silicon burning in stars. However, it is not binding energy per defined nucleon (as defined above), which controls which exact nuclei are made, because within stars, neutrons are free to convert to protons to release even more energy, per generic nucleon, if the result is a stable nucleus with a larger fraction of protons. In fact, it has been argued that photodisintegration of ^{62}Ni to form ^{56}Fe may be energetically possible in an extremely hot star core, due to this beta decay conversion of neutrons to protons.[10] The conclusion is that at the pressure and temperature conditions in the cores of large stars, energy is released by converting all matter into ^{56}Fe nuclei (ionized atoms). (However, at high temperatures not all matter will be in the lowest energy state.) This energetic maximum should also hold for ambient conditions, say $T = 298$ K and $p = 1$ atm, for neutral condensed matter consisting of ^{56}Fe atoms—however, in these conditions nuclei of atoms are inhibited from fusing into the most stable and low energy state of matter.

It is generally believed that iron-56 is more common than nickel isotopes in the universe for mechanistic reasons, because its unstable progenitor nickel-56 is copiously made by staged build-up of 14 helium nuclei inside supernovas, where it has no time to decay to iron before being released into the interstellar medium in a matter of a few minutes, as the supernova explodes. However, nickel-56 then decays to cobalt-56 within a few weeks, then this radioisotope finally decays to iron-56 with a half life of about 77.3 days. The radioactive decay-powered light curve of such a process has been observed to happen in type II supernovae, such as SN 1987A. In a star, there are no good ways to create nickel-62 by alpha-addition processes, or else there would presumably be more of this highly stable nuclide in the universe.

1.5.1 Measuring the binding energy

The fact that the maximum binding energy is found in medium-sized nuclei is a consequence of the trade-off in the effects of two opposing forces that have different range characteristics. The attractive nuclear force (strong nuclear force), which binds protons and neutrons equally to each other, has a limited range due to a rapid exponential decrease in this force with distance. However, the repelling electromagnetic force, which acts between protons to force nuclei apart, falls off with distance much more slowly (as the inverse square of distance). For nuclei larger than about four nucleons in diameter, the additional repelling force of additional protons more than offsets any binding energy that results between further added nucleons as a result of additional strong force interactions. Such nuclei become increasingly less tightly bound as their size increases, though most of them are still stable. Finally, nuclei containing more than 209 nucleons (larger than about 6 nucleons in diameter) are all too large to be stable, and are subject to spontaneous decay to smaller nuclei.

Nuclear fusion produces energy by combining the very lightest elements into more tightly bound elements (such as hydrogen into helium), and nuclear fission produces energy by splitting the heaviest elements (such as uranium and plutonium) into more tightly bound elements (such as barium and krypton). Both processes produce energy, because middle-sized nuclei are the most tightly bound of all.

As seen above in the example of deuterium, nuclear binding energies are large enough that they may be easily measured as fractional mass deficits, according to the equivalence of mass and energy. The atomic binding energy is simply the amount of energy (and mass) released, when a collection of free nucleons are joined together to form a nucleus.

Nuclear binding energy can be computed from the difference in mass of a nucleus, and the sum of the masses of the number of free neutrons and protons that make up the nucleus. Once this mass difference, called the mass defect or mass deficiency, is known, Einstein's mass-energy equivalence formula $E = mc^2$ can be used to compute the binding energy of any nucleus. Early nuclear physicists used to refer to computing this value as a "packing fraction" calculation.

For example, the atomic mass unit (1 u) is defined as 1/12 of the mass of a ^{12}C atom—but the atomic mass of a ^{1}H atom (which is a proton plus electron) is 1.007825 u, so each nucleon in ^{12}C has lost, on average, about 0.8% of its mass in the form of binding energy.

1.5.2 Semiempirical formula for nuclear binding energy

Main article: Semi-empirical mass formula

For a nucleus with A nucleons, including Z protons and N neutrons, a semi-empirical formula for the binding energy (BE) per nucleon is:

$$\frac{\text{BE}}{A \cdot \text{MeV}} = a - \frac{b}{A^{1/3}} - \frac{cZ^2}{A^{4/3}} - \frac{d\left(N - Z\right)^2}{A^2} \pm \frac{e}{A^{7/4}}$$

where the coefficients are given by: $a = 14.0$; $b = 13.0$; $c = 0.585$; $d = 19.3$; $e = 33$.

The first term a is called the saturation contribution and ensures that the binding energy per nucleon is the same for all nuclei to a first approximation. The term $-b/A^{1/3}$ is a surface tension effect and is proportional to the number of nucleons that are situated on the nuclear surface; it is largest for light nuclei. The term $-cZ^2/A^{4/3}$ is the Coulomb electrostatic repulsion; this becomes more important as Z increases. The symmetry correction term $-d(N - Z)^2/A^2$ takes into account the fact that in the absence of other effects the most stable arrangement has equal numbers of protons and neutrons; this is because the n-p interaction in a nucleus is stronger than either the n-n or p-p interaction. The pairing term $\pm e/A^{7/4}$ is purely empirical; it is + for even-even nuclei and - for odd-odd nuclei.

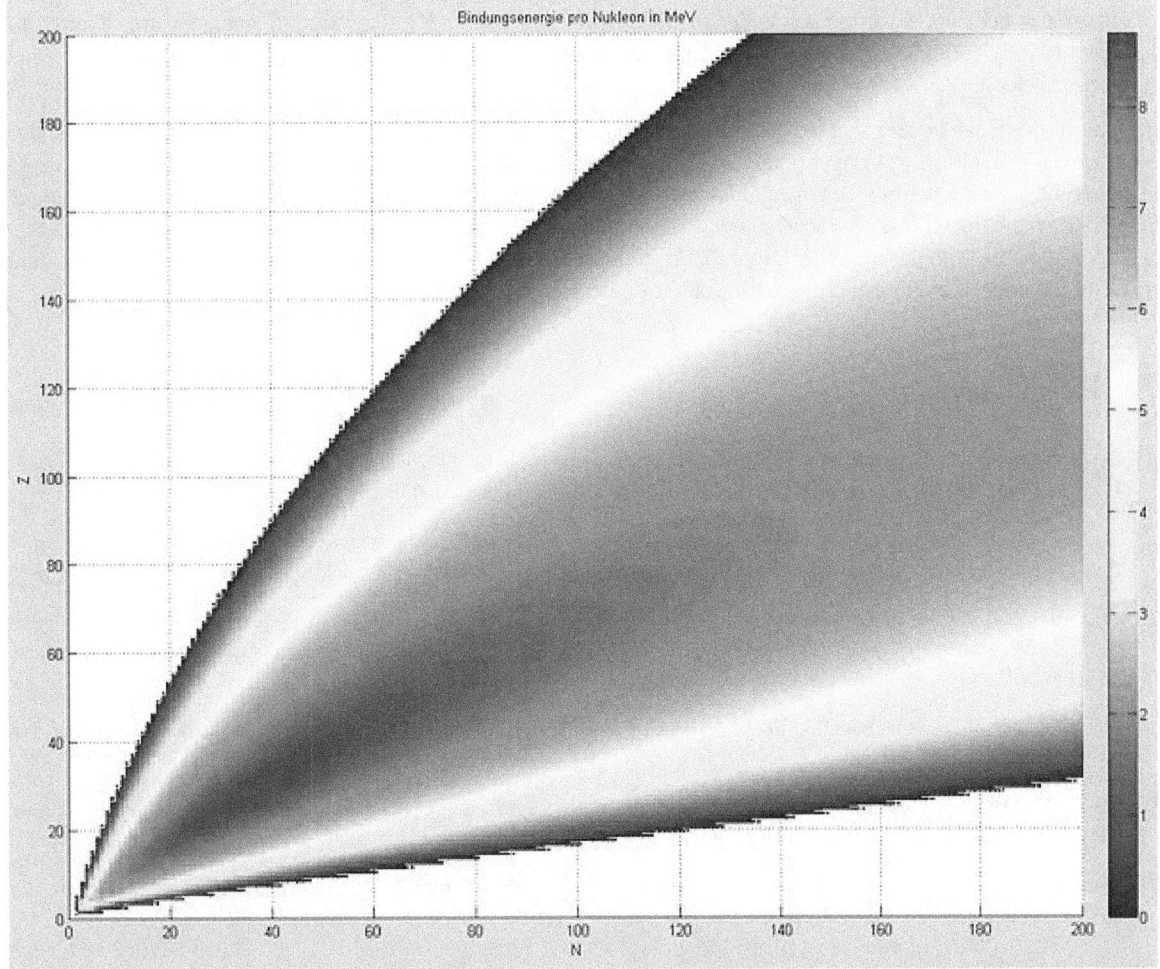

A graphical representation of the semi-empirical binding energy formula. The binding energy per nucleon in MeV (highest numbers in dark red, in excess of 8.5 MeV per nucleon) is plotted for various nuclides as a function of Z, the atomic number (y-axis), vs. N, the number of neutrons (x-axis). The highest numbers are seen for Z = 26 (iron).

1.6   Example values deduced from experimentally measured atom nuclide masses

The following table lists some binding energies and mass defect values.[11] Notice also that we use 1 u = (931.494028 ± 0.000023) MeV. To calculate the binding energy we use the formula $Z (m_p + m_e) + N m_n - m_{nuclide}$ where Z denotes the number of protons in the nuclides and N their number of neutrons. We take $m_p = 938.2723$ MeV, $m_e = 0.5110$ MeV and $m_n = 939.5656$ MeV. The letter A denotes the sum of Z and N (number of nucleons in the nuclide). If we assume the reference nucleon has the mass of a neutron (so that all "total" binding energies calculated are maximal) we could define the total binding energy as the difference from the mass of the nucleus, and the mass of a collection of A free neutrons. In other words, it would be $(Z + N) m_n - m_{nuclide}$. The "*total* binding energy per nucleon" would be this value divided by A.

^{56}Fe has the lowest nucleon-specific mass of the four nuclides listed in this table, but this does not imply it is the strongest bound atom per hadron, unless the choice of beginning hadrons is completely free. Iron releases the largest energy if any 56 nucleons are allowed to build a nuclide—changing one to another if necessary, The highest binding energy per hadron, with the hadrons starting as the same number of protons Z and total nucleons A as in the bound nucleus, is ^{62}Ni. Thus, the true absolute value of the total binding energy of a nucleus depends on what we are allowed to construct the nucleus out of. If all nuclei of mass number A were to be allowed to be constructed of A neutrons, then ^{56}Fe would release the most energy per nucleon, since it has a larger fraction of protons than ^{62}Ni. However, if nucleons are required to be constructed of only the same number of protons and neutrons that they contain, then nickel-62 is the most tightly bound nucleus, per nucleon.

In the table above it can be seen that the decay of a neutron, as well as the transformation of tritium into helium-3, releases energy; hence, it manifests a stronger bound new state when measured against the mass of an equal number of neutrons (and also a lighter state per number of total hadrons). Such reactions are not driven by changes in binding energies as calculated from previously fixed N and Z numbers of neutrons and protons, but rather in decreases in the total mass of the nuclide/per nucleon, with the reaction. (Note that the Binding Energy given above for hydrogen-1 is the atomic binding energy, not the nuclear binding energy which would be zero.)

1.7   References

[1] Dr. Rod Nave of the Department of Physics and Astronomy (July 2010). "Nuclear Binding Energy". *Hyperphysics - a free web resource from GSU*. Georgia State University. Retrieved 2010-07-11. External link in |work= (help)

[2] "Nuclear binding energy". *How to solve for nuclear binding energy. Guides to solving many of the types of quantitative problems found in Chemistry 116. See: Guides*. Purdue University. July 2010. Retrieved 2010-07-10. External link in |work= (help)

[3] "Nuclear Energy". *Energy Education is an interactive curriculum supplement for secondary-school science students, funded by the U. S. Department of Energy and the Texas State Energy Conservation Office (SECO)*. U. S. Department of Energy and the Texas State Energy Conservation Office (SECO). July 2010. Retrieved 2010-07-10.

[4] Stern, Dr. David P. (September 23, 2004). "Nuclear Physics". *"From Stargazers to Starships" Public domain content*. NASA website. Retrieved 2010-07-11.

[5] Stern, Dr. David P. (November 15, 2004). "A Review of Nuclear Structure". *"From Stargazers to Starships" Public domain content*. NASA website. Retrieved 2010-07-11.

[6] Stern, Dr. David P. (February 11, 2009). "Nuclear Binding Energy". *"From Stargazers to Starships" Public domain content*. NASA website. Retrieved 2010-07-11.

[7] Turning Lead into Gold

[8] "Marie Curie - X-rays and Uranium Rays". aip.org. Retrieved 2006-04-10.

[9] Fewell, M. P. (1995). "The atomic nuclide with the highest mean binding energy". *American Journal of Physics* **63** (7): 653–658. Bibcode:1995AmJPh..63..653F. doi:10.1119/1.17828.

[10] M.P. Fewell, 1995

[11] Jagdish K. Tuli, Nuclear Wallet Cards, 7th edition, April 2005, Brookhaven National Laboratory, US National Nuclear Data Center

Chapter 2

Atomic nucleus

The **nucleus** is the small, dense region consisting of protons and neutrons at the center of an atom. The atomic nucleus was discovered in 1911 by Ernest Rutherford based on the 1909 Geiger–Marsden gold foil experiment. After the discovery of the neutron in 1932, models for a nucleus composed of protons and neutrons were quickly developed by Dmitri Ivanenko[1] and Werner Heisenberg.[2][3][4][5][6] Almost all of the mass of an atom is located in the nucleus, with a very small contribution from the electron cloud. Protons and neutrons are bound together to form a nucleus by the nuclear force.

The diameter of the nucleus is in the range of 1.75 fm (1.75×10^{-15} m) for hydrogen (the diameter of a single proton)[7] to about 15 fm for the heaviest atoms, such as uranium. These dimensions are much smaller than the diameter of the atom itself (nucleus + electron cloud), by a factor of about 23,000 (uranium) to about 145,000 (hydrogen).

The branch of physics concerned with the study and understanding of the atomic nucleus, including its composition and the forces which bind it together, is called nuclear physics.

2.1 Introduction

2.1.1 History

Main article: Rutherford model

The nucleus was discovered in 1911, as a result of Ernest Rutherford's efforts to test Thomson's "plum pudding model" of the atom.[8] The electron had already been discovered earlier by J.J. Thomson himself. Knowing that atoms are neutral, Thomson postulated that there must be a positive charge as well. In his plum pudding model, Thomson stated that an atom consisted of negative electrons randomly scattered within a sphere of positive charge. Ernest Rutherford later devised an experiment, performed by Hans Geiger and Ernest Marsden under Rutherford's direction, that involved the deflection of alpha particles directed at a thin sheet of metal foil. He reasoned that if Thomson's model were correct, the positively charged alpha nuclei would easily pass through the foil with very little deviation in their paths as the foil should act in a manner as to be neutrally charged if the negative and positive charges are so intimately mixed as to make it appear neutral. To his surprise, many of the particles were deflected at very large angles. Because the mass of alpha particles is about 8000 times that of an electron, it became apparent that a very strong force must be present if it could deflect the massive and fast moving helium nuclei. He realized that the plum pudding model could not be accurate and that the deflections of the alpha particles could only be explained if the positive and negatives charges were in fact separated from each other and that the mass of the atom was a concentrated point of positive charge. Thus, the idea of a nuclear atom with a dense center of positive charge and mass became justified.

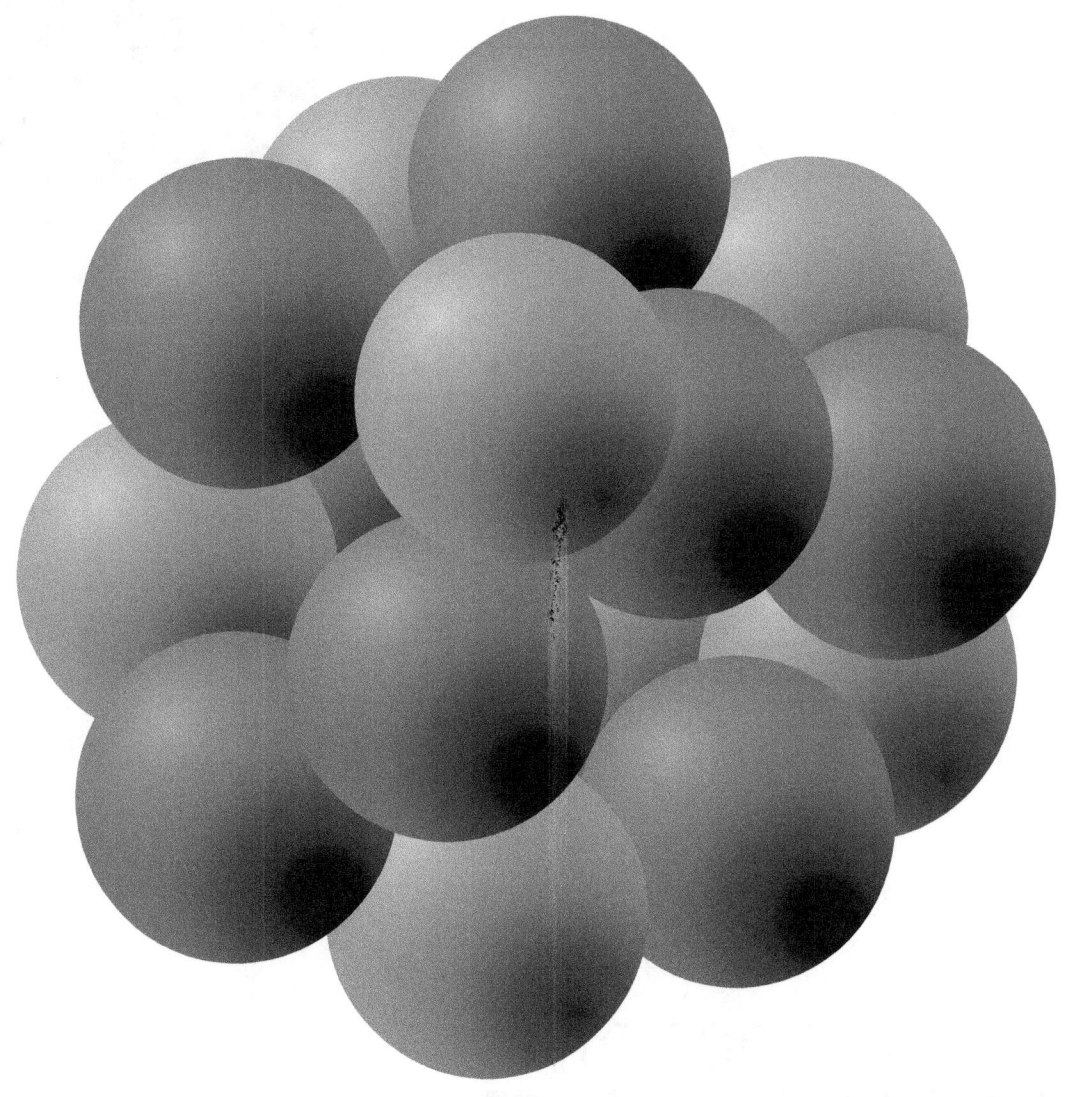

A model of the atomic nucleus showing it as a compact bundle of the two types of nucleons: protons (red) and neutrons (blue). In this diagram, protons and neutrons look like little balls stuck together, but an actual nucleus (as understood by modern nuclear physics) cannot be explained like this, but only by using quantum mechanics. In a nucleus which occupies a certain energy level (for example, the ground state), each nucleon can be said to occupy a range of locations.

2.1.2 Etymology

The term **nucleus** is from the Latin word *nucleus*, a diminutive of *nux* ("nut"), meaning the kernel (i.e., the "small nut") inside a watery type of fruit (like a peach). In 1844, Michael Faraday used the term to refer to the "central point of an atom". The modern atomic meaning was proposed by Ernest Rutherford in 1912.[9] The adoption of the term "nucleus" to atomic theory, however, was not immediate. In 1916, for example, Gilbert N. Lewis stated, in his famous article *The Atom and the Molecule*, that "the atom is composed of the *kernel* and an outer atom or *shell*"[10]

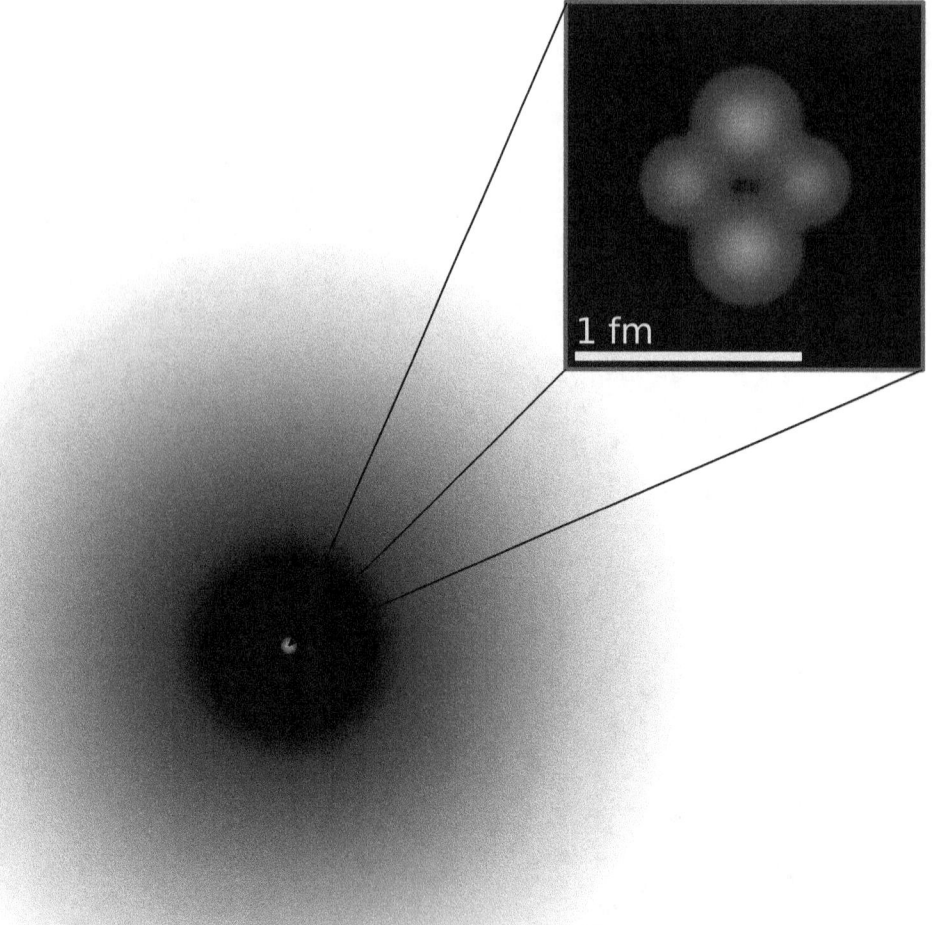

$$1 \text{ Å} = 100{,}000 \text{ fm}$$

*A figurative depiction of the helium−4 atom with the electron cloud in shades of gray. In the nucleus, the two protons and two neutrons are depicted in red and blue. This depiction shows the particles as separate, whereas in an actual helium atom, the protons are superimposed in space and most likely found at the very center of the nucleus, and the same is true of the two neutrons. Thus, all four particles are most likely found in exactly the same space, at the central point. Classical images of separate particles fail to model known charge distributions in very small nuclei. A more accurate image is that the spatial distribution of nucleons in helium's nucleus, although on a far smaller scale, is much closer to the helium **electron cloud** shown here, than to the fanciful nucleus image.*

2.1.3   Nuclear makeup

The nucleus of an atom consists of neutrons and protons, which in turn are the manifestation of fundamental particles, called quarks, that are held in association by the nuclear strong force in certain stable combinations of hadrons, called baryons. The nuclear strong force extends far enough from each baryon so as to bind the neutrons and protons together against the repulsive force of the positively charged protons. The nuclear strong force has a very short range and essentially drops to zero just beyond the edge of the nucleus. The collective action of the positively charged nucleus is to hold the electrically negative charged electrons in their orbits about the nucleus. The collection of negatively charged electrons orbiting the nucleus display an affinity for certain configurations and numbers of electrons that make their orbits stable. Which chemical element an atom represents is determined by the number of protons in the nucleus; the atom will have an equal number of electrons orbiting that nucleus. Individual chemical elements can create more stable electron configurations by combining to share their electrons. It is that sharing of electrons to create stable electronic orbits about the nucleus that appears to us as the chemistry of our macro world.

While protons define the entire charge of a nucleus and, hence, its chemical identity, neutrons are electrically neutral, but contribute to the mass of a nucleus to nearly the same extent as the protons. Neutrons explain the phenomenon of isotopes – varieties of the same chemical element which differ only in their atomic mass, not their chemical action.

2.2 Protons and neutrons

Protons and neutrons are fermions, with different values of the strong isospin quantum number, so two protons and two neutrons can share the same space wave function since they are not identical quantum entities. They sometimes are viewed as two different quantum states of the same particle, the *nucleon*.[11][12] Two fermions, such as two protons, or two neutrons, or a proton + neutron (the deuteron) can exhibit bosonic behavior when they become loosely bound in pairs.

In the rare case of a hypernucleus, a third baryon called a hyperon, with a different value of the strangeness quantum number, can also share the wave function. However, the latter type of nuclei is extremely unstable and not found on Earth except in high energy physics experiments.

The neutron has a positively charged core of radius $\approx$ 0.3 fm surrounded by a compensating negative charge of radius between 0.3 fm and 2 fm. The proton has an approximately exponentially decaying positive charge distribution with a mean square radius of about 0.8 fm.[13]

2.3 Forces

Nuclei are bound together by the residual strong force (nuclear force). The residual strong force is a minor residuum of the strong interaction which binds quarks together to form protons and neutrons. This force is much weaker *between* neutrons and protons because it is mostly neutralized within them, in the same way that electromagnetic forces *between* neutral atoms (such as van der Waals forces that act between two inert gas atoms) are much weaker than the electromagnetic forces that hold the parts of the atoms internally together (for example, the forces that hold the electrons in an inert gas atom bound to its nucleus).

The nuclear force is highly attractive at the distance of typical nucleon separation, and this overwhelms the repulsion between protons which is due to the electromagnetic force, thus allowing nuclei to exist. However, because the residual strong force has a limited range because it decays quickly with distance (see Yukawa potential), only nuclei smaller than a certain size can be completely stable. The largest known completely stable (e.g., stable to alpha, beta, and gamma decay) nucleus is lead-208 which contains a total of 208 nucleons (126 neutrons and 82 protons). Nuclei larger than this maximal size of 208 particles are unstable and (as a trend) become increasingly short-lived with larger size, as the number of neutrons and protons which compose them increases beyond this number. However, bismuth-209 is also stable to beta decay and has the longest half-life to alpha decay of any known isotope, estimated at a billion times longer than the age of the universe.

The residual strong force is effective over a very short range (usually only a few fermis; roughly one or two nucleon diameters) and causes an attraction between any pair of nucleons. For example, between protons and neutrons to form [NP] deuteron, and also between protons and protons, and neutrons and neutrons.

2.4 Halo nuclei and strong force range limits

The effective absolute limit of the range of the strong force is represented by halo nuclei such as lithium-11 or boron-14, in which dineutrons, or other collections of neutrons, orbit at distances of about ten fermis (roughly similar to the 8 fermi radius of the nucleus of uranium-238). These nuclei are not maximally dense. Halo nuclei form at the extreme edges of the chart of the nuclides—the neutron drip line and proton drip line—and are all unstable with short half-lives, measured in milliseconds; for example, lithium-11 has a half-life of 8.8 milliseconds.

Halos in effect represent an excited state with nucleons in an outer quantum shell which has unfilled energy levels "below" it (both in terms of radius and energy). The halo may be made of either neutrons [NN, NNN] or protons [PP, PPP]. Nuclei which have a single neutron halo include ^{11}Be and ^{19}C. A two-neutron halo is exhibited by ^{6}He, ^{11}Li, ^{17}B, ^{19}B and ^{22}C. Two-neutron halo nuclei break into three fragments, never two, and are called *Borromean nuclei* because of this behavior (referring to a system of three interlocked rings in which breaking any ring frees both of the others). ^{8}He and ^{14}Be both exhibit a four-neutron halo. Nuclei which have a proton halo include ^{8}B and ^{26}P. A two-proton halo is exhibited by ^{17}Ne and ^{27}S. Proton halos are expected to be more rare and unstable than the neutron examples, because of the repulsive electromagnetic forces of the excess proton(s).

2.5 Nuclear models

Although the standard model of physics is widely believed to completely describe the composition and behavior of the nucleus, generating predictions from theory is much more difficult than for most other areas of particle physics. This is due to two reasons:

- In principle, the physics within a nuclei can be derived entirely from quantum chromodynamics (QCD). In practice however, current computational and mathematical approaches for solving QCD in low-energy systems such as the nuclei are extremely limited. This is due to the phase transition that occurs between high-energy quark matter and low-energy hadronic matter, which renders perturbative techniques unusable, making it difficult to construct an accurate QCD-derived model of the forces between nucleons. Current approaches are limited to either phenomenological models such as the Argonne v18 potential or chiral effective field theory.[14]

- Even if the nuclear force is well constrained, a significant amount of computational power is required to accurately compute the properties of nuclei *ab initio*. Developments in many-body theory have made this possible for many low mass and relatively stable nuclei, but further improvements in both computational power and mathematical approaches are required before heavy nuclei or highly unstable nuclei can be tackled.

Historically, experiments have been compared to relatively crude models that are necessarily imperfect. None of these models can completely explain experimental data on nuclear structure.[15]

The nuclear radius (R) is considered to be one of the basic quantities that any model must predict. For stable nuclei (not halo nuclei or other unstable distorted nuclei) the nuclear radius is roughly proportional to the cube root of the mass number (A) of the nucleus, and particularly in nuclei containing many nucleons, as they arrange in more spherical configurations:

The stable nucleus has approximately a constant density and therefore the nuclear radius R can be approximated by the following formula,

$$R = r_0 A^{1/3}$$

where A = Atomic mass number (the number of protons Z, plus the number of neutrons N) and r_0 = 1.25 fm = 1.25 × 10^{-15} m. In this equation, the constant r_0 varies by 0.2 fm, depending on the nucleus in question, but this is less than 20% change from a constant.[16]

In other words, packing protons and neutrons in the nucleus gives *approximately* the same total size result as packing hard spheres of a constant size (like marbles) into a tight spherical or almost spherical bag (some stable nuclei are not quite spherical, but are known to be prolate).

2.5.1 Liquid drop model

Main article: Semi-empirical mass formula

Early models of the nucleus viewed the nucleus as a rotating liquid drop. In this model, the trade-off of long-range electromagnetic forces and relatively short-range nuclear forces, together cause behavior which resembled surface tension forces in liquid drops of different sizes. This formula is successful at explaining many important phenomena of nuclei, such as their changing amounts of binding energy as their size and composition changes (see semi-empirical mass formula), but it does not explain the special stability which occurs when nuclei have special "magic numbers" of protons or neutrons.

The terms in the semi-empirical mass formula, which can be used to approximate the binding energy of many nuclei, are considered as the sum of five types of energies (see below). Then the picture of a nucleus as a drop of incompressible liquid roughly accounts for the observed variation of binding energy of the nucleus:

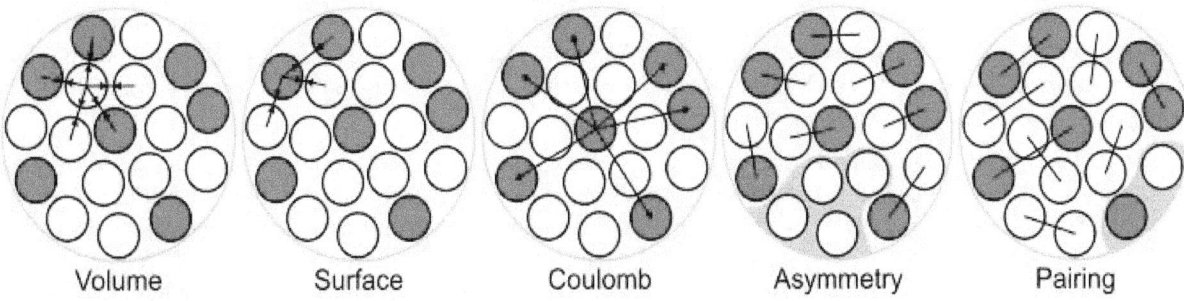

Volume   Surface   Coulomb   Asymmetry   Pairing

Volume energy. When an assembly of nucleons of the same size is packed together into the smallest volume, each interior nucleon has a certain number of other nucleons in contact with it. So, this nuclear energy is proportional to the volume.

Surface energy. A nucleon at the surface of a nucleus interacts with fewer other nucleons than one in the interior of the nucleus and hence its binding energy is less. This surface energy term takes that into account and is therefore negative and is proportional to the surface area.

Coulomb Energy. The electric repulsion between each pair of protons in a nucleus contributes toward decreasing its binding energy.

Asymmetry energy (also called Pauli Energy). An energy associated with the Pauli exclusion principle. Were it not for the Coulomb energy, the most stable form of nuclear matter would have the same number of neutrons as protons, since unequal numbers of neutrons and protons imply filling higher energy levels for one type of particle, while leaving lower energy levels vacant for the other type.

Pairing energy. An energy which is a correction term that arises from the tendency of proton pairs and neutron pairs to occur. An even number of particles is more stable than an odd number.

2.5.2 Shell models and other quantum models

Main article: Nuclear shell model

A number of models for the nucleus have also been proposed in which nucleons occupy orbitals, much like the atomic orbitals in atomic physics theory. These wave models imagine nucleons to be either sizeless point particles in potential wells, or else probability waves as in the "optical model", frictionlessly orbiting at high speed in potential wells.

In the above models, the nucleons may occupy orbitals in pairs, due to being fermions, which allows to explain even/odd Z and N effects well-known from experiments. The exact nature and capacity of nuclear shells differs from those of electrons in atomic orbitals, primarily because the potential well in which the nucleons move (especially in larger nuclei) is quite different from the central electromagnetic potential well which binds electrons in atoms. Some resemblance to atomic orbital models may be seen in a small atomic nucleus like that of helium-4, in which the two protons and two neutrons separately occupy 1s orbitals analogous to the 1s orbital for the two electrons in the helium atom, and achieve unusual stability for the same reason. Nuclei with 5 nucleons are all extremely unstable and short-lived, yet, helium-3, with 3 nucleons, is very stable even with lack of a closed 1s orbital shell. Another nucleus with 3 nucleons, the triton hydrogen-3 is unstable and will decay into helium-3 when isolated. Weak nuclear stability with 2 nucleons {NP} in the 1s orbital is found in the deuteron hydrogen-2, with only one nucleon in each of the proton and neutron potential wells. While each nucleon is a fermion, the {NP} deuteron is a boson and thus does not follow Pauli Exclusion for close packing within shells. Lithium-6 with 6 nucleons is highly stable without a closed second 1p shell orbital. For light nuclei with total nucleon numbers 1 to 6 only those with 5 do not show some evidence of stability. Observations of beta-stability of light nuclei outside closed shells indicate that nuclear stability is much more complex than simple closure of shell orbitals with magic numbers of protons and neutrons.

For larger nuclei, the shells occupied by nucleons begin to differ significantly from electron shells, but nevertheless, present nuclear theory does predict the magic numbers of filled nuclear shells for both protons and neutrons. The closure of the stable shells predicts unusually stable configurations, analogous to the noble group of nearly-inert gases in chemistry. An example is the stability of the closed shell of 50 protons, which allows tin to have 10 stable isotopes, more than any other element. Similarly, the distance from shell-closure explains the unusual instability of isotopes which have far from stable numbers of these particles, such as the radioactive elements 43 (technetium) and 61 (promethium), each of which is preceded and followed by 17 or more stable elements.

There are however problems with the shell model when an attempt is made to account for nuclear properties well away from closed shells. This has led to complex *post hoc* distortions of the shape of the potential well to fit experimental data, but the question remains whether these mathematical manipulations actually correspond to the spatial deformations in real nuclei. Problems with the shell model have led some to propose realistic two-body and three-body nuclear force effects involving nucleon clusters and then build the nucleus on this basis. Two such cluster models are the Close-Packed Spheron Model of Linus Pauling and the 2D Ising Model of MacGregor.[15]

2.5.3   Consistency between models

Main article: Nuclear structure

As with the case of superfluid liquid helium, atomic nuclei are an example of a state in which both (1) "ordinary" particle physical rules for volume and (2) non-intuitive quantum mechanical rules for a wave-like nature apply. In superfluid helium, the helium atoms have volume, and essentially "touch" each other, yet at the same time exhibit strange bulk properties, consistent with a Bose–Einstein condensation. The latter reveals that they also have a wave-like nature and do not exhibit standard fluid properties, such as friction. For nuclei made of hadrons which are fermions, the same type of condensation does not occur, yet nevertheless, many nuclear properties can only be explained similarly by a combination of properties of particles with volume, in addition to the frictionless motion characteristic of the wave-like behavior of objects trapped in Erwin Schrödinger's quantum orbitals.

2.6   See also

- Giant resonance

- List of particles

- Nuclear medicine

- Radioactivity

- Semi-empirical mass formula

2.7 References

[1] Iwanenko, D.D., The neutron hypothesis, Nature **129** (1932) 798.

[2] Heisenberg, W. (1932). "Über den Bau der Atomkerne. I". *Z. Phys.* **77**: 1–11. Bibcode:19:[3] Heisenberg, W.

(1932). "Über den Bau der Atomkerne. II". *Z. Phys.* **78** (3–4): 156–164. Bibcode:1932ZPhy...78..156H.
doi:10.1007/BF01337585.

[4] Heisenberg, W. (1933). "Über den Bau der Atomkerne. III". *Z. Phys.* **80** (9–10): 587–596. Bibcode:1933ZPhy...80..587H.
doi:10.1007/BF01335696.

[5] Miller A. I. *Early Quantum Electrodynamics: A Sourcebook*, Cambridge University Press, Cambridge, 1995, ISBN 0521568919,
pp. 84–88.

[6] Bernard Fernandez and Georges Ripka (2012). "Nuclear Theory After the Discovery of the Neutron". *Unravelling the Mystery of the Atomic Nucleus: A Sixty Year Journey 1896 — 1956*. Springer. p. 263. ISBN 9781461441809. Retrieved 15 February 2013.

[7] Geoff Brumfiel (July 7, 2010). "The proton shrinks in size". *Nature*. doi:10.1038/news.2010.337.

[8] *Rutgers University*. "The Rutherford Experiment". physics.rutgers.edu. Retrieved February 26, 2013.

[9] D. Harper. "Nucleus". *Online Etymology Dictionary*. Retrieved 2010-03-06.

[10] G.N. Lewis (1916). "The Atom and the Molecule". *Journal o f the American Chemical Society* **38** (4): 4. doi:10.
1021/ja02261a

[11] A.G. Sitenko, V.K. Tartakovskiĭ (1997). *Theory of Nucleus: Nuclear Structure and Nuclear Interaction*. Kluwer Academic. p. 3. ISBN 0-7923-4423-5.

[12] M.A. Srednicki (2007). *Quantum Field Theory*. Cambridge University Press. pp. 522–523. ISBN 978-0-521-86449-7.

[13] J.-L. Basdevant, J. Rich, M. Spiro (2005). *Fundamentals in Nuclear Physics*. Springer. p. 155. ISBN 0-387-01672-4.

[14] Machleidt, R.; Entem, D.R. (2011). "Chiral eff ective field theory and nuclear forces". *Physics Reports* **503** (1): 1–75. arXiv:1105.2675. Bibcode:2011PhR...503....1M. doi:10.1016/j.physrep.2011.02.001.

[15] N.D. Cook (2010). *Models of the Atomic Nucleus* (2nd ed.). Springer. p. 57 ff. ISBN 978-3-642-14736-4.

[16] K.S. Krane (1987). *Introductory Nuclear Physics*. Wiley-VCH. ISBN 0-471-80553-X.

2.8 External links

- The Nucleus – a chapter from an online textbook

- The LIVEChart of Nuclides – IAEA in Java or HTML

- Article on the "nuclear shell model," giving nuclear shell filling for the various elements. Accessed Sept. 16, 2009.

- Timeline: Subatomic Concepts, Nuclear Science & Technology.

Chapter 3

Nucleon

For the Ford concept car, see Ford Nucleon. For the fictional power source in the Transformers universe, see Nucleon (Transformers).

In chemistry and physics, a **nucleon** is one of the particles that makes up the atomic nucleus. Each atomic nucleus consists of one or more nucleons, and each atom in turn consists of a cluster of nucleons surrounded by one or more electrons. There are two known kinds of nucleon: the neutron and the proton. The mass number of a given atomic isotope is identical to its number of nucleons. Thus the term nucleon number may be used in place of the more common terms mass number or atomic mass number.

Until the 1960s, nucleons were thought to be elementary particles, each of which would not then have been made up of smaller parts. Now they are known to be composite particles, made of three quarks bound together by the so-called strong interaction. The interaction between two or more nucleons is called internucleon interactions or nuclear force, which is also ultimately caused by the strong interaction. (Before the discovery of quarks, the term "strong interaction" referred to just internucleon interactions.)

Nucleons sit at the boundary where particle physics and nuclear physics overlap. Particle physics, particularly quantum chromodynamics, provides the fundamental equations that explain the properties of quarks and of the strong interaction. These equations explain quantitatively how quarks can bind together into protons and neutrons (and all the other hadrons). However, when multiple nucleons are assembled into an atomic nucleus (nuclide), these fundamental equations become too difficult to solve directly (see lattice QCD). Instead, nuclides are studied within nuclear physics, which studies nucleons and their interactions by approximations and models, such as the nuclear shell model. These models can successfully explain nuclide properties, for example, whether or not a certain nuclide undergoes radioactive decay.

The proton and neutron are both baryons and both fermions. They are quite similar. One carries a non-zero net charge and the other carries a zero net charge; the proton's mass is only 0.1% less than the neutron's. Thus, they can be viewed as two states of the same nucleon. They together form the isospin doublet ($\mathbf{I} = {}^1/_2$). In isospin space, neutrons can be rotationally transformed into protons, and vice versa. These nucleons are acted upon equally by the strong interaction. This implies that strong interaction is invariant when doing rotation transformation in isospin space. According to the Noether theorem, isospin is conserved with respect to the strong interaction.[1]:129–130

3.1   Overview

Main articles: Proton and Neutron

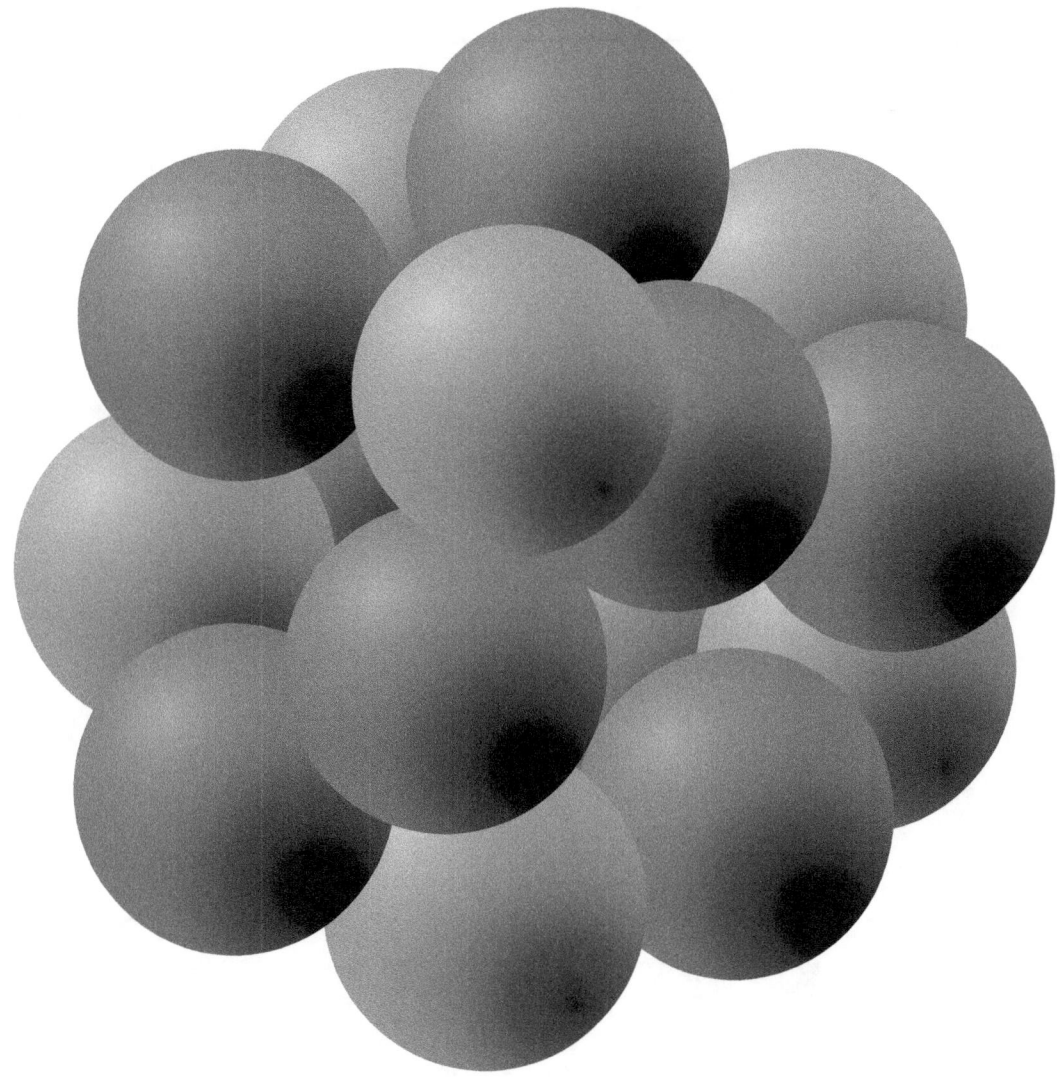

An atomic nucleus is a compact bundle of the two types of nucleons: Protons (red) and neutrons (blue). In this picture, the protons and neutrons look like little balls stuck together, but an actual nucleus, as understood by modern nuclear physics, does not look like this. An actual nucleus can only be accurately described using quantum mechanics. For example, in a real nucleus, each nucleon is in multiple locations at once, spread throughout the nucleus.

3.1.1 Properties

Nucleon quark composition

Proton (p): uud

Neutron (n): udd

Antiproton (p̄): ūūd̄

Antineutron (n̄): ūd̄d̄

Protons and neutrons are most important and best known for constituting atomic nuclei, but they can also be found on their own, not part of a larger nucleus. A proton on its own is the nucleus of the hydrogen-1 atom (^{1}H). A neutron on its own is unstable (see below), but they can be found in nuclear reactions (see neutron radiation) and are used in scientific analysis (see neutron scattering).

Both the proton and neutron are made of three quarks. The proton is made of two up quarks and one down quark, while the neutron is one up quark and two down quarks. The quarks are held together by the strong force. It is also said that the quarks are held together by gluons, but this is just a different way to say the same thing (gluons mediate the strong force).

An up quark has electric charge $+\frac{2}{3}\,e$, and a down quark has charge $-\frac{1}{3}\,e$, so the total electric charge of the proton and neutron are $+e$ and 0, respectively. The word "neutron" comes from the fact that it is electrically "neutral".

The mass of the proton and neutron is quite similar: The proton is 1.6726×10^{-27} kg or 938.27 MeV/c^2, while the neutron is 1.6749×10^{-27} kg or 939.57 MeV/c^2. The neutron is roughly 0.1% heavier. The similarity in mass can be explained roughly by the slight difference in mass of up quark and down quark composing the nucleons. However, detailed explanation remains an unsolved problem in particle physics.[1]:135–136

The spin of both protons and neutrons is $\frac{1}{2}$. This means they are fermions not bosons, and therefore, like electrons, they are subject to the Pauli exclusion principle. This is a very important fact in nuclear physics: Protons and neutrons in an atomic nucleus cannot all be in the same quantum state, but instead they spread out into nuclear shells analogous to electron shells in chemistry. Another reason that the spin of the proton and neutron is important is because it is the source of nuclear spin in larger nuclei. Nuclear spin is best known for its crucial role in the NMR/MRI technique for chemistry and biochemistry analysis.

The magnetic moment of a proton, denoted μ_p, is 2.79 nuclear magnetons (μN), while the magnetic moment of a neutron is $\mu_n = -1.91$ μN. These parameters are also important in NMR/MRI.

3.1.2   Stability

A neutron by itself is an unstable particle: It undergoes β– decay (a type of radioactive decay) by turning into a proton, electron, and electron antineutrino, with a half-life around ten minutes. (See the Neutron article for further discussion of neutron decay.) A proton by itself is thought to be stable, or at least its lifetime is too long to measure. (This is an important issue in particle physics, see Proton decay.)

Inside a nucleus, on the other hand, both protons and neutrons can be stable or unstable, depending on the nuclide. Inside some nuclides, a neutron can turn into a proton (plus other particles) as described above; inside other nuclides the reverse can happen, where a proton turns into a neutron (plus other particles) through β+ decay or electron capture; and inside still other nuclides, both protons and neutrons are stable and do not change form.

3.1.3   Antinucleons

Main articles: Antineutron, Antiproton and Antimatter

Both of the nucleons have corresponding antiparticles: The antiproton and the antineutron. These antimatter particles have the same mass and opposite charge as the proton and neutron respectively, and they interact in the same way. (This is generally believed to be *exactly* true, due to CPT symmetry. If there is a difference, it is too small to measure in all experiments to date.) In particular, antinucleons can bind into an "antinucleus". So far, scientists have created antideuterium[2][3] and antihelium-3[4] nuclei.

3.2   Tables of detailed properties

3.2.1   Nucleons

^**a** The masses of the proton and neutron are known with far greater precision in atomic mass units (u) than in MeV/c^2, due to the relatively poorly known value of the elementary charge. The conversion factor used is 1 u = 931.494028±0.000023 MeV/c^2. The masses of their antiparticles are assumed to be identical, and no experiments have refuted this to date. Current experiments show any percent difference between the masses of the proton and antiproton must be less than 2×10^{-9}[PDG 1] and the difference between the neutron and antineutron masses is on the order of $(9\pm6) \times 10^{-5}$ MeV/c^2.[PDG 2]

^**b** At least 10^{35} years. See proton decay.

^**c** For free neutrons; in most common nuclei, neutrons are stable.

3.2.2 Nucleon resonances

Nucleon resonances are excited states of nucleon particles, often corresponding to one of the quarks having a flipped spin state, or with different orbital angular momentum when the particle decays. Only resonances with a 3 or 4 star rating at the Particle Data Group (PDG) are included in this table. Due to their extraordinarily short lifetimes, many properties of these particles are still under investigation.

The symbol format is given as N(M) L_2I_2J, where M is the particle's approximate mass, L is the orbital angular momentum of the Nucleon-meson pair produced when it decays, and I and J are the particle's isospin and total angular momentum respectively. Since nucleons are defined as having $\frac{1}{2}$ isospin, the first number will always be 1, and the second number will always be odd. When discussing nucleon resonances, sometimes the N is omitted and the order is reversed, giving L_2I_2J (M). For example, a proton can be symbolized as "N(939) S_{11}" or "S_{11} (939)".

The table below lists only the base resonance; each individual entry represents 4 baryons: 2 nucleon resonances particles, as well as their 2 antiparticles. Each resonance exists in a form with a positive electric charge (Q), with a quark composition of uud like the proton, and a neutral form, with a quark composition of udd like the neutron, as well as the corresponding antiparticles with antiquark compositions of uud and udd respectively. Since they contain no strange, charm, bottom, or top quarks, these particles do not possess strangeness, etc. The table only lists the resonances with an isospin of $\frac{1}{2}$. For resonances with $\frac{3}{2}$ isospin, see the Delta baryon article.

† *The $P_{11}(939)$ nucleon represents the excited state of a normal proton or neutron, for example, within the nucleus of an atom. Such particles are usually stable within the nucleus, i.e. Lithium-6.*

3.3 Quark model classification

In the quark model with SU(2) flavour, the two nucleons are part of the ground state doublet. The proton has quark content of *uud*, and the neutron, *udd*. In SU(3) flavour, they are part of the ground state octet (**8**) of spin $\frac{1}{2}$ baryons, known as the Eightfold way. The other members of this octet are the hyperons strange isotriplet Σ+, Σ0, Σ−, the Λ and the strange isodoublet Ξ0, Ξ−. One can extend this multiplet in SU(4) flavour (with the inclusion of the charm quark) to the ground state **20**-plet, or to SU(6) flavour (with the inclusion of the top and bottom quarks) to the ground state **56**-plet.

The article on isospin provides an explicit expression for the nucleon wave functions in terms of the quark flavour eigenstates.

3.4 Models

Although it is known that the nucleon is made from three quarks, as of 2006, it is not known how to solve the equations of motion for quantum chromodynamics. Thus, the study of the low-energy properties of the nucleon are performed by means of models. The only first-principles approach available is to attempt to solve the equations of QCD numerically, using lattice QCD. This requires complicated algorithms and very powerful supercomputers. However, several analytic models also exist:

The Skyrmion models the nucleon as a topological soliton in a non-linear SU(2) pion field. The topological stability of the Skyrmion is interpreted as the conservation of baryon number, that is, the non-decay of the nucleon. The local topological winding number density is identified with the local baryon number density of the nucleon. With the pion isospin vector field oriented in the shape of a hedgehog space, the model is readily solvable, and is thus sometimes called the **hedgehog model**. The hedgehog model is able to predict low-energy parameters, such as the nucleon mass, radius and axial coupling constant, to approximately 30% of experimental values.

The MIT bag model confines three non-interacting quarks to a spherical cavity, with the boundary condition that the quark vector current vanish on the boundary. The non-interacting treatment of the quarks is justified by appealing to the idea of asymptotic freedom, whereas the hard boundary condition is justified by quark confinement. Mathematically, the model vaguely resembles that of a radar cavity, with solutions to the Dirac equation standing in for solutions to the Maxwell equations and the vanishing vector current boundary condition standing for the conducting metal walls of the radar cavity. If the radius of the bag is set to the radius of the nucleon, the bag model predicts a nucleon mass that is within 30% of the actual mass. Although the basic bag model does not provide a pion-mediated interaction, it describes excellently the nucleon-nucleon forces through the 6-quark bag s-channel mechanism using the P matrix.[5] [6]

The **chiral bag model**[7] merges the MIT bag model and the Skyrmion model. In this model, a hole is punched out of the middle of the Skyrmion, and replaced with a bag model. The boundary condition is provided by the requirement of continuity of the axial vector current across the bag boundary. Very curiously, the missing part of the topological winding number (the baryon number) of the hole punched into the Skyrmion is exactly made up by the non-zero vacuum expectation value (or spectral asymmetry) of the quark fields inside the bag. As of 2006, this remarkable trade-off between topology and the spectrum of an operator does not have any grounding or explanation in the mathematical theory of Hilbert spaces and their relationship to geometry. Several other properties of the chiral bag are notable: it provides a better fit to the low energy nucleon properties, to within 5–10%, and these are almost completely independent of the chiral bag radius (as long as the radius is less than the nucleon radius). This independence of radius is referred to as the **Cheshire Cat principle**, after the fading to a smile of Lewis Carroll's Cheshire Cat. It is expected that a first-principles solution of the equations of QCD will demonstrate a similar duality of quark-pion descriptions.

3.5   See also

- Hadrons

- Electroweak interaction

3.6   Further reading

- A.W. Thomas and W.Weise, *The Structure of the Nucleon*, (2001) Wiley-WCH, Berlin, ISBN ISBN 3-527-40297-7

- YAN Kun. Equation of average binding energy per nucleon. doi:10.3969/j.issn.1004-2903.2011.01.018

- Brown, G. E.; Jackson, A. D. (1976). *The Nucleon–Nucleon Interaction*. North-Holland Publishing. ISBN 0-7204-0335-9.

- Vepstas, L.; Jackson, A.D.; Goldhaber, A.S. (1984). "Two-phase models of baryons and the chiral Casimir effect". *Physics Letters B* **140** (5–6): 280–284. Bibcode:1984PhLB..140..280V. doi:10.1016/0370-2693(84)90753-6.

- Vepstas, L.; Jackson, A. D. (1990). "Justifying the chiral bag". *Physics Reports* **187** (3): 109–143 doi:10.1016/0370-1573(90)90056-8.

- Nakamura, N.; Particle Data Group; et al. (2011). "Review of Particle Physics". *Journal of Physics G* **37** (7): 075021. Bibcode:2010JPhG...37g5021N. doi:10.1088/0954-3899/37/7A/075021.

3.7   References

[1] Griffiths, David J. (2008), *Introduction to Elementary Particles* (2nd revised ed.), WILEY-VCH, ISBN 978-3-527-40601-2

[2] Massam, T; Muller, Th.; Righini, B.; Schneegans, M.; Zichichi, A. (1965). "Experimental observation of antideuteron production". *Il Nuovo Cimento* **39**: 10–14. Bibcode:1965NCimS..39...10M. doi:10.1007/BF02814251.

[3] Dorfan, D. E; Eades, J.; Lederman, L. M.; Lee, W.; Ting, C. C. (June 1965). "Observation of Antideuterons". *Phys. Rev. Lett.* **14** (24): 1003–1006. Bibcode:1965PhRvL..14.1003D. doi:10.1103/PhysRevLett.14.1003.

[4] R. Arsenescu; et al. (2003). "Antihelium-3 production in lead-lead collisions at 158 *A* GeV/*c*". *New Journal of Physics* **5**: 1. Bibcode:2003NJPh....5....1A. doi:10.1088/1367-2630/5/1/301.

[5] Jaffe, R. L.; Low, F. E. (1979). "Connection between quark-model eigenstates and low-energy scattering". *Phys. Rev. D* **19**: 2105. Bibcode:1979PhRvD..19.2105J. doi:10.1103/PhysRevD.19.2105.

[6] Yu; Simonov, A. (1981). "The quark compound bag model and the Jaff e-Low P matrix". *Phys. Lett. B* **107**: 1. doi:10.1016/0370-2693(81)91133-3.

[7] Gerald E. Brown and Mannque Rho (March 1979). "The little bag". *Phys. Lett. B* **82** (2): 177–180. Bibcode: doi:10.1016/0370-2693(79)90729-9.

3.7.1 Particle listings

[1] Particle listings – p

[2] Particle listings – n

[3] Particle listings — Note on N and Delta Resonances

[4] Particle listings — N(1440)

[5] Particle listings — N(1520)

[6] Particle listings — N(1535)

[7] Particle listings — N(1650)

[8] Particle listings — N(1675)

[9] Particle listings — N(1680)

[10] Particle listings — N(1700)

[11] Particle listings — N(1710)

[12] Particle listings — N(1720)

[13] Particle listings — N(2190)

[14] Particle listings — N(2220)

[15] Particle listings — N(2250)

Chapter 4

Nuclear force

This article is about the force that holds nucleons together in a nucleus. For the force that holds quarks together in a nucleon, see Strong interaction.

The **nuclear force** (or **nucleon–nucleon interaction** or **residual strong force**) is the force between protons and

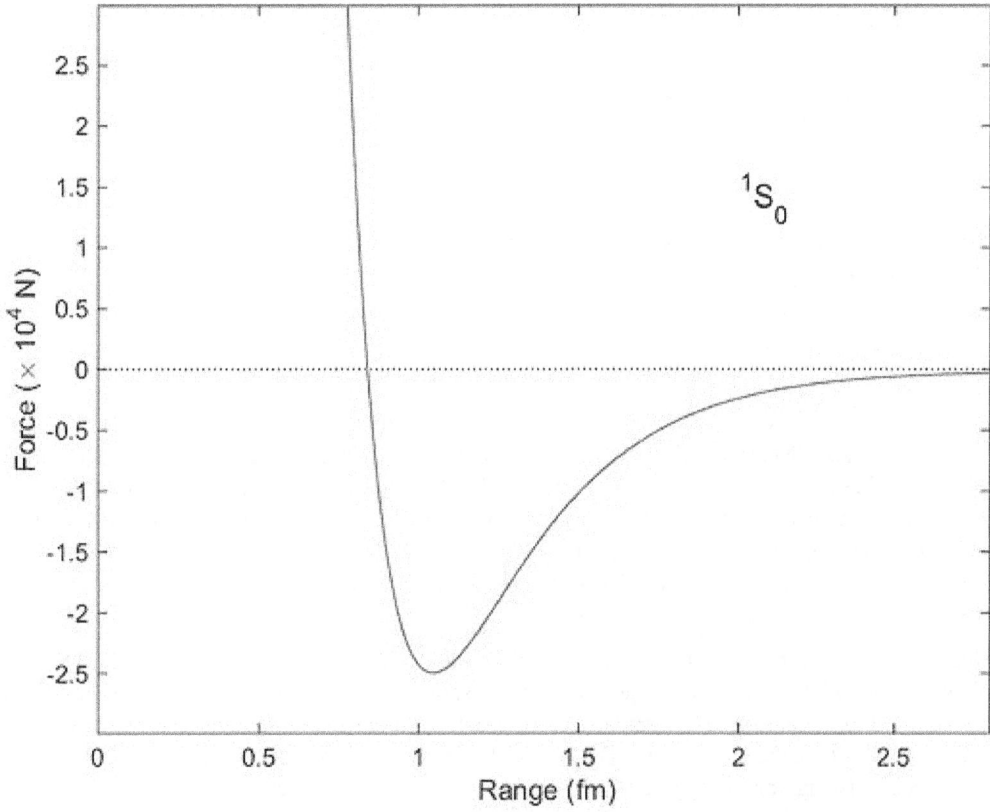

Force (in units of 10,000 N) between two nucleons as a function of distance as computed from the Reid potential (1968).[1] The spins of the neutron and proton are aligned, and they are in the S angular momentum state. The attractive (negative) force has a maximum at a distance of about 1 fm with a force of about 25,000 N. Particles much closer than a distance of 0.8 fm experience a large repulsive (positive) force. Particles separated by a distance greater than 1 fm are still attracted (Yukawa potential), but the force falls as an exponential function of distance.

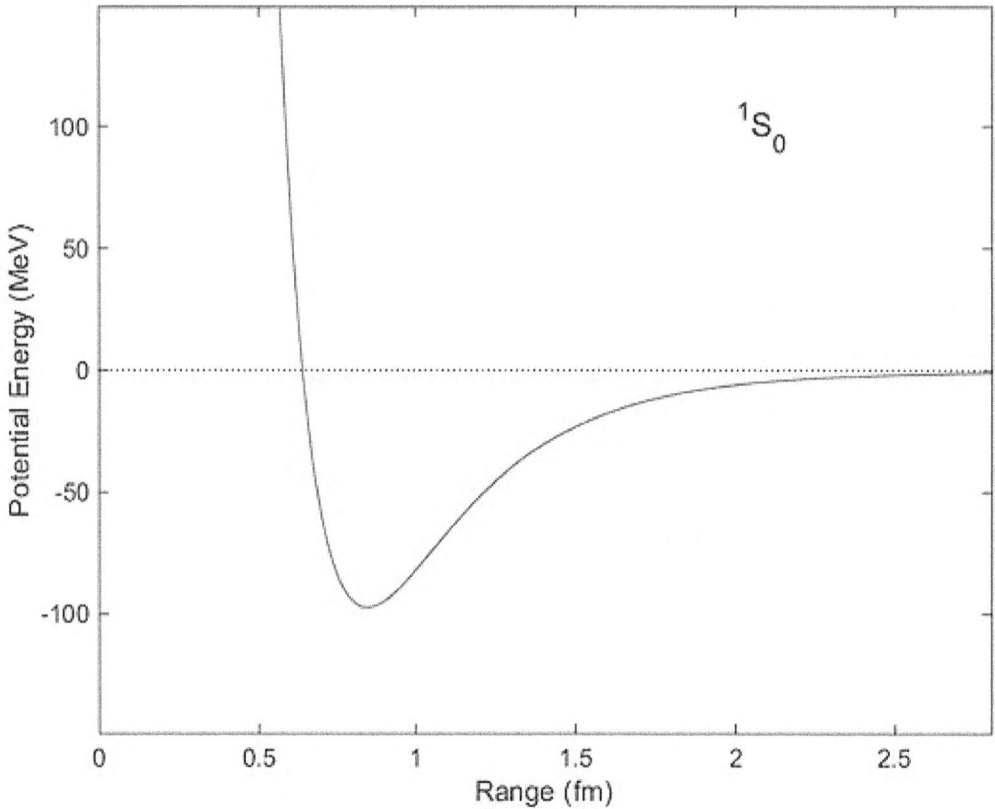

Corresponding potential energy (in units of MeV) of two nucleons as a function of distance as computed from the Reid potential. The potential well is a minimum at a distance of about 0.8 fm. With this potential nucleons can become bound with a negative "binding energy."

neutrons, subatomic particles that are collectively called nucleons. The nuclear force is responsible for binding protons and neutrons into atomic nuclei. Neutrons and protons are affected by the nuclear force almost identically. Since protons have charge +1 *e*, they experience a Coulomb repulsion that tends to push them apart, but at short range the nuclear force is sufficiently attractive as to overcome the electromagnetic repulsive force. The mass of a nucleus is less than the sum total of the individual masses of the protons and neutrons which form it. The difference in mass between bound and unbound nucleons is known as the mass defect. Energy is released when some large nuclei break apart, and it is this energy that used in nuclear power and nuclear weapons.[2][3]

The nuclear force is powerfully attractive between nucleons at distances of about 1 femtometer (fm, or 1.0×10^{-15} metres) between their centers, but rapidly decreases to insignificance at distances beyond about 2.5 fm. At distances less than 0.7 fm, the nuclear force becomes repulsive. This repulsive component is responsible for the physical size of nuclei, since the nucleons can come no closer than the force allows. By comparison, the size of an atom, measured in angstroms (Å, or 1.0×10^{-10} m), is five orders of magnitude larger. The nuclear force is not simple, however, since it depends on the nucleon spins, has a tensor component, and may depend on the relative momentum of the nucleons.[4]

A quantitative description of the nuclear force relies on partially empirical equations that model the internucleon potential energies, or potentials. (Generally, forces within a system of particles can be more simply modeled by describing the system's potential energy; the negative gradient of a potential is equal to the vector force.) The constants for the equations are phenomenological, that is, determined by fitting the equations to experimental data. The internucleon potentials attempt to describe the properties of nucleon–nucleon interaction. Once determined, any given potential can be used in, e.g., the Schrödinger equation to determine the quantum mechanical properties of the nucleon system.

The discovery of the neutron in 1932 revealed that atomic nuclei were made of protons and neutrons, held together by an attractive force. By 1935 the nuclear force was conceived to be transmitted by particles called mesons. This theoretical development included a description of the Yukawa potential, an early example of a nuclear potential. Mesons, predicted by theory, were discovered experimentally in 1947. By the 1970s, the quark model had been developed, which showed that the mesons and nucleons were composed of quarks and gluons. By this new model, the nuclear force, resulting from the exchange of mesons between neighboring nucleons, is a residual effect of the strong force.

4.1 Description

The nuclear force is only felt between particles composed of quarks, or hadrons. At small separations between nucleons (less than ~ 0.7 fm between their centers, depending upon spin alignment) the force becomes repulsive, which keeps the nucleons at a certain average separation, even if they are of different types. This repulsion arises from the Pauli exclusion force for identical nucleons (such as two neutrons or two protons). A Pauli exclusion force also occurs between quarks of the same type within nucleons, when the nucleons are different (a proton and a neutron, for example). The nuclear force also has a "tensor" component which depends on whether or not the spins (angular momentum vectors) of the nucleons are aligned (point in the same direction) or anti-aligned (i.e., point in opposite directions in space).

At distances larger than 0.7 fm the force becomes attractive between spin-aligned nucleons, becoming maximal at a center–center distance of about 0.9 fm. Beyond this distance the force drops exponentially, until beyond about 2.0 fm separation, the force is negligible. Nucleons have a radius of about 0.8 fm.[5]

At short distances (less than 1.7 fm or so), the nuclear force is stronger than the Coulomb force between protons; it thus overcomes the repulsion of protons inside the nucleus. However, the Coulomb force between protons has a much larger range due to its decay as the inverse square of charge separation, and Coulomb repulsion thus becomes the only significant force between protons when their separation exceeds about 2 to 2.5 fm.

For two particles that are the same (such as two neutrons or two protons) the force is not enough to bind the particles, since the spin vectors of two particles of the same type must point in opposite directions when the particles are near each other and are (save for spin) in the same quantum state. This requirement for fermions stems from the Pauli exclusion principle. For fermion particles of different types (such as a proton and neutron), particles may be close to each other and have aligned spins without violating the Pauli exclusion principle, and the nuclear force may bind them (in this case, into a deuteron), since the nuclear force is much stronger for spin-aligned particles. But if the particles' spins are anti-aligned the nuclear force is too weak to bind them, even if they are of different types.

To disassemble a nucleus into unbound protons and neutrons requires work against the nuclear force. Conversely, energy is released when a nucleus is created from free nucleons or other nuclei: the nuclear binding energy. Because of mass–energy equivalence (i.e. Einstein's famous formula $E = mc^2$), releasing this energy causes the mass of the nucleus to be lower than the total mass of the individual nucleons, leading to the so-called "mass defect".[6]

The nuclear force is nearly independent of whether the nucleons are neutrons or protons. This property is called *charge independence*. The force depends on whether the spins of the nucleons are parallel or antiparallel, and it has a noncentral or *tensor* component. This part of the force does not conserve orbital angular momentum, which is a constant of motion under central forces.

The symmetry resulting in the strong force, proposed by Werner Heisenberg, is that protons and neutrons are identical in every respect, other than their charge. This is not completely true, because neutrons are a tiny bit heavier, but it is an approximate symmetry. Protons and neutrons are therefore viewed as the same particle, but with different isospin quantum number. The strong force is invariant under SU(2) transformations, just as particles with "regular spin" are. Isospin and "regular" spin are related under this SU(2) symmetry group. There are only strong attractions when the total isospin is 0, as is confirmed by experiment.[7]

The information on nuclear force are obtained by scattering experiments and the study of light nuclei binding energy.

The nuclear force occurs by the exchange of virtual light mesons, such as the virtual pions, as well as two types of virtual mesons with spin (vector mesons), the rho mesons and the omega mesons. The vector mesons account for the spin-dependence of the nuclear force in this "virtual meson" picture.

The nuclear force is separate from what historically was known as the weak nuclear force. The weak interaction is one

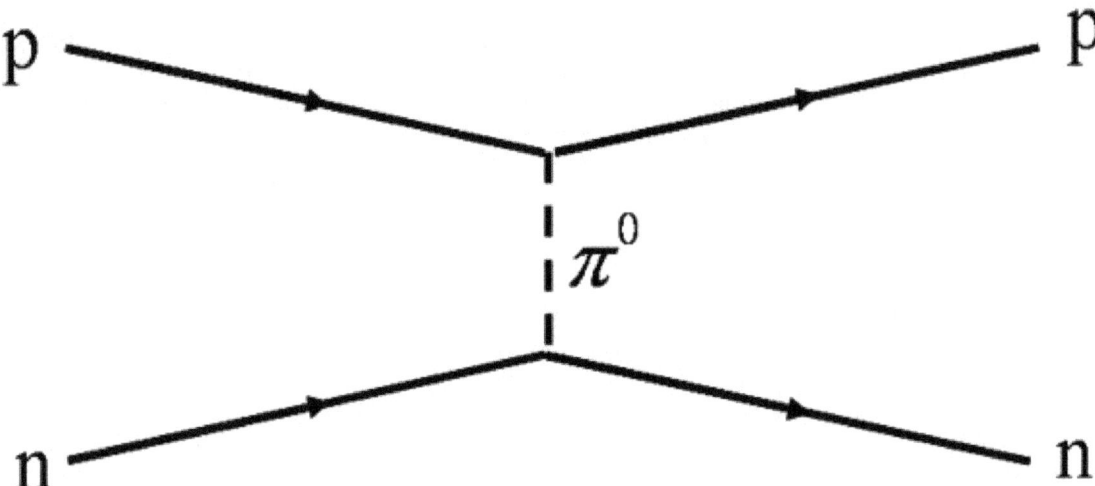

A Feynman diagram of a strong proton–neutron interaction mediated by a neutral pion. Time proceeds from left to right.

of the four fundamental interactions, and it refers to such processes as beta decay. The weak force plays no role in the interaction of nucleons, though it is responsible for the decay of neutrons to protons and vice versa.

4.2 History

The nuclear force has been at the heart of nuclear physics ever since the field was born in 1932 with the discovery of the neutron by James Chadwick. The traditional goal of nuclear physics is to understand the properties of atomic nuclei in terms of the 'bare' interaction between pairs of nucleons, or nucleon–nucleon forces (NN forces).

Within months after the discovery of the neutron, Werner Heisenberg[8][9][10] and Dmitri Ivanenko[11] had proposed proton–neutron models for the nucleus.[12] Heisenberg approached the description of protons and neutrons in the nucleus through quantum mechanics, an approach that was not at all obvious at the time. Heisenberg's theory for protons and neutrons in the nucleus was a "major step toward understanding the nucleus as a quantum mechanical system."[13] Heisenberg introduced the first theory of nuclear exchange forces that bind the nucleons. He considered protons and neutrons to be different quantum states of the same particle, i.e., nucleons distinguished by the value of their nuclear isospin quantum numbers.

One of the earliest models for the nucleus was the liquid drop model developed in the 1930s. One property of nuclei is that the average binding energy per nucleon is approximately the same for all stable nuclei, which is similar to a liquid drop. The liquid drop model treated the nucleus as a drop of incompressible nuclear fluid, with nucleons behaving like molecules in a liquid. The model was first proposed by George Gamow and then developed by Niels Bohr, Werner Heisenberg and Carl Friedrich von Weizsäcker. This crude model did not explain all the properties of the nucleus, but it did explain the spherical shape of most nuclei. The model also gave good predictions for the nuclear binding energy of nuclei.

In 1934, Hideki Yukawa made the earliest attempt to explain the nature of the nuclear force. According to his theory, massive bosons (mesons) mediate the interaction between two nucleons. Although, in light of quantum chromodynamics (QCD), meson theory is no longer perceived as fundamental, the meson-exchange concept (where hadrons are treated as elementary particles) continues to represent the best working model for a quantitative *NN* potential. The Yukawa potential (also called a screened Coulomb potential) is a potential of the form

$$V_{\text{Yukawa}}(r) = -g^2 \frac{e^{-\mu r}}{r},$$

where g is a magnitude scaling constant, i.e., the amplitude of potential, μ is the Yukawa particle mass, r is the radial distance to the particle. The potential is monotone increasing, implying that the force is always attractive. The constants are determined empirically. The Yukawa potential depends only on the distance between particles, r, hence it models a central force.

Throughout the 1930s a group at Columbia University led by I. I. Rabi developed magnetic resonance techniques to determine the magnetic moments of nuclei. These measurements led to the discovery in 1939 that the deuteron also possessed an electric quadrupole moment.[14][15] This electrical property of the deuteron had been interfering with the measurements by the Rabi group. The deuteron, composed of a proton and a neutron, is one of the simplest nuclear systems. The discovery meant that the physical shape of the deuteron was not symmetric, which provided valuable insight into the nature of the nuclear force binding nucleons. In particular, the result showed that the nuclear force was not a central force, but had a tensor character.[1] Hans Bethe identified the discovery of the deuteron's quadrupole moment as one of the important events during the formative years of nuclear physics.[14]

Historically, the task of describing the nuclear force phenomenologically was formidable. The first semi-empirical quantitative models came in the mid-1950s,[1] such as the Woods–Saxon potential (1954). There was substantial progress in experiment and theory related to the nuclear force in the 1960s and 1970s. One influential model was the Reid potential (1968).[1] In recent years, experimenters have concentrated on the subtleties of the nuclear force, such as its charge dependence, the precise value of the πNN coupling constant, improved phase shift analysis, high-precision NN data, high-precision NN potentials, NN scattering at intermediate and high energies, and attempts to derive the nuclear force from QCD.

4.3 The nuclear force as a residual of the strong force

The nuclear force is a residual effect of the more fundamental strong force, or strong interaction. The strong interaction is the attractive force that binds the elementary particles called quarks together to form the nucleons themselves. This more powerful force is mediated by particles called gluons. Gluons hold quarks together with a force like that of electric charge, but of far greater strength. Quarks, gluons and their dynamics are mostly confined within nucleons, but residual influences extend slightly beyond nucleon boundaries to give rise to the nuclear force.

The nuclear forces arising between nucleons are analogous to the forces in chemistry between neutral atoms or molecules called London forces. Such forces between atoms are much weaker than the attractive electrical forces that hold the atoms themselves together (i.e., that bind electrons to the nucleus), and their range between atoms is shorter, because they arise from small separation of charges inside the neutral atom. Similarly, even though nucleons are made of quarks in combinations which cancel most gluon forces (they are "color neutral"), some combinations of quarks and gluons nevertheless leak away from nucleons, in the form of short-range nuclear force fields that extend from one nucleon to another nearby nucleon. These nuclear forces are very weak compared to direct gluon forces ("color forces" or strong forces) inside nucleons, and the nuclear forces extend only over a few nuclear diameters, falling exponentially with distance. Nevertheless, they are strong enough to bind neutrons and protons over short distances, and overcome the electrical repulsion between protons in the nucleus.

Sometimes, the nuclear force is called the **residual strong force**, in contrast to the strong interactions which arise from QCD. This phrasing arose during the 1970s when QCD was being established. Before that time, the *strong nuclear force* referred to the inter-nucleon potential. After the verification of the quark model, *strong interaction* has come to mean QCD.

4.4 Nucleon–nucleon potentials

Two-nucleon systems such as the deuteron, the nucleus of a deuterium atom, as well as proton–proton or neutron–proton scattering are ideal for studying the NN force. Such systems can be described by attributing a *potential* (such as the Yukawa potential) to the nucleons and using the potentials in a Schrödinger equation. The form of the potential is derived phenomenologically, although for the long-range interaction, meson-exchange theories help to construct the potential. The parameters of the potential are determined by fitting to experimental data such as the deuteron binding energy or NN

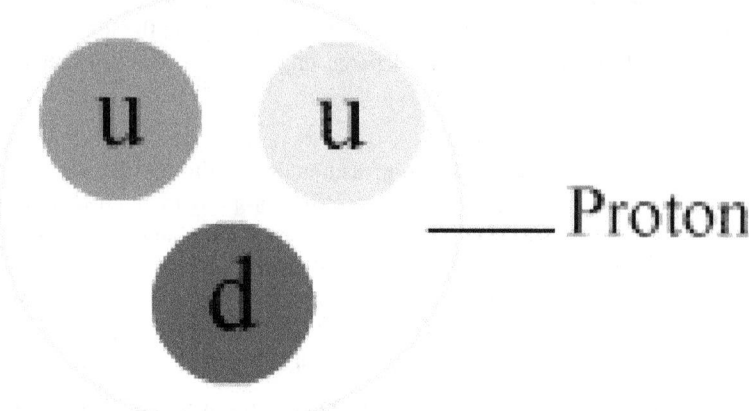

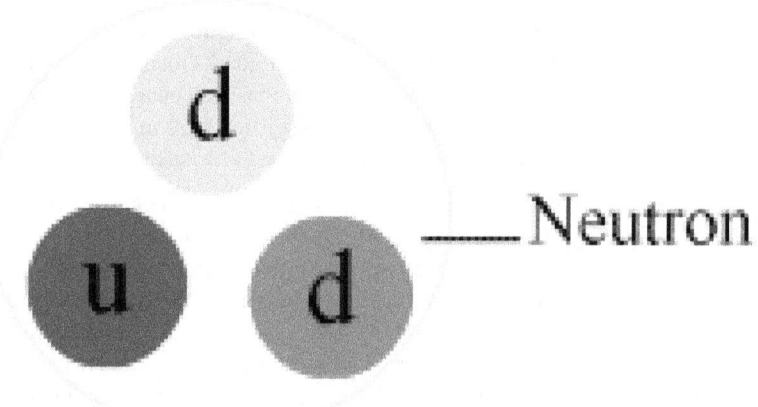

An animation of the interaction. The colored double circles are gluons. Anticolors are shown as per this diagram (larger version).

elastic scattering cross sections (or, equivalently in this context, so-called *NN* phase shifts).

The most widely used *NN* potentials are the Paris potential, the Argonne AV18 potential ,[16] the CD-Bonn potential and the Nijmegen potentials.

A more recent approach is to develop effective field theories for a consistent description of nucleon–nucleon and three-nucleon forces. In particular, chiral symmetry breaking can be analyzed in terms of an effective field theory (called chiral perturbation theory) which allows perturbative calculations of the interactions between nucleons with pions as exchange particles.

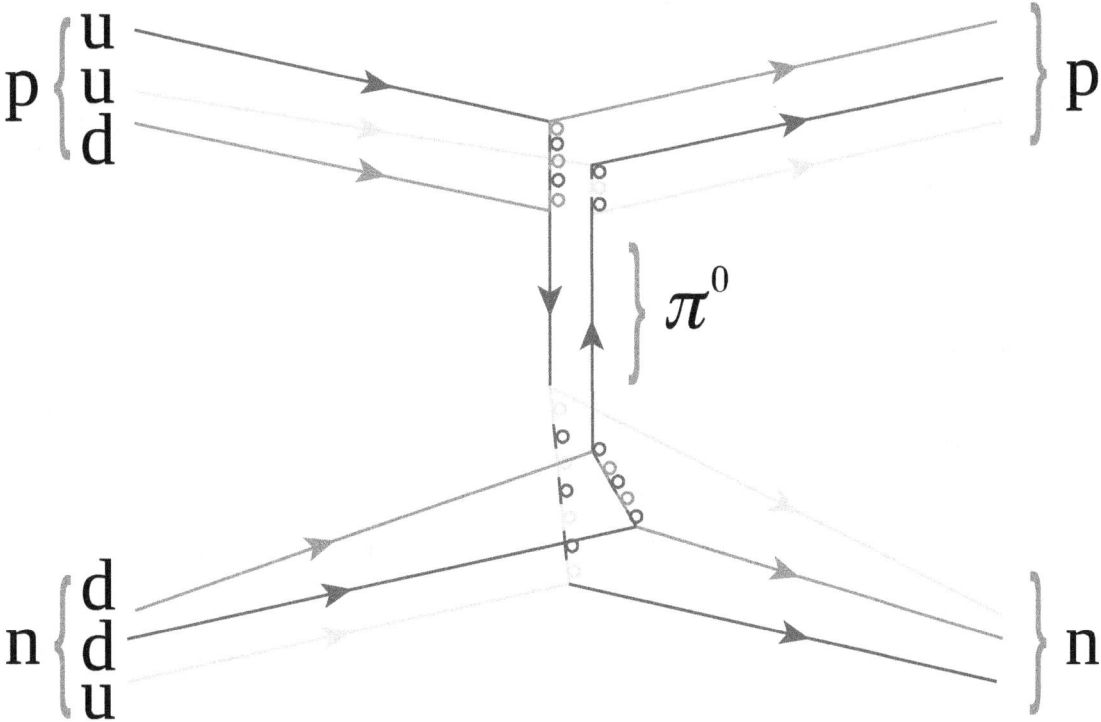

The same diagram as that above with the individual quark constituents shown, to illustrate how the fundamental *strong interaction gives rise to the* **nuclear force***. Straight lines are quarks, while multi-colored loops are gluons (the carriers of the fundamental force). Other gluons, which bind together the proton, neutron, and pion "in-flight," are not shown.*

4.4.1 From nucleons to nuclei

The ultimate goal of nuclear physics would be to describe all nuclear interactions from the basic interactions between nucleons. This is called the *microscopic* or *ab initio* approach of nuclear physics. There are two major obstacles to overcome before this dream can become reality:

- Calculations in many-body systems are difficult and require advanced computation techniques.

- There is evidence that three-nucleon forces (and possibly higher multi-particle interactions) play a significant role. This means that three-nucleon potentials must be included into the model.

This is an active area of research with ongoing advances in computational techniques leading to better first-principles calculations of the nuclear shell structure. Two- and three-nucleon potentials have been implemented for nuclides up to $A = 12$.

4.4.2 Nuclear potentials

A successful way of describing nuclear interactions is to construct one potential for the whole nucleus instead of considering all its nucleon components. This is called the *macroscopic* approach. For example, scattering of neutrons from nuclei can be described by considering a plane wave in the potential of the nucleus, which comprises a real part and an imaginary part. This model is often called the **optical model** since it resembles the case of light scattered by an opaque glass sphere.

Nuclear potentials can be *local* or *global*: local potentials are limited to a narrow energy range and/or a narrow nuclear mass range, while global potentials, which have more parameters and are usually less accurate, are functions of the energy and the nuclear mass and can therefore be used in a wider range of applications.

4.5 See also

- Strong interaction

- Standard Model

4.6 References

[1] Reid, R.V. (1968). "Local phenomenological nucleon–nucleon potentials". *Annals o f Physics* **50**: 411–448 doi:10.1016/0003-4916(68)90126-7.

[2] Binding Energy, Mass Defect, Furry Elephant physics educational site, retr 2012 7 1

[3] Chapter 4 NUCLEAR PROCESSES, THE STRONG FORCE, M. Ragheb 1/30/2013, University of Illinois

[4] Kenneth S. Krane (1988). *Introductory Nuclear Physics*. Wiley & Sons. ISBN 0-471-80553-X.

[5] Povh, B.; Rith, K.; Scholz, C.; Zetsche, F. (2002). *Particles and Nuclei: An Introduction to the* Springer-Verlag. p. 73. ISBN 978-3-540-43823-6.

[6] Stern, Dr. Swapnil Nikam (February 11, 2009). "Nuclear Binding Energy". " From Stargazers to Retrieved 2010-12-30.

[7] Griffiths, David, Introduction to Elementary Particles

[8] Heisenberg, W. (1932). "Über den Bau der Atomkerne. I". *Z. Phys.* **77**: 1–11. doi:10.1007/BF01342433.

[9] Heisenberg, W. (1932). "Über den Bau der Atomkerne. II". *Z. Phys.* **78** (3–4): 156–164. doi:10.1007/BF01337585.

[10] Heisenberg, W. (1933). "Über den Bau der Atomkerne. III". *Z. Phys.* **80** (9–10): 587–596. doi:10.1007/BF01335696.

[11] Iwanenko, D.D., The neutron hypothesis, Nature **129** (1932) 798.

[12] Miller A. I. *Earl y Quantum Electrodynamics: A Sourcebook*, Cambridge University Press, Cambridge, 1995, pp. 84–88.

[13] Brown, L.M.; Rechenberg, H. (1996). *The Origin of the Concept of Nuclear Forces*. Bristol and Philadelphia: Institute of Physics Publishing. ISBN 0750303735.

[14] John S. Rigden (1987). *Rabi, Scientist and Citizen*. New York: Basic Books, Inc. pp. 99–114. ISBN 9780674004351. Retrieved May 9, 2015.

[15] Kellogg, J.M.; Rabi, I.I.; Ramsey, N.F.; Zacharias, J.R. (1939). "An electrical quadrupole moment of the deuteron". *Physical Review* **55**: 318–319. Bibcode:1939PhRv...55..318K. doi:10.1103/physrev.55.318. Retrieved May 9, 2015.

[16] Wiringa, R. B.; Stoks, V. G. J.; Schiavilla, R. (1995). "Accurate nucleon–nucleon potential with charge-independence breaking". *Physical Review C* **51**: 38. arXiv:nucl-th/9408016. Bibcode:1995PhRvC..51...38W. doi:10.1103/PhysRevC.51.38.

4.7 Bibliography

- Gerald Edward Brown and A. D. Jackson, *The Nucleon–Nucleon Interaction*, (1976) North-Holland Publishing, Amsterdam ISBN 0-7204-0335-9

- R. Machleidt and I. Slaus, "The nucleon–nucleon interaction", *J. Phys.* G **27** (2001) R69 *(topical review)*.

- E.A. Nersesov, *Fundamentals of atomic and nuclear physics*, (1990), Mir Publishers, Moscow, ISBN 5-06-001249-2

- P. Navrátil and W.E. Ormand, "Ab initio shell model with a genuine three-nucleon force for the p-shell nuclei", Phys. Rev. C **68**, 034305 (2003).

Chapter 5

Nuclear structure

Understanding the structure of the atomic nucleus is one of the central challenges in nuclear physics.

5.1 Models

5.1.1 The liquid drop model

Main article: Semi-empirical mass formula

The liquid drop model is one of the first models of **nuclear structure**, proposed by Carl Friedrich von Weizsäcker in 1935.[1] It describes the nucleus as a semiclassical fluid made up of neutrons and protons, with an internal repulsive electrostatic force proportional to the number of protons. The quantum mechanical nature of these particles appears via the Pauli exclusion principle, which states that no two nucleons of the same kind can be at the same state. Thus the fluid is actually what is known as a Fermi liquid. In this model, the binding energy of a nucleus with Z protons and N neutrons is given by

$$E_B = a_V A - a_S A^{2/3} - a_C \frac{Z^2}{A^{1/3}} - a_A \frac{(N-Z)^2}{A} - \delta(A, Z)$$

where $A = Z + N$ is the total number of nucleons. The terms proportional to A and $A^{2/3}$ represent the volume and surface energy of the liquid drop, the term proportional to Z^2 represents the electrostatic energy, the term proportional to $(N - Z)^2$ represents the Pauli exclusion principle and the last term $\delta(A, Z)$ is the pairing term, which lowers the energy for even numbers of protons or neutrons. The coefficients a_V, a_S, a_C, a_A and the strength of the pairing term may be estimated theoretically, or fit to data. This simple model reproduces the main features of the binding energy of nuclei.

The assumption of nucleus as a drop of Fermi liquid is still widely used in the form of Finite Range Droplet Model (FRDM), due to the possible good reproduction of nuclear binding energy on the whole chart, with the necessary accuracy for predictions of unknown nuclei.[2]

5.1.2 The shell model

Main article: Nuclear shell model

The expression "shell model" is ambiguous in that it refers to two different eras in the state of the art. It was previously used to describe the existence of nucleon shells in the nucleus according to an approach closer to what is now called mean field theory. Nowadays, it refers to a formalism analogous to the configuration interaction formalism used in quantum chemistry. We shall introduce the latter here.

5.1.3 Introduction to the shell concept

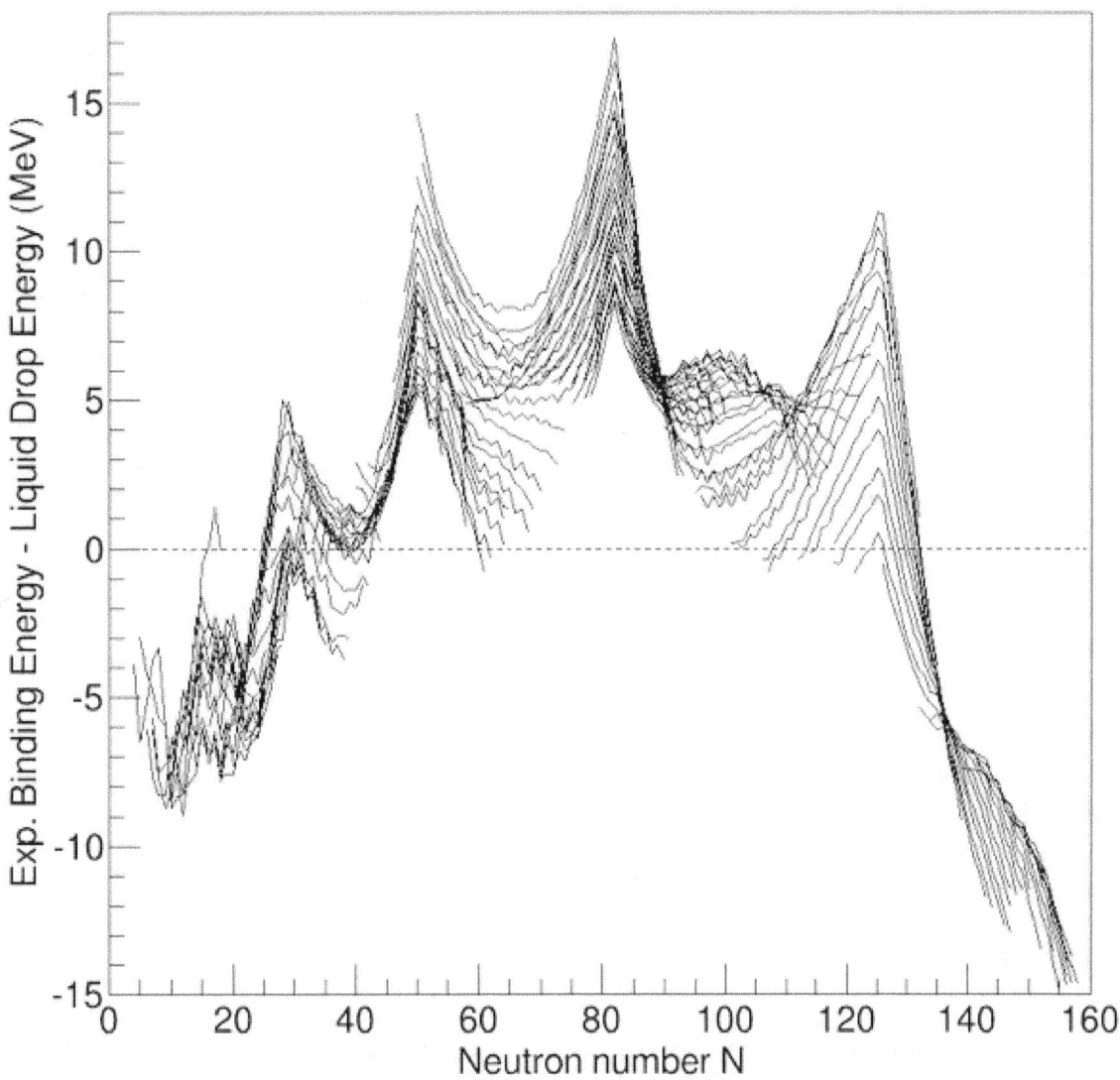

Difference between experimental binding energies and the liquid drop model prediction as a function of neutron number for Z>7

Systematic measurements of the binding energy of atomic nuclei show systematic deviations with respect to those estimated from the liquid drop model. In particular, some nuclei having certain values for the number of protons and/or neutrons are bound more tightly together than predicted by the liquid drop model. These nuclei are called singly/doubly magic. This observation led scientists to assume the existence of a shell structure of nucleons (protons and neutrons) within the nucleus, like that of electrons within atoms.

Indeed, nucleons are quantum objects. Strictly speaking, one should not speak of energies of individual nucleons, because they are all correlated with each other. However, as an approximation one may envision an average nucleus, within which nucleons propagate individually. Owing to their quantum character, they may only occupy *discrete* energy levels. These levels are by no means uniformly distributed; some intervals of energy are crowded, and some are empty, generating a gap in possible energies. A shell is such a set of levels separated from the other ones by a wide empty gap.

The energy levels are found by solving the Schrödinger equation for a single nucleon moving in the average potential generated by all other nucleons. Each level may be occupied by a nucleon, or empty. Some levels accommodate several different quantum states with the same energy; they are said to be *degenerate*. This occurs in particular if the average nucleus has some symmetry.

The concept of shells allows one to understand why some nuclei are bound more tightly than others. This is because two nucleons of the same kind cannot be in the same state (Pauli exclusion principle). So the lowest-energy state of the nucleus is one where nucleons fill all energy levels from the bottom up to some level. A nucleus with full shells is exceptionally stable, as will be explained.

As with electrons in the electron shell model, protons in the outermost shell are relatively loosely bound to the nucleus if there are only few protons in that shell, because they are farthest from the center of the nucleus. Therefore, nuclei which have a full outer proton shell will be more tightly bound and have a higher binding energy than other nuclei with a similar total number of protons. All this is also true for neutrons.

Furthermore, the energy needed to excite the nucleus (i.e. moving a nucleon to a higher, previously unoccupied level) is exceptionally high in such nuclei. Whenever this unoccupied level is the next after a full shell, the only way to excite the nucleus is to raise one nucleon *across the gap*, thus spending a large amount of energy. Otherwise, if the highest occupied energy level lies in a partly filled shell, much less energy is required to raise a nucleon to a higher state in the same shell.

Some evolution of the shell structure observed in stable nuclei is expected away from the valley of stability. For example, observations of unstable isotopes have shown shifting and even a reordering of the single particle levels of which the shell structure is composed.[3] This is sometimes observed as the creation of an island of inversion or in the reduction of excitation energy gaps above the traditional magic numbers.

Basic hypotheses

Some basic hypotheses are made in order to give a precise conceptual framework to the shell model :

- The atomic nucleus is a quantum n-body system.

- The internal motion of nucleons within the nucleus is non-relativistic, and their behavior is governed by the Schrödinger equation.

- Nucleons are considered to be pointlike, without any internal structure.

Brief description of the formalism

The general process used in the shell model calculations is the following. First a Hamiltonian for the nucleus is defined. Usually, for computational practicality, only one- and two-body terms are taken into account in this definition. The interaction is an effective theory : it contains free parameters which have to be fitted with experimental data.

The next step consists in defining a basis of single-particle states, i.e. a set of wavefunctions describing all possible nucleon states. Most of the time, this basis is obtained via a Hartree–Fock computation. With this set of one-particle states, Slater determinants are built, that is, wavefunctions for Z proton variables or N neutron variables, which are antisymmetrized products of single-particle wavefunctions (antisymmetrized meaning that under exchange of variables for any pair of nucleons, the wavefunction only changes sign).

In principle, the number of quantum states available for a single nucleon at a finite energy is finite, say n. The number of nucleons in the nucleus must be smaller than the number of available states, otherwise the nucleus cannot hold all of its nucleons. There are thus several ways to choose Z (or N) states among the n possible. In combinatorial mathematics, the number of choices of Z objects among n is the binomial coefficient C^Z_n. If n is much larger than Z (or N), this increases roughly like n^Z. Practically, this number becomes so large that every computation is impossible for $A=N+Z$ larger than 8.

To obviate this difficulty, the space of possible single-particle states is divided into a core and a valence shell, by analogy with chemistry. The core is a set of single-particles which are assumed to be inactive, in the sense that they are the well bound lowest-energy states, and that there is no need to reexamine their situation. They do not appear in the Slater determinants, contrary to the states in the valence space, which is the space of all single-particle states *not in the core*, but possibly to be considered in the choice of the build of the (Z-) N-body wavefunction. The set of all possible Slater determinants in the valence space defines a basis for (Z-) N-body states.

The last step consists in computing the matrix of the Hamiltonian within this basis, and to diagonalize it. In spite of the reduction of the dimension of the basis owing to the fixation of the core, the matrices to be diagonalized reach easily dimensions of the order of 10^9, and demand specific diagonalization techniques.

The shell model calculations give in general an excellent fit with experimental data. They depend however strongly on two main factors :

- The way to divide the single-particle space into core and valence.

- The effective nucleon–nucleon interaction.

5.2 Mean field theories

5.2.1 The independent-particle model

The interaction between nucleons, which is a consequence of strong interactions and binds the nucleons within the nucleus, exhibits the peculiar behaviour of having a finite range: it vanishes when the distance between two nucleons becomes too large; it is attractive at medium range, and repulsive at very small range. This last property correlates with the Pauli exclusion principle according to which two fermions (nucleons are fermions) cannot be in the same quantum state. This results in a very large mean free path predicted for a nucleon within the nucleus.[4]

The main idea of the Independent Particle approach is that a nucleon moves inside a certain potential well (which keeps it bound to the nucleus) independently from the other nucleons. This amounts to replacing a N-body problem (N particles interacting) by N single-body problems. This essential simplification of the problem is the cornerstone of mean field theories. These are also widely used in atomic physics, where electrons move in a mean field due to the central nucleus and the electron cloud itself.

The independent particle model and mean field theories (we shall see that there exist several variants) have a great success in describing the properties o the nucleus starting from an effective interaction or an effective potential, thus are a basic part of atomic nucleus theory. One should also notice that they are modular enough, in that it is quite easy to extend the model to introduce effects such as nuclear pairing, or collective motions of the nucleon like rotation, or vibration, adding the corresponding energy terms in the formalism. This implies that in many representation the mean field is only a starting point for a more complete description which introduces correlations reproducing properties like collective excitations and nucleon transfer.[5][6]

5.2.2 Nuclear potential and effective interaction

A large part of the practical difficulties met in mean field theories is the definition (or calculation) of the potential of the mean field itself. One can very roughly distinguish between two approaches :

- The **phenomenological** approach is a parameterization of the nuclear potential by an appropriate mathematical function. Historically, this procedure was applied with the greatest success by Sven Gösta Nilsson, who used as a potential a (deformed) harmonic oscillator potential. The most recent parameterizations are based on more realistic functions, which account more accurately for scattering experiments, for example. In particular the form known as the Woods Saxon potential can be mentioned.

- The **self-consistent** or Hartree–Fock approach aims to deduce mathematically the nuclear potential from an effective nucleon–nucleon interaction. This technique implies a resolution of the Schrödinger equation in an iterative fashion, starting from an ansatz wavefunction and improving it variationally, since the potential depends there upon the wavefunctions to be determined. The latter are written as Slater determinants.

In the case of the Hartree–Fock approaches, the trouble is not to find the mathematical function which describes best the nuclear potential, but that which describes best the nucleon–nucleon interaction. Indeed, in contrast with atomic physics where the interaction is known (it is the Coulomb interaction), the nucleon–nucleon interaction within the nucleus is not known analytically.

There are two main reasons for this fact. First, the strong interaction acts essentially among the quarks forming the nucleons. The nucleon–nucleon interaction in vacuum is a mere *consequence* of the quark–quark interaction. While the latter is well understood in the framework of the Standard Model at high energies, it is much more complicated in low energies due to color confinement and asymptotic freedom. Thus there is yet no fundamental theory allowing one to deduce the nucleon–nucleon interaction from the quark–quark interaction. Furthermore, even if this problem were solved, there would remain a large difference between the ideal (and conceptually simpler) case of two nucleons interacting in vacuum, and that of these nucleons interacting in the nuclear matter. To go further, it was necessary to invent the concept of effective interaction. The latter is basically a mathematical function with several arbitrary parameters, which are adjusted to agree with experimental data.

Most modern interaction are zero-range so they act only when the two nucleons are in contact, as introduced by Tony Skyrme.[7]

5.2.3   The self-consistent approaches of the Hartree–Fock type

In the Hartree–Fock approach of the n-body problem, the starting point is a Hamiltonian containing n kinetic energy terms, and potential terms. As mentioned before, one of the mean field theory hypotheses is that only the two-body interaction is to be taken into account. The potential term of the Hamiltonian represents all possible two-body interactions in the set of n fermions. It is the first hypothesis.

The second step consists in assuming that the wavefunction of the system can be written as a Slater determinant of one-particle spin-orbitals. This statement is the mathematical translation of the independent-particle model. This is the second hypothesis.

There remains now to determine the components of this Slater determinant, that is, the individual wavefunctions of the nucleons. To this end, it is assumed that the total wavefunction (the Slater determinant) is such that the energy is minimum. This is the third hypothesis.

Technically, it means that one must compute the mean value of the (known) two-body Hamiltonian on the (unknown) Slater determinant, and impose that its mathematical variation vanishes. This leads to a set of equations where the unknowns are the individual wavefunctions: the Hartree–Fock equations. Solving these equations gives the wavefunctions and individual energy levels of nucleons, and so the total energy of the nucleus and its wavefunction.

This short account of the Hartree–Fock method explains why it is called also the variational approach. At the beginning of the calculation, the total energy is a "function of the individual wavefunctions" (a so-called functional), and everything is then made in order to optimize the choice of these wavefunctions so that the functional has a minimum – hopefully absolute, and not only local. To be more precise, there should be mentioned that the energy is a functional of the density, defined as the sum of the individual squared wavefunctions. Let us note also that the Hartree–Fock method is also used in atomic physics and condensed matter physics as Density Functional Theory, DFT.

The process of solving the Hartree–Fock equations can only be iterative, since these are in fact a Schrödinger equation in which the potential depends on the density, that is, precisely on the wavefunctions to be determined. Practically, the algorithm is started with a set of individual grossly reasonable wavefunctions (in general the eigenfunctions of a harmonic oscillator). These allow to compute the density, and therefrom the Hartree–Fock potential. Once this done, the Schrödinger equation is solved anew, and so on. The calculation stops – convergence is reached – when the difference among wavefunctions, or energy levels, for two successive iterations is less than a fixed value. Then the mean field potential is completely determined, and the Hartree–Fock equations become standard Schrödinger equations. The corresponding

Hamiltonian is then called the Hartree–Fock Hamiltonian.

5.2.4   The relativistic mean field approaches

Born first in the 1970s with the works of John Dirk Walecka on quantum hadrodynamics, the relativistic models of the nucleus were sharpened up towards the end of the 1980s by P. Ring and coworkers. The starting point of these approaches is the relativistic quantum field theory. In this context, the nucleon interactions occur via the exchange of virtual particles called mesons. The idea is, in a first step, to build a Lagrangian containing these interaction terms. Second, by an application of the least action principle, one gets a set of equations of motion. The real particles (here the nucleons) obey the Dirac equation, whilst the virtual ones (here the mesons) obey the Klein–Gordon equations.

In view of the non-perturbative nature of strong interaction, and also in view of the fact that the exact potential form of this interaction between groups of nucleons is relatively badly known, the use of such an approach in the case of atomic nuclei requires drastic approximations. The main simplification consists in replacing in the equations all field terms (which are operators in the mathematical sense) by their mean value (which are functions). In this way, one gets a system of coupled integro-differential equations, which can be solved numerically, if not analytically.

5.2.5   The interacting boson model

The interacting boson model (IBM) is a model in nuclear physics in which nucleons are represented as pairs, each of them acting as a boson particle, with integral spin of 0, 2 or 4. There are several branches of this model - in one of them (IBM-1) one can group all types of nucleons in pairs, in others (for instance - IBM-2) one considers protons and neutrons in pairs separately. See also: Interacting boson model.

5.2.6   Spontaneous breaking of symmetry in nuclear physics

One of the focal points of all physics is symmetry. The nucleon–nucleon interaction and all effective interactions used in practice have certain symmetries. They are invariant by translation (changing the frame of reference so that directions are not altered), by rotation (turning the frame of reference around some axis), or parity (changing the sense of axes) in the sense that the interaction does not change under any of these operations. Nevertheless, in the Hartree–Fock approach, solutions which are not invariant under such a symmetry can appear. One speaks then of spontaneous symmetry breaking.

Qualitatively, these spontaneous symmetry breakings can be explained in the following way : in the mean field theory, the nucleus is described as a set of independent particles. Most additional correlations among nucleons which do not enter the mean field are neglected. They can appear however by a breaking of the symmetry of the mean field Hamiltonian, which is only approximate. If the density used to start the iterations of the Hartree–Fock process breaks certain symmetries, the final Hartree–Fock Hamiltonian may break these symmetries, if it is advantageous to keep these broken from the point of view of the total energy.

It may also converge towards a symmetric solution. In any case, if the final solution breaks the symmetry, for example, the rotational symmetry, so that the nucleus appears not to be spherical, but elliptic, all configurations deduced from this deformed nucleus by a rotation are just as good solutions for the Hartree–Fock problem. The ground state of the nucleus is then *degenerate*.

A similar phenomenon happens with the nuclear pairing, which violates the conservation of the number of baryons (see below).

5.3   Extensions of the mean field theories

5.3.1   Nuclear pairing phenomenon

The most common extension to mean field theory is the nuclear pairing. Nuclei with an even number of nucleons are systematically more bound than those with an odd one. This implies that each nucleon binds with another one to form a pair, consequently the system cannot be described as independent particles subjected to a common a mean field. When the nucleus has an even number of protons and neutrons, each one of them finds a partner. To excite such a system, one must at least use such an energy as to break a pair. Conversely, in the case of odd number of protons or neutrons, there exists an unpaired nucleon, which needs less energy to be excited.

This phenomenon is closely analogous to that of Type 1 superconductivity in solid state physics. The first theoretical description of nuclear pairing was proposed at the end of the 1950s by Aage Bohr, Ben Mottelson, and David Pines (which contributed to the reception of the Nobel Prize in Physics in 1975 by Bohr and Mottelson).[8] It was close to the BCS theory of Bardeen, Cooper and Schrieffer, which accounts for metal superconductivity. Theoretically, the pairing phenomenon as described by the BCS theory combines with the mean field theory: nucleons are both subject to the mean field potential and to the pairing interaction.

Hartree–Fock–Bogolyubov (HFB) approach is a more sophisticated approach,[9] that enable to consider the pairing and mean field interactions consistently on equal footing, and is now the de facto standard in the mean field treatment of nuclear systems.

5.3.2 Symmetry Restoration

Peculiarity of mean field methods is the calculation of nuclear property by explicit symmetry breaking. The calculation of the mean field with self-consistent methods (e.g. Hartree-Fock), breaks rotational symmetry, and the calculation of pairing property breaks particle-number.

Several techniques for symmetry restoration by projecting on good quantum numbers have been developed.[10]

5.3.3 Particle Vibrations Coupling

Mean field methods (eventually considering symmetry restoration) are a good approximation for the ground state of the system, even postulating a system of independent particles. Higher-order corrections consider the fact that the particles interact together by the means of correlation. These correlations can be introduced taking into account the coupling of independent particle degrees of freedom, low-energy collective excitation of systems with even number of protons and neutrons.

In these way excited states can be reproduced by the means of Random Phase Approximation (RPA), and eventually consistently calculating also corrections to the ground state (e.g. by the means of Nuclear Field Theory [6]).

5.4   See also

Nuclear magnetic moment

5.5   References

[1] von Weizsäcker, C. F. (1935). "Zur Theorie der Kernmassen". *Zeitschrift für Physik* (in German) **96** (7–8): 431–458. Bibcode:1935ZPhy...96..431W. doi:10.1007/BF01337700.

[2] Moeller, P.; Myers, W. D.; Swiatecki, W. J.; Treiner, J. (3 Sep 1984). "Finite Range Droplet Model". *Conference: 7. international conference on atomic masses and fundamental constants (AMCO-7), Darmstadt-Seeheim, F.R. Germany.*

[3] Sorlin, O.; Porquet, M.-G. (2008). "Nuclear magic numbers: New features far from stability". *Progress in Particle and Nuclear Physics* **61** (2): 602–673. arXiv:0805.2561. Bibcode:2008PrPNP..61..602S. doi:10.1016/j.ppnp.2008.05.001.

[4] Brink, David; Broglia, Ricardo A. (2005). *Nuclear Superfluidity.* Cambridge University Press.

[5] Ring, P.; Schuck, P. (1980). *The nuclear many-body problem.* Springer Verlag. ISBN 3-540-21206-X.

[6] http://arxiv.org/abs/1504.05335

[7] http://www.sciencedirect.com/science/article/pii/0375947475903383

[8] Broglia, Ricardo A.; Zelevinsky, Vladimir (2013). *Fifty Years of Nuclear BCS: Pairing in Finite Systems*. World Scientific. ISBN 978-981-4412-48-3.

[9] http://www.fuw.edu.pl/~{}dobaczew/hfbtho16w/node2.html

[10] B. F. Bayman, Nucl. Phys. 15, 33 (1960)

See also: Nuclear physics § References

5.5.1 General audience

- James M. Cork ; *Radioactivité & physique nucléaire*, Dunod (1949).

5.5.2 Introductory texts

- Luc Valentin ; *Le monde subatomique - Des quarks aux centrales nucléaires*, Hermann (1986).

- Luc Valentin ; *Noyaux et particules - Modèles et symétries*, Hermann (1997).

- David Halliday ; *Introductory Nuclear Physics*, Wiley & Sons (1957).

- Kenneth Krane ; *Introductory Nuclear Physics*, Wiley & Sons (1987).

- Carlos Bertulani ; *Nuclear Physics in a Nutshell*, Princeton University Press (2007).

5.5.3 Fundamental texts

- Peter E. Hodgson; *Nuclear Reactions and Nuclear Structure*. Oxford University Press (1971).

- Irving Kaplan; *Nuclear physics*, the Addison-Wesley Series in Nuclear Science & Engineering, Addison-Wesley (1956). 2nd edition (1962).

- A. Bohr & B. Mottelson ; *Nuclear Structure*, 2 vol., Benjamin (1969–1975). Volume 1 : *Single Particle Motion* ; Volume 2 : *Nuclear Deformations*. Réédité par World Scientific Publishing Company (1998), ISBN 981-02-3197-0.

- P. Ring & P. Schuck; *The nuclear many-body problem*, Springer Verlag (1980), ISBN 3-540-21206-X

- A. de Shalit & H. Feshbach; *Theoretical Nuclear Physics*, 2 vol., John Wiley & Sons (1974). Volume 1: *Nuclear Structure*; Volume 2: *Nuclear Reactions*, ISBN 0-471-20385-8

5.6 External links

English

- (English) Institut de Physique Nucléaire (IPN), France

- (English) Facility for Antiproton and Ion Research (FAIR), Germany

- (English) Gesellschaft für Schwerionenforschung (GSI), Germany

- (English) Joint Institute for Nuclear Research (JINR), Russia

- (English) Argonne National Laboratory (ANL), USA

- (English) Riken, Japan

- (English) National Superconducting Cyclotron Laboratory, Michigan State University, USA

- (English) Facility for Rare Isotope Beams, Michigan State University, USA

French

- (French) Institut de Physique Nucléaire (IPN), France

- (French) Centre de Spectrométrie Nucléaire et de Spectrométrie de Masse (CSNSM), France

- (French) Service de Physique Nucléaire CEA/DAM, France

- (French) Institut National de Physique Nucléaire et de Physique des Particules (In2p3), France

- (French) Grand Accélérateur National d'Ions Lourds (GANIL), France

- (French) Commissariat à l'Energie Atomique (CEA), France

- (French) Centre Européen de Recherches Nucléaires, Suisse

-  **The LIVEChart of Nuclides - IAEA**

See also: Nuclear physics § External links

Chapter 6

Nuclear reaction

In nuclear physics and nuclear chemistry, a **nuclear reaction** is semantically considered to be the process in which two nuclei, or else a nucleus of an atom and a subatomic particle (such as a proton, neutron, or high energy electron) from outside the atom, collide to produce one or more nuclides that are different from the nuclide(s) that began the process. Thus, a nuclear reaction must cause a transformation of at least one nuclide to another. If a nucleus interacts with another nucleus or particle and they then separate without changing the nature of any nuclide, the process is simply referred to as a type of nuclear scattering, rather than a nuclear reaction.

In principle, a reaction can involve more than two particles colliding, but because the probability of three or more nuclei to meet at the same time at the same place is much less than for two nuclei, such an event is exceptionally rare (see triple alpha process for an example very close to a three-body nuclear reaction). "Nuclear reaction" is a term implying an **induced** change in a nuclide, and thus it does not apply to any type of radioactive decay (which by definition is a spontaneous process).

Natural nuclear reactions occur in the interaction between cosmic rays and matter, and nuclear reactions can be employed artificially to obtain nuclear energy, at an adjustable rate, on demand. Perhaps the most notable nuclear reactions are the nuclear chain reactions in fissionable materials that produces induced nuclear fission, and the various nuclear fusion reactions of light elements that power the energy production of the Sun and stars. Both of these types of reactions are employed in nuclear weapons.

6.1   Notation

Nuclear reactions may be shown in a form similar to chemical equations, for which invariant mass must balance for each side of the equation, and in which transformations of particles must follow certain conservation laws, such as conservation of charge and baryon number (total atomic mass number). An example of this notation follows:

To balance the equation above for mass, charge and mass number, the second nucleus to the right must have atomic number 2 and mass number 4; it is therefore also helium-4. The complete equation therefore reads:

or more simply:

Instead of using the full equations in the style above, in many situations a compact notation is used to describe nuclear reactions. This style of the form A(b,c)D is equivalent to A + b producing c + D. Common light particles are often

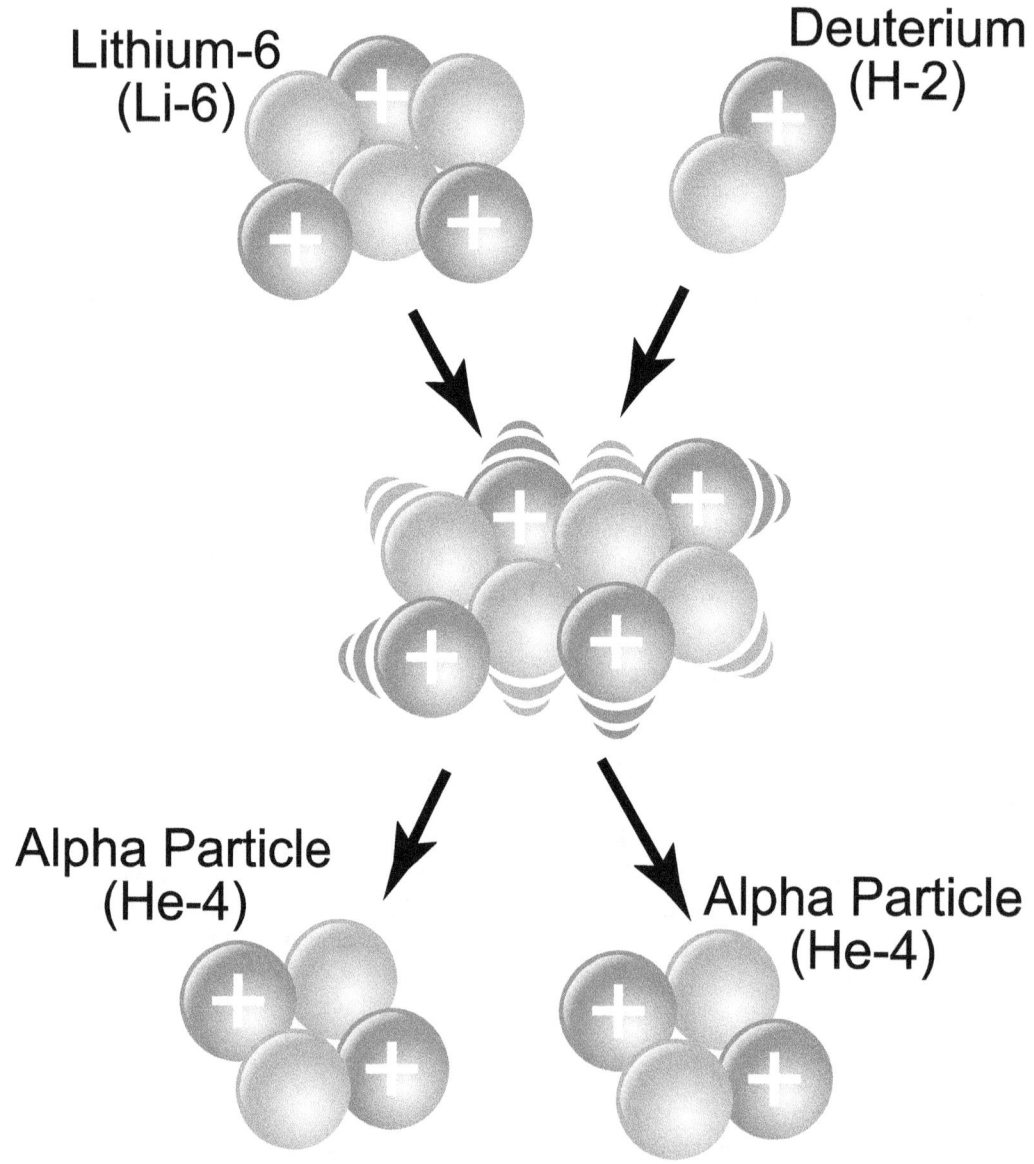

Lithium-6 – Deuterium Reaction

*In this symbolic representing of a nuclear reaction, lithium-6 (6
3Li) and deuterium (2
1H) react to form the highly excited intermediate nucleus 8
4Be which then decays immediately into two alpha particles of helium-4 (4
2He). Protons are symbolically represented by red spheres, and neutrons by blue spheres.*

abbreviated in this shorthand, typically p for proton, n for neutron, d for deuteron, α representing an alpha particle or helium-4, β for beta particle or electron, γ for gamma photon, etc. The reaction above would be written as Li-6(d,α)α.[1][2]

6.2 History

In 1917, Ernest Rutherford was able to accomplish transmutation of nitrogen into oxygen at the University of Manchester, using alpha particles directed at nitrogen $^{14}N + \alpha \rightarrow {}^{17}O + p$. This was the first observation of an induced nuclear reaction, that is, a reaction in which particles from one decay are used to transform another atomic nucleus. Eventually, in 1932 at Cambridge University, a fully artificial nuclear reaction and nuclear transmutation was achieved by Rutherford's colleagues John Cockcroft and Ernest Walton, who used artificially accelerated protons against lithium-7, to split the nucleus into two alpha particles. The feat was popularly known as "splitting the atom", although it was not the modern nuclear fission reaction later discovered in heavy elements, in 1938 by the German scientists Otto Hahn and Fritz Straßmann.[3]

6.3 Energy conservation

Kinetic energy may be released during the course of a reaction (exothermic reaction) or kinetic energy may have to be supplied for the reaction to take place (endothermic reaction). This can be calculated by reference to a table of very accurate particle rest masses,[4] as follows: according to the reference tables, the 6
3Li nucleus has a relative atomic mass of 6.015 atomic mass units (abbreviated u), the deuterium has 2.014 u, and the helium-4 nucleus has 4.0026 u Thus:

- Total rest mass on left side = 6.015 + 2.014 = 8.029 u

- Total rest mass on right side = 2 × 4.0026 = 8.0052 u

- Missing rest mass = 8.029 − 8.0052 = 0.0238 atomic mass units.

In a nuclear reaction, the total (relativistic) energy is conserved. The "missing" rest mass must therefore reappear as kinetic energy released in the reaction; its source is the nuclear binding energy. Using Einstein's mass-energy equivalence formula $E = mc^2$, the amount of energy released can be determined. We first need the energy equivalent of one atomic mass unit:

$$1 \text{ u } c^2 = (1.66054 \times 10^{-27} \text{ kg}) \times (2.99792 \times 10^8 \text{ m/s})^2$$
$$= 1.49242 \times 10^{-10} \text{ kg (m/s)}^2 = 1.49242 \times 10^{-10} \text{ J (Joule)}$$
$$\times (1 \text{ MeV} / 1.60218 \times 10^{-13} \text{ J})$$
$$= 931.49 \text{ MeV},$$

so $1 \text{ u } c^2 = 931.49$ MeV.

Hence, the energy released is 0.0238 × 931 MeV = 22.2 MeV.

Expressed differently: the mass is reduced by 0.3%, corresponding to 0.3% of 90 PJ/kg is 270 TJ/kg.

This is a large amount of energy for a nuclear reaction; the amount is so high because the binding energy per nucleon of the helium-4 nucleus is unusually high, because the He-4 nucleus is "doubly magic". (The He-4 nucleus is unusually stable and tightly bound for the same reason that the helium atom is inert: each pair of protons and neutrons in He-4 occupies a filled **1s** nuclear orbital in the same way that the pair of electrons in the helium atom occupy a filled **1s** electron orbital). Consequently, alpha particles appear frequently on the right hand side of nuclear reactions.

The energy released in a nuclear reaction can appear mainly in one of three ways:

- kinetic energy of the product particles

- emission of very high energy photons, called gamma rays

- some energy may remain in the nucleus, as a metastable energy level.

When the product nucleus is metastable, this is indicated by placing an asterisk ("*") next to its atomic number. This energy is eventually released through nuclear decay.

A small amount of energy may also emerge in the form of X-rays. Generally, the product nucleus has a different atomic number, and thus the configuration of its electron shells is wrong. As the electrons rearrange themselves and drop to lower energy levels, internal transition X-rays (X-rays with precisely defined emission lines) may be emitted.

6.4   Q-value and energy balance

In writing down the reaction equation, in a way analogous to a chemical equation, one may in addition give the reaction energy on the right side:

$$\text{Target nucleus} + \text{projectile} \rightarrow \text{Final nucleus} + \text{ejectile} + Q.$$

For the particular case discussed above, the reaction energy has already been calculated as $Q = 22.2$ MeV. Hence:

The reaction energy (the "Q-value") is positive for exothermal reactions and negative for endothermal reactions. On the one hand, it is the difference between the sums of kinetic energies on the final side and on the initial side. But on the other hand, it is also the difference between the nuclear rest masses on the initial side and on the final side (in this way, we have calculated the Q-value above).

6.5   Reaction rates

If the reaction equation is balanced, that does not mean that the reaction really occurs. The rate at which reactions occur depends on the particle energy, the particle flux and the reaction cross section. An example of a large repository of reaction rates is the REACLIB database, as maintained by the Joint Institute for Nuclear Astrophysics.

6.6   Neutrons vs. ions

In the initial collision which begins the reaction, the particles must approach closely enough so that the short range strong force can affect them. As most common nuclear particles are positively charged, this means they must overcome considerable electrostatic repulsion before the reaction can begin. Even if the target nucleus is part of a neutral atom, the other particle must penetrate well beyond the electron cloud and closely approach the nucleus, which is positively charged. Thus, such particles must be first accelerated to high energy, for example by:

- particle accelerators

- nuclear decay (alpha particles are the main type of interest here, since beta and gamma rays are rarely involved in nuclear reactions)

- very high temperatures, on the order of millions of degrees, producing thermonuclear reactions

- cosmic rays

Also, since the force of repulsion is proportional to the product of the two charges, reactions between heavy nuclei are rarer, and require higher initiating energy, than those between a heavy and light nucleus; while reactions between two light nuclei are the most common ones.

Neutrons, on the other hand, have no electric charge to cause repulsion, and are able to initiate a nuclear reaction at very low energies. In fact, at extremely low particle energies (corresponding, say, to thermal equilibrium at room temperature), the neutron's de Broglie wavelength is greatly increased, possibly greatly increasing its capture cross section, at energies close to resonances of the nuclei involved. Thus low energy neutrons *may* be even more reactive than high energy neutrons.

6.7 Notable types

While the number of possible nuclear reactions is immense, there are several types which are more common, or otherwise notable. Some examples include:

- Fusion reactions — two light nuclei join to form a heavier one, with additional particles (usually protons or neutrons) thrown off to conserve momentum.

- Spallation — a nucleus is hit by a particle with sufficient energy and momentum to knock out several small fragments or smash it into many fragments.

- Induced gamma emission belongs to a class in which only photons were involved in creating and destroying states of nuclear excitation.

- Alpha decay - Though driven by the same underlying forces as spontaneous fission, α decay is usually considered to be separate from the latter. The often-quoted idea that "nuclear reactions" are confined to induced processes is incorrect. "Radioactive decays" are a subgroup of "nuclear reactions" that are spontaneous rather than induced. For example, so-called "hot alpha particles" with unusually high energies may actually be produced in induced ternary fission, which is an induced nuclear reaction (contrasting with spontaneous fission). Such alphas occur from spontaneous ternary fission as well.

- Neutron-induced nuclear fission reactions – a very heavy nucleus, spontaneously or after absorbing additional light particles (usually neutrons), splits into two or sometimes three pieces. This is an induced nuclear reaction. Spontaneous fission, which occurs without assistance of the neutron, is usually not considered a nuclear reaction. At most, it is not an *induced* nuclear reaction.

6.7.1 Direct reactions

An intermediate energy projectile transfers energy or picks up or loses nucleons to the nucleus in a single quick (10^{-21} second) event. Energy and momentum transfer are relatively small. These are particularly useful in experimental nuclear physics, because the reaction mechanisms are often simple enough to calculate with sufficient accuracy to probe the structure of the target nucleus.

Inelastic scattering

Main article: Inelastic scattering

Only energy and momentum are transferred.

- (p,p') tests differences between nuclear states

- (α,α') measures nuclear surface shapes and sizes. Since α particles that hit the nucleus react more violently, elastic and shallow inelastic α scattering are sensitive to the shapes and sizes of the targets, like light scattered from a small black object.

- (e,e') is useful for probing the interior structure. Since electrons interact less strongly than do protons and neutrons, they reach to the centers of the targets and their wave functions are less distorted by passing through the nucleus.

Transfer reactions

Usually at moderately low energy, one or more nucleons are transferred between the projectile and target. These are useful in studying outer shell structure of nuclei.

- (α,n) and (α,p) reactions. Some of the earliest nuclear reactions studied involved an alpha particle produced by alpha decay, knocking a nucleon from a target nucleus.

- (d,n) and (d,p) reactions. A deuteron beam impinges on a target; the target nuclei absorb either the neutron or proton from the deuteron. The deuteron is so loosely bound that this is almost the same as proton or neutron capture. A compound nucleus may be formed, leading to additional neutrons being emitted more slowly. (d,n) reactions are used to generate energetic neutrons.

- The strangeness exchange reaction (K, π) has been used to study hypernuclei.

- The reaction ^{14}N(α,p)^{17}O performed by Rutherford in 1917 (reported 1919), is generally regarded as the first nuclear transmutation experiment.

Reactions with neutrons

Reactions with neutrons are important in nuclear reactors and nuclear weapons. While the best known neutron reactions are neutron scattering, neutron capture, and nuclear fission, for some light nuclei (especially odd-odd nuclei) the most probable reaction with a thermal neutron is a transfer reaction:

Some reactions are only possible with fast neutrons:

- (n,2n) reactions produce small amounts of protactinium-231 and uranium-232 in the thorium cycle which is otherwise relatively free of highly radioactive actinide products.

- ^{9}Be + n $\rightarrow$ 2α + 2n can contribute some additional neutrons in the beryllium neutron reflector of a nuclear weapon.

- ^{7}Li + n $\rightarrow$ T + α + n unexpectedly contributed additional yield in Castle Bravo, Castle Romeo, and Castle Yankee, the three highest-yield nuclear tests conducted by the U.S.

6.7.2 Compound nuclear reactions

Either a low energy projectile is absorbed or a higher energy particle transfers energy to the nucleus, leaving it with too much energy to be fully bound together. On a time scale of about 10^{-19} seconds, particles, usually neutrons, are "boiled" off. That is, it remains together until enough energy happens to be concentrated in one neutron to escape the mutual attraction. Charged particles rarely boil off because of the coulomb barrier. The excited quasi-bound nucleus is called a **compound nucleus**.

- Low energy (e, e' xn), (γ, xn) (the xn indicating one or more neutrons), where the gamma or virtual gamma energy is near the giant dipole resonance. These increase the need for radiation shielding around electron accelerators.

Further information: Spallation § Nuclear spallation

6.8   See also

- nuclear chain reaction

- Oppenheimer-Phillips process

- atomic nucleus

- atomic number

- atomic mass

- carbon-nitrogen cycle

- acoplanarity

6.9 References

[1] The Astrophysics Spectator: Hydrogen Fusion Rates in Stars

[2] R. J. D. Tilley Understanding solids: the science of materials, John Wiley and Sons, 2004, ISBN 0-470-85275-5, p. 495

[3] Cockcroft and Walton split lithium with high energy protons April 1932.

[4] a table of atomic masses

- Direct Reactions

- Compound Nucleus Reactions

6.10 Sources

- M.G. Bowler, Nuclear Physics, Pergamon Press 1973. ISBN 0-08-016983-X

Chapter 7

Nuclear fission

"Splitting the atom" redirects here. For the EP, see Splitting the Atom.

In nuclear physics and nuclear chemistry, **nuclear fission** is either a nuclear reaction or a radioactive decay process in which the nucleus of an atom splits into smaller parts (lighter nuclei). The fission process often produces free neutrons and photons (in the form of gamma rays), and releases a very large amount of energy even by the energetic standards of radioactive decay.

Nuclear fission of heavy elements was discovered on December 17, 1938 by German Otto Hahn and his assistant Fritz Strassmann, and explained theoretically in January 1939 by Lise Meitner and her nephew Otto Robert Frisch. Frisch named the process by analogy with biological fission of living cells. It is an exothermic reaction which can release large amounts of energy both as electromagnetic radiation and as kinetic energy of the fragments (heating the bulk material where fission takes place). In order for fission to produce energy, the total binding energy of the resulting elements must be less negative (higher energy) than that of the starting element.

Fission is a form of nuclear transmutation because the resulting fragments are not the same element as the original atom. The two nuclei produced are most often of comparable but slightly different sizes, typically with a mass ratio of products of about 3 to 2, for common fissile isotopes.[1][2] Most fissions are binary fissions (producing two charged fragments), but occasionally (2 to 4 times per 1000 events), *three* positively charged fragments are produced, in a ternary fission. The smallest of these fragments in ternary processes ranges in size from a proton to an argon nucleus.

Apart from fission induced by a neutron, harnessed and exploited by humans, a natural form of spontaneous radioactive decay (not requiring a neutron) is also referred to as fission, and occurs especially in very high-mass-number isotopes. Spontaneous fission was discovered in 1940 by Flyorov, Petrzhak and Kurchatov[3] in Moscow, when they decided to confirm that, without bombardment by neutrons, the fission rate of uranium was indeed negligible, as predicted by Niels Bohr; it wasn't.[3]

The unpredictable composition of the products (which vary in a broad probabilistic and somewhat chaotic manner) distinguishes fission from purely quantum-tunnelling processes such as proton emission, alpha decay and cluster decay, which give the same products each time. Nuclear fission produces energy for nuclear power and drives the explosion of nuclear weapons. Both uses are possible because certain substances called nuclear fuels undergo fission when struck by fission neutrons, and in turn emit neutrons when they break apart. This makes possible a self-sustaining nuclear chain reaction that releases energy at a controlled rate in a nuclear reactor or at a very rapid uncontrolled rate in a nuclear weapon.

The amount of free energy contained in nuclear fuel is millions of times the amount of free energy contained in a similar mass of chemical fuel such as gasoline, making nuclear fission a very dense source of energy. The products of nuclear fission, however, are on average far more radioactive than the heavy elements which are normally fissioned as fuel, and remain so for significant amounts of time, giving rise to a nuclear waste problem. Concerns over nuclear waste accumulation and over the destructive potential of nuclear weapons may counterbalance the desirable qualities of fission as an energy source, and give rise to ongoing political debate over nuclear power.

7.1 Physical overview

7.1.1 Mechanism

Nuclear fission can occur without neutron bombardment as a type of radioactive decay. This type of fission (called spontaneous fission) is rare except in a few heavy isotopes. In engineered nuclear devices, essentially all nuclear fission occurs as a "nuclear reaction" — a bombardment-driven process that results from the collision of two subatomic particles. In nuclear reactions, a subatomic particle collides with an atomic nucleus and causes changes to it. Nuclear reactions are thus driven by the mechanics of bombardment, not by the relatively constant exponential decay and half-life characteristic of spontaneous radioactive processes.

Many types of nuclear reactions are currently known. Nuclear fission differs importantly from other types of nuclear reactions, in that it can be amplified and sometimes controlled via a nuclear chain reaction (one type of general chain reaction). In such a reaction, free neutrons released by each fission event can trigger yet more events, which in turn release more neutrons and cause more fissions.

The chemical element isotopes that can sustain a fission chain reaction are called nuclear fuels, and are said to be *fissile*. The most common nuclear fuels are ^{235}U (the isotope of uranium with an atomic mass of 235 and of use in nuclear reactors) and ^{239}Pu (the isotope of plutonium with an atomic mass of 239). These fuels break apart into a bimodal range of chemical elements with atomic masses centering near 95 and 135 **u** (fission products). Most nuclear fuels undergo spontaneous fission only very slowly, decaying instead mainly via an alpha/beta decay chain over periods of millennia to eons. In a nuclear reactor or nuclear weapon, the overwhelming majority of fission events are induced by bombardment with another particle, a neutron, which is itself produced by prior fission events.

Nuclear fissions in fissile fuels are the result of the nuclear excitation energy produced when a fissile nucleus captures a neutron. This energy, resulting from the neutron capture, is a result of the attractive nuclear force acting between the neutron and nucleus. It is enough to deform the nucleus into a double-lobed "drop," to the point that nuclear fragments exceed the distances at which the nuclear force can hold two groups of charged nucleons together, and when this happens, the two fragments complete their separation and then are driven further apart by their mutually repulsive charges, in a process which becomes irreversible with greater and greater distance. A similar process occurs in fissionable isotopes (such as uranium-238), but in order to fission, these isotopes require additional energy provided by fast neutrons (such as those produced by nuclear fusion in thermonuclear weapons).

The liquid drop model of the atomic nucleus predicts equal-sized fission products as an outcome of nuclear deformation. The more sophisticated nuclear shell model is needed to mechanistically explain the route to the more energetically favorable outcome, in which one fission product is slightly smaller than the other. A theory of the fission based on shell model has been formulated by Maria Goeppert Mayer.

The most common fission process is binary fission, and it produces the fission products noted above, at 95±15 and 135±15 **u**. However, the binary process happens merely because it is the most probable. In anywhere from 2 to 4 fissions per 1000 in a nuclear reactor, a process called ternary fission produces three positively charged fragments (plus neutrons) and the smallest of these may range from so small a charge and mass as a proton (Z=1), to as large a fragment as argon (Z=18). The most common small fragments, however, are composed of 90% helium-4 nuclei with more energy than alpha particles from alpha decay (so-called "long range alphas" at ~ 16 MeV), plus helium-6 nuclei, and tritons (the nuclei of tritium). The ternary process is less common, but still ends up producing significant helium-4 and tritium gas buildup in the fuel rods of modern nuclear reactors.[4]

7.1.2 Energetics

Input

The fission of a heavy nucleus requires a total input energy of about 7 to 8 million electron volts (MeV) to initially overcome the nuclear force which holds the nucleus into a spherical or nearly spherical shape, and from there, deform it into a two-lobed ("peanut") shape in which the lobes are able to continue to separate from each other, pushed by their mutual positive charge, in the most common process of binary fission (two positively charged fission products + neutrons). Once the nuclear lobes have been pushed to a critical distance, beyond which the short range strong force can no longer hold them together, the process of their separation proceeds from the energy of the (longer range) electromagnetic repulsion between the fragments. The result is two fission fragments moving away from each other, at high energy.

About 6 MeV of the fission-input energy is supplied by the simple binding of an extra neutron to the heavy nucleus via the strong force; however, in many fissionable isotopes, this amount of energy is not enough for fission. Uranium-238, for example, has a near-zero fission cross section for neutrons of less than one MeV energy. If no additional energy is supplied by any other mechanism, the nucleus will not fission, but will merely absorb the neutron, as happens when U-238 absorbs slow and even some fraction of fast neutrons, to become U-239. The remaining energy to initiate fission can be supplied by two other mechanisms: one of these is more kinetic energy of the incoming neutron, which is increasingly able to fission a fissionable heavy nucleus as it exceeds a kinetic energy of one MeV or more (so-called fast neutrons). Such high energy neutrons are able to fission U-238 directly (see thermonuclear weapon for application, where the fast neutrons are supplied by nuclear fusion). However, this process cannot happen to a great extent in a nuclear reactor, as too small a fraction of the fission neutrons produced by any type of fission have enough energy to efficiently fission U-238 (fission neutrons have a mode energy of 2 MeV, but a median of only 0.75 MeV, meaning half of them have less than this insufficient energy).[5]

Among the heavy actinide elements, however, those isotopes that have an odd number of neutrons (such as U-235 with 143 neutrons) bind an extra neutron with an additional 1 to 2 MeV of energy over an isotope of the same element with an even number of neutrons (such as U-238 with 146 neutrons). This extra binding energy is made available as a result of the mechanism of neutron pairing effects. This extra energy results from the Pauli exclusion principle allowing an extra neutron to occupy the same nuclear orbital as the last neutron in the nucleus, so that the two form a pair. In such isotopes, therefore, no neutron kinetic energy is needed, for all the necessary energy is supplied by absorption of any neutron, either of the slow or fast variety (the former are used in moderated nuclear reactors, and the latter are used in fast neutron reactors, and in weapons). As noted above, the subgroup of fissionable elements that may be fissioned efficiently with their own fission neutrons (thus potentially causing a nuclear chain reaction in relatively small amounts of the pure material) are termed "fissile." Examples of fissile isotopes are U-235 and plutonium-239.

Output

Typical fission events release about two hundred million eV (200 MeV) of energy for each fission event. The exact isotope which is fissioned, and whether or not it is fissionable or fissile, has only a small impact on the amount of energy released. This can be easily seen by examining the curve of binding energy (image below), and noting that the average binding energy of the actinide nuclides beginning with uranium is around 7.6 MeV per nucleon. Looking further left on the curve of binding energy, where the fission products cluster, it is easily observed that the binding energy of the fission products tends to center around 8.5 MeV per nucleon. Thus, in any fission event of an isotope in the actinide's range of mass, roughly 0.9 MeV is released per nucleon of the starting element. The fission of U235 by a slow neutron yields nearly identical energy to the fission of U238 by a fast neutron. This energy release profile holds true for thorium and the various minor actinides as well.[6]

By contrast, most chemical oxidation reactions (such as burning coal or TNT) release at most a few eV per event. So, nuclear fuel contains at least ten million times more usable energy per unit mass than does chemical fuel. The energy of nuclear fission is released as kinetic energy of the fission products and fragments, and as electromagnetic radiation in the form of gamma rays; in a nuclear reactor, the energy is converted to heat as the particles and gamma rays collide with the atoms that make up the reactor and its working fluid, usually water or occasionally heavy water or molten salts.

When a uranium nucleus fissions into two daughter nuclei fragments, about 0.1 percent of the mass of the uranium nucleus[7] appears as the fission energy of ~200 MeV. For uranium-235 (total mean fission energy 202.5 MeV), typically ~169 MeV appears as the kinetic energy of the daughter nuclei, which fly apart at about 3% of the speed of light, due to Coulomb repulsion. Also, an average of 2.5 neutrons are emitted, with a mean kinetic energy per neutron of ~2 MeV (total of 4.8 MeV).[8] The fission reaction also releases ~7 MeV in prompt gamma ray photons. The latter figure means that a nuclear fission explosion or criticality accident emits about 3.5% of its energy as gamma rays, less than 2.5% of its energy as fast neutrons (total of both types of radiation ~ 6%), and the rest as kinetic energy of fission fragments (this appears almost immediately when the fragments impact surrounding matter, as simple heat). In an atomic bomb, this heat may serve to raise the temperature of the bomb core to 100 million kelvin and cause secondary emission of soft X-rays, which convert some of this energy to ionizing radiation. However, in nuclear reactors, the fission fragment kinetic energy remains as low-temperature heat, which itself causes little or no ionization.

So-called neutron bombs (enhanced radiation weapons) have been constructed which release a larger fraction of their energy as ionizing radiation (specifically, neutrons), but these are all thermonuclear devices which rely on the nuclear fusion stage to produce the extra radiation. The energy dynamics of pure fission bombs always remain at about 6% yield of the total in radiation, as a prompt result of fission.

The total *prompt fission* energy amounts to about 181 MeV, or ~ 89% of the total energy which is eventually released by fission over time. The remaining ~ 11% is released in beta decays which have various half-lives, but begin as a process in the fission products immediately; and in delayed gamma emissions associated with these beta decays. For example, in uranium-235 this delayed energy is divided into about 6.5 MeV in betas, 8.8 MeV in antineutrinos (released at the same time as the betas), and finally, an additional 6.3 MeV in delayed gamma emission from the excited beta-decay products (for a mean total of ~10 gamma ray emissions per fission, in all). Thus, about 6.5% of the total energy of fission is released some time after the event, as non-prompt or delayed ionizing radiation, and the delayed ionizing energy is about evenly divided between gamma and beta ray energy.

In a reactor that has been operating for some time, the radioactive fission products will have built up to steady state concentrations such that their rate of decay is equal to their rate of formation, so that their fractional total contribution to reactor heat (via beta decay) is the same as these radioisotopic fractional contributions to the energy of fission. Under these conditions, the 6.5% of fission which appears as delayed ionizing radiation (delayed gammas and betas from radioactive fission products) contributes to the steady-state reactor heat production under power. It is this output fraction which remains when the reactor is suddenly shut down (undergoes scram). For this reason, the reactor decay heat output begins at 6.5% of the full reactor steady state fission power, once the reactor is shut down. However, within hours, due to decay of these isotopes, the decay power output is far less. See decay heat for detail.

The remainder of the delayed energy (8.8 MeV/202.5 MeV = 4.3% of total fission energy) is emitted as antineutrinos, which as a practical matter, are not considered "ionizing radiation." The reason is that energy released as antineutrinos is not captured by the reactor material as heat, and escapes directly through all materials (including the Earth) at nearly the speed of light, and into interplanetary space (the amount absorbed is minuscule). Neutrino radiation is ordinarily not classed as ionizing radiation, because it is almost entirely not absorbed and therefore does not produce effects (although the very rare neutrino event is ionizing). Almost all of the rest of the radiation (6.5% delayed beta and gamma radiation) is eventually converted to heat in a reactor core or its shielding.

Some processes involving neutrons are notable for absorbing or finally yielding energy — for example neutron kinetic energy does not yield heat immediately if the neutron is captured by a uranium-238 atom to breed plutonium-239, but this energy is emitted if the plutonium-239 is later fissioned. On the other hand, so-called delayed neutrons emitted as radioactive decay products with half-lives up to several minutes, from fission-daughters, are very important to reactor control, because they give a characteristic "reaction" time for the total nuclear reaction to double in size, if the reaction is run in a "delayed-critical" zone which deliberately relies on these neutrons for a supercritical chain-reaction (one in which each fission cycle yields more neutrons than it absorbs). Without their existence, the nuclear chain-reaction would be prompt critical and increase in size faster than it could be controlled by human intervention. In this case, the first experimental atomic reactors would have run away to a dangerous and messy "prompt critical reaction" before their operators could have manually shut them down (for this reason, designer Enrico Fermi included radiation-counter-triggered control rods, suspended by electromagnets, which could automatically drop into the center of Chicago Pile-1). If these delayed neutrons are captured without producing fissions, they produce heat as well.[9]

7.1.3 Product nuclei and binding energy

Main articles: fission product and fission product yield

In fission there is a preference to yield fragments with even proton numbers, which is called the odd-even effect on the fragments charge distribution. However, no odd-even effect is observed on fragment **mass number** distribution. This result is attributed to nucleon pair breaking.

In nuclear fission events the nuclei may break into any combination of lighter nuclei, but the most common event is not fission to equal mass nuclei of about mass 120; the most common event (depending on isotope and process) is a slightly unequal fission in which one daughter nucleus has a mass of about 90 to 100 **u** and the other the remaining 130 to 140 **u**.[10] Unequal fissions are energetically more favorable because this allows one product to be closer to the energetic minimum near mass 60 **u** (only a quarter of the average fissionable mass), while the other nucleus with mass 135 **u** is still not far out of the range of the most tightly bound nuclei (another statement of this, is that the atomic binding energy curve is slightly steeper to the left of mass 120 **u** than to the right of it).

7.1.4   Origin of the active energy and the curve of binding energy

Nuclear fission of heavy elements produces energy because the specific binding energy (binding energy per mass) of intermediate-mass nuclei with atomic numbers and atomic masses close to ^{62}Ni and ^{56}Fe is greater than the nucleon-specific binding energy of very heavy nuclei, so that energy is released when heavy nuclei are broken apart. The total rest masses of the fission products (**Mp**) from a single reaction is less than the mass of the original fuel nucleus (**M**). The excess mass Δ**m** = **M** – **Mp** is the invariant mass of the energy that is released as photons (gamma rays) and kinetic energy of the fission fragments, according to the mass-energy equivalence formula $E = mc^2$.

The variation in specific binding energy with atomic number is due to the interplay of the two fundamental forces acting on the component nucleons (protons and neutrons) that make up the nucleus. Nuclei are bound by an attractive nuclear force between nucleons, which overcomes the electrostatic repulsion between protons. However, the nuclear force acts only over relatively short ranges (a few nucleon diameters), since it follows an exponentially decaying Yukawa potential which makes it insignificant at longer distances. The electrostatic repulsion is of longer range, since it decays by an inverse-square rule, so that nuclei larger than about 12 nucleons in diameter reach a point that the total electrostatic repulsion overcomes the nuclear force and causes them to be spontaneously unstable. For the same reason, larger nuclei (more than about eight nucleons in diameter) are less tightly bound per unit mass than are smaller nuclei; breaking a large nucleus into two or more intermediate-sized nuclei releases energy. The origin of this energy is the nuclear force, which intermediate-sized nuclei allows to act more efficiently, because each nucleon has more neighbors which are within the short range attraction of this force. Thus less energy is needed in the smaller nuclei and the difference to the state before is set free.

Also because of the short range of the strong binding force, large stable nuclei must contain proportionally more neutrons than do the lightest elements, which are most stable with a **1 to 1 ratio** of protons and neutrons. Nuclei which have more than 20 protons cannot be stable unless they have more than an equal number of neutrons. Extra neutrons stabilize heavy elements because they add to strong-force binding (which acts between all nucleons) without adding to proton–proton repulsion. Fission products have, on average, about the same ratio of neutrons and protons as their parent nucleus, and are therefore usually unstable to beta decay (which changes neutrons to protons) because they have proportionally too many neutrons compared to stable isotopes of similar mass.

This tendency for fission product nuclei to beta-decay is the fundamental cause of the problem of radioactive high level waste from nuclear reactors. Fission products tend to be beta emitters, emitting fast-moving electrons to conserve electric charge, as excess neutrons convert to protons in the fission-product atoms. See Fission products (by element) for a description of fission products sorted by element.

7.1.5   Chain reactions

Main article: Nuclear chain reaction

Several heavy elements, such as uranium, thorium, and plutonium, undergo both spontaneous fission, a form of radioactive decay and *induced fission*, a form of nuclear reaction. Elemental isotopes that undergo induced fission when struck by a free neutron are called fissionable; isotopes that undergo fission when struck by a thermal, slow moving neutron are also called fissile. A few particularly fissile and readily obtainable isotopes (notably ^{233}U, ^{235}U and ^{239}Pu) are called nuclear fuels because they can sustain a chain reaction and can be obtained in large enough quantities to be useful.

All fissionable and fissile isotopes undergo a small amount of spontaneous fission which releases a few free neutrons into any sample of nuclear fuel. Such neutrons would escape rapidly from the fuel and become a free neutron, with a mean lifetime of about 15 minutes before decaying to protons and beta particles. However, neutrons almost invariably impact and are absorbed by other nuclei in the vicinity long before this happens (newly created fission neutrons move at about 7% of the speed of light, and even moderated neutrons move at about 8 times the speed of sound). Some neutrons will impact fuel nuclei and induce further fissions, releasing yet more neutrons. If enough nuclear fuel is assembled in one place, or if the escaping neutrons are sufficiently contained, then these freshly emitted neutrons outnumber the neutrons that escape from the assembly, and a *sustained nuclear chain reaction* will take place.

An assembly that supports a sustained nuclear chain reaction is called a critical assembly or, if the assembly is almost entirely made of a nuclear fuel, a critical mass. The word "critical" refers to a cusp in the behavior of the differential equation that governs the number of free neutrons present in the fuel: if less than a critical mass is present, then the amount of neutrons is determined by radioactive decay, but if a critical mass or more is present, then the amount of neutrons is controlled instead by the physics of the chain reaction. The actual mass of a *critical mass* of nuclear fuel depends strongly on the geometry and surrounding materials.

Not all fissionable isotopes can sustain a chain reaction. For example, ^{238}U, the most abundant form of uranium, is fissionable but not fissile: it undergoes induced fission when impacted by an energetic neutron with over 1 MeV of kinetic energy. However, too few of the neutrons produced by ^{238}U fission are energetic enough to induce further fissions in ^{238}U, so no chain reaction is possible with this isotope. Instead, bombarding ^{238}U with slow neutrons causes it to absorb them (becoming ^{239}U) and decay by beta emission to ^{239}Np which then decays again by the same process to ^{239}Pu; that process is used to manufacture ^{239}Pu in breeder reactors. In-situ plutonium production also contributes to the neutron chain reaction in other types of reactors after sufficient plutonium-239 has been produced, since plutonium-239 is also a fissile element which serves as fuel. It is estimated that up to half of the power produced by a standard "non-breeder" reactor is produced by the fission of plutonium-239 produced in place, over the total life-cycle of a fuel load.

Fissionable, non-fissile isotopes can be used as fission energy source even without a chain reaction. Bombarding ^{238}U with fast neutrons induces fissions, releasing energy as long as the external neutron source is present. This is an important effect in all reactors where fast neutrons from the fissile isotope can cause the fission of nearby ^{238}U nuclei, which means that some small part of the ^{238}U is "burned-up" in all nuclear fuels, especially in fast breeder reactors that operate with higher-energy neutrons. That same fast-fission effect is used to augment the energy released by modern thermonuclear weapons, by jacketing the weapon with ^{238}U to react with neutrons released by nuclear fusion at the center of the device. But the explosive effects of nuclear fission chain reactions can be reduced by using substances like moderators which slow down the speed of secondary neutrons.[11]

7.1.6   Fission reactors

Critical fission reactors are the most common type of nuclear reactor. In a critical fission reactor, neutrons produced by fission of fuel atoms are used to induce yet more fissions, to sustain a controllable amount of energy release. Devices that produce engineered but non-self-sustaining fission reactions are subcritical fission reactors. Such devices use radioactive decay or particle accelerators to trigger fissions.

Critical fission reactors are built for three primary purposes, which typically involve different engineering trade-offs to take advantage of either the heat or the neutrons produced by the fission chain reaction:

- *power reactors* are intended to produce heat for nuclear power, either as part of a generating station or a local power system such as a nuclear submarine.

- *research reactors* are intended to produce neutrons and/or activate radioactive sources for scientific, medical, engineering, or other research purposes.

- *breeder reactors* are intended to produce nuclear fuels in bulk from more abundant isotopes. The better known fast breeder reactor makes ^{239}Pu (a nuclear fuel) from the naturally very abundant ^{238}U (not a nuclear fuel). Thermal breeder reactors previously tested using ^{232}Th to breed the fissile isotope ^{233}U (thorium fuel cycle) continue to be studied and developed.

While, in principle, all fission reactors can act in all three capacities, in practice the tasks lead to conflicting engineering goals and most reactors have been built with only one of the above tasks in mind. (There are several early counter-examples, such as the Hanford N reactor, now decommissioned). Power reactors generally convert the kinetic energy of fission products into heat, which is used to heat a working fluid and drive a heat engine that generates mechanical or electrical power. The working fluid is usually water with a steam turbine, but some designs use other materials such as gaseous helium. Research reactors produce neutrons that are used in various ways, with the heat of fission being treated as an unavoidable waste product. Breeder reactors are a specialized form of research reactor, with the caveat that the sample being irradiated is usually the fuel itself, a mixture of ^{238}U and ^{235}U. For a more detailed description of the physics and operating principles of critical fission reactors, see nuclear reactor physics. For a description of their social, political, and environmental aspects, see nuclear power.

7.1.7   Fission bombs

One class of nuclear weapon, a *fission bomb* (not to be confused with the *fusion bomb*), otherwise known as an *atomic bomb* or *atom bomb*, is a fission reactor designed to liberate as much energy as possible as rapidly as possible, before the released energy causes the reactor to explode (and the chain reaction to stop). Development of nuclear weapons was the motivation behind early research into nuclear fission: the Manhattan Project of the U.S. military during World War II carried out most of the early scientific work on fission chain reactions, culminating in the Trinity test bomb and the Little Boy and Fat Man bombs that were exploded over the cities Hiroshima, and Nagasaki, Japan in August 1945.

Even the first fission bombs were thousands of times more explosive than a comparable mass of chemical explosive. For example, Little Boy weighed a total of about four tons (of which 60 kg was nuclear fuel) and was 11 feet (3.4 m) long; it also yielded an explosion equivalent to about 15 kilotons of TNT, destroying a large part of the city of Hiroshima. Modern nuclear weapons (which include a thermonuclear *fusion* as well as one or more fission stages) are hundreds of times more energetic for their weight than the first pure fission atomic bombs (see nuclear weapon yield), so that a modern single missile warhead bomb weighing less than 1/8 as much as Little Boy (see for example W88) has a yield of 475,000 tons of TNT, and could bring destruction to about 10 times the city area.

While the fundamental physics of the fission chain reaction in a nuclear weapon is similar to the physics of a controlled nuclear reactor, the two types of device must be engineered quite differently (see nuclear reactor physics). A nuclear bomb is designed to release all its energy at once, while a reactor is designed to generate a steady supply of useful power. While overheating of a reactor can lead to, and has led to, meltdown and steam explosions, the much lower uranium enrichment makes it impossible for a nuclear reactor to explode with the same destructive power as a nuclear weapon. It is also difficult to extract useful power from a nuclear bomb, although at least one rocket propulsion system, Project Orion, was intended to work by exploding fission bombs behind a massively padded and shielded spacecraft.

The strategic importance of nuclear weapons is a major reason why the technology of nuclear fission is politically sensitive. Viable fission bomb designs are, arguably, within the capabilities of many being relatively simple from an engineering viewpoint. However, the difficulty of obtaining fissile nuclear material to realize the designs, is the key to the relative unavailability of nuclear weapons to all but modern industrialized governments with special programs to produce fissile materials (see uranium enrichment and nuclear fuel cycle).

7.2   History

7.2.1   Discovery of nuclear fission

The discovery of nuclear fission occurred in 1938 in the buildings of Kaiser Wilhelm Society for Chemistry, today part of the Free University of Berlin, following nearly five decades of work on the science of radioactivity and the elaboration of new nuclear physics that described the components of atoms. In 1911, Ernest Rutherford proposed a model of the atom in which a very small, dense and positively charged nucleus of protons (the neutron had not yet been discovered) was surrounded by orbiting, negatively charged electrons (the Rutherford model).[13] Niels Bohr improved upon this in 1913 by reconciling the quantum behavior of electrons (the Bohr model). Work by Henri Becquerel, Marie Curie, Pierre Curie, and Rutherford further elaborated that the nucleus, though tightly bound, could undergo different forms of radioactive decay, and thereby transmute into other elements. (For example, by alpha decay: the emission of an alpha particle—two protons and two neutrons bound together into a particle identical to a helium nucleus.)

Some work in nuclear transmutation had been done. In 1917, Rutherford was able to accomplish transmutation of nitrogen into oxygen, using alpha particles directed at nitrogen $^{14}N + \alpha \rightarrow {}^{17}O + p$. This was the first observation of a nuclear reaction, that is, a reaction in which particles from one decay are used to transform another atomic nucleus. Eventually, in 1932, a fully artificial nuclear reaction and nuclear transmutation was achieved by Rutherford's colleagues Ernest

Walton and John Cockcroft, who used artificially accelerated protons against lithium-7, to split this nucleus into two alpha particles. The feat was popularly known as "splitting the atom", although it was not the modern nuclear fission reaction later discovered in heavy elements, which is discussed below.[14] Meanwhile, the possibility of *combining* nuclei—nuclear fusion—had been studied in connection with understanding the processes which power stars. The first artificial fusion reaction had been achieved by Mark Oliphant in 1932, using two accelerated deuterium nuclei (each consisting of a single proton bound to a single neutron) to create a helium nucleus.[15]

After English physicist James Chadwick discovered the neutron in 1932,[16] Enrico Fermi and his colleagues in Rome studied the results of bombarding uranium with neutrons in 1934.[17] Fermi concluded that his experiments had created new elements with 93 and 94 protons, which the group dubbed ausonium and hesperium. However, not all were convinced by Fermi's analysis of his results. The German chemist Ida Noddack notably suggested in print in 1934 that instead of creating a new, heavier element 93, that "it is conceivable that the nucleus breaks up into several large fragments."[18][19] However, Noddack's conclusion was not pursued at the time.

After the Fermi publication, Otto Hahn, Lise Meitner, and Fritz Strassmann began performing similar experiments in Berlin. Meitner, an Austrian Jew, lost her citizenship with the "Anschluss", the occupation and annexation of Austria into Nazi Germany in March 1938, but she fled in July 1938 to Sweden and started a correspondence by mail with Hahn in Berlin. By coincidence, her nephew Otto Robert Frisch, also a refugee, was also in Sweden when Meitner received a letter from Hahn dated 19 December describing his chemical proof that some of the product of the bombardment of uranium with neutrons was barium. Hahn suggested a *bursting* of the nucleus, but he was unsure of what the physical basis for the results were. Barium had an atomic mass 40% less than uranium, and no previously known methods of radioactive decay could account for such a large difference in the mass of the nucleus. Frisch was skeptical, but Meitner trusted Hahn's ability as a chemist. Marie Curie had been separating barium from radium for many years, and the techniques were well-known. According to Frisch:

> Was it a mistake? No, said Lise Meitner; Hahn was too good a chemist for that. But how could barium be formed from uranium? No larger fragments than protons or helium nuclei (alpha particles) had ever been chipped away from nuclei, and to chip off a large number not nearly enough energy was available. Nor was it possible that the uranium nucleus could have been cleaved right across. A nucleus was not like a brittle solid that can be cleaved or broken; George Gamow had suggested early on, and Bohr had given good arguments that a nucleus was much more like a liquid drop. Perhaps a drop could divide itself into two smaller drops in a more gradual manner, by first becoming elongated, then constricted, and finally being torn rather than broken in two? We knew that there were strong forces that would resist such a process, just as the surface tension of an ordinary liquid drop tends to resist its division into two smaller ones. But nuclei differed from ordinary drops in one important way: they were electrically charged, and that was known to counteract the surface tension.

> The charge of a uranium nucleus, we found, was indeed large enough to overcome the effect of the surface tension almost completely; so the uranium nucleus might indeed resemble a very wobbly unstable drop, ready to divide itself at the slightest provocation, such as the impact of a single neutron. But there was another problem. After separation, the two drops would be driven apart by their mutual electric repulsion and would acquire high speed and hence a very large energy, about 200 MeV in all; where could that energy come from? ...Lise Meitner... worked out that the two nuclei formed by the division of a uranium nucleus together would be lighter than the original uranium nucleus by about one-fifth the mass of a proton. Now whenever mass disappears energy is created, according to Einstein's formula $E = mc^2$, and one-fifth of a proton mass was just equivalent to 200 MeV. So here was the source for that energy; it all fitted![20]

In short, Meitner and Frisch had correctly interpreted Hahn's results to mean that the nucleus of uranium had split roughly in half. Frisch suggested the process be named "nuclear fission," by analogy to the process of living cell division into two cells, which was then called binary fission. Just as the term nuclear "chain reaction" would later be borrowed from chemistry, so the term "fission" was borrowed from biology.

On 22 December 1938, Hahn and Strassmann sent a manuscript to *Naturwissenschaften* reporting that they had discovered the element barium after bombarding uranium with neutrons.[21] Simultaneously, they communicated these results to Meitner in Sweden. She and Frisch correctly interpreted the results as evidence of nuclear fission.[22] Frisch confirmed this experimentally on 13 January 1939.[23][24] For proving that the barium resulting from his bombardment of uranium with neutrons was the product of nuclear fission, Hahn was awarded the Nobel Prize for Chemistry in 1944 (the sole recipient) "for his discovery of the fission of heavy nuclei". (The award was actually given to Hahn in 1945, as "the Nobel Committee for Chemistry decided that none of the year's nominations met the criteria as outlined in the will of Alfred Nobel." In such cases, the Nobel Foundation's statutes permit that year's prize be reserved until the following year.)[25]

News spread quickly of the new discovery, which was correctly seen as an entirely novel physical effect with great scientific—and potentially practical—possibilities. Meitner's and Frisch's interpretation of the discovery of Hahn and Strassmann crossed the Atlantic Ocean with Niels Bohr, who was to lecture at Princeton University. I.I. Rabi and Willis Lamb, two Columbia University physicists working at Princeton, heard the news and carried it back to Columbia. Rabi said he told Enrico Fermi; Fermi gave credit to Lamb. Bohr soon thereafter went from Princeton to Columbia to see Fermi. Not finding Fermi in his office, Bohr went down to the cyclotron area and found Herbert L. Anderson. Bohr grabbed him by the shoulder and said: "Young man, let me explain to you about something new and exciting in physics."[26] It was clear to a number of scientists at Columbia that they should try to detect the energy released in the nuclear fission of uranium from neutron bombardment. On 25 January 1939, a Columbia University team conducted the first nuclear fission experiment in the United States,[27] which was done in the basement of Pupin Hall; the members of the team were Herbert L. Anderson, Eugene T. Booth, John R. Dunning, Enrico Fermi, G. Norris Glasoe, and Francis G. Slack. The experiment involved placing uranium oxide inside of an ionization chamber and irradiating it with neutrons, and measuring the energy thus released. The results confirmed that fission was occurring and hinted strongly that it was the isotope uranium 235 in particular that was fissioning. The next day, the Fifth Washington Conference on Theoretical Physics began in Washington, D.C. under the joint auspices of the George Washington University and the Carnegie Institution of Washington. There, the news on nuclear fission was spread even further, which fostered many more experimental demonstrations.[28]

During this period the Hungarian physicist Leó Szilárd, who was residing in the United States at the time, realized that the neutron-driven fission of heavy atoms could be used to create a nuclear chain reaction. Such a reaction using neutrons was an idea he had first formulated in 1933, upon reading Rutherford's disparaging remarks about generating power from his team's 1932 experiment using protons to split lithium. However, Szilárd had not been able to achieve a neutron-driven chain reaction with neutron-rich light atoms. In theory, if in a neutron-driven chain reaction the number of secondary neutrons produced was greater than one, then each such reaction could trigger multiple additional reactions, producing an exponentially increasing number of reactions. It was thus a possibility that the fission of uranium could yield vast amounts of energy for civilian or military purposes (i.e., electric power generation or atomic bombs).

Szilard now urged Fermi (in New York) and Frédéric Joliot-Curie (in Paris) to refrain from publishing on the possibility of a chain reaction, lest the Nazi government become aware of the possibilities on the eve of what would later be known as World War II. With some hesitation Fermi agreed to self-censor. But Joliot-Curie did not, and in April 1939 his team in Paris, including Hans von Halban and Lew Kowarski, reported in the journal *Nature* that the number of neutrons emitted with nuclear fission of ^{235}U was then reported at 3.5 per fission.[29] (They later corrected this to 2.6 per fission.) Simultaneous work by Szilard and Walter Zinn confirmed these results. The results suggested the possibility of building nuclear reactors (first called "neutronic reactors" by Szilard and Fermi) and even nuclear bombs. However, much was still unknown about fission and chain reaction systems.

7.2.2   Fission chain reaction realized

"Chain reactions" at that time were a known phenomenon in *chemistry*, but the analogous process in nuclear physics, using neutrons, had been foreseen as early as 1933 by Szilárd, although Szilárd at that time had no idea with what materials the process might be initiated. Szilárd considered that neutrons would be ideal for such a situation, since they lacked an electrostatic charge.

With the news of fission neutrons from uranium fission, Szilárd immediately understood the possibility of a nuclear chain reaction using uranium. In the summer, Fermi and Szilard proposed the idea of a nuclear reactor (pile) to mediate this process. The pile would use natural uranium as fuel. Fermi had shown much earlier that neutrons were far more effectively captured by atoms if they were of low energy (so-called "slow" or "thermal" neutrons), because for quantum reasons it made the atoms look like much larger targets to the neutrons. Thus to slow down the secondary neutrons released by the fissioning uranium nuclei, Fermi and Szilard proposed a graphite "moderator," against which the fast, high-energy secondary neutrons would collide, effectively slowing them down. With enough uranium, and with pure-enough graphite, their "pile" could theoretically sustain a slow-neutron chain reaction. This would result in the production of heat, as well as the creation of radioactive fission products.

In August 1939, Szilard and fellow Hungarian refugees physicists Teller and Wigner thought that the Germans might make use of the fission chain reaction and were spurred to attempt to attract the attention of the United States government to the issue. Towards this, they persuaded German-Jewish refugee Albert Einstein to lend his name to a letter directed to President Franklin Roosevelt. The Einstein–Szilárd letter suggested the possibility of a uranium bomb deliverable by ship, which would destroy "an entire harbor and much of the surrounding countryside." The President received the letter on 11 October 1939 — shortly after World War II began in Europe, but two years before U.S. entry into it. Roosevelt ordered that a scientific committee be authorized for overseeing uranium work and allocated a small sum of money for pile research.

In England, James Chadwick proposed an atomic bomb utilizing natural uranium, based on a paper by Rudolf Peierls with the mass needed for critical state being 30–40 tons. In America, J. Robert Oppenheimer thought that a cube of uranium deuteride 10 cm on a side (about 11 kg of uranium) might "blow itself to hell." In this design it was still thought that a moderator would need to be used for nuclear bomb fission (this turned out not to be the case if the fissile isotope was separated). In December, Werner Heisenberg delivered a report to the German Ministry of War on the possibility of a uranium bomb. Most of these models were still under the assumption that the bombs would be powered by slow neutron reactions—and thus be similar to a reactor undergoing a meltdown.

In Birmingham, England, Frisch teamed up with Peierls, a fellow German-Jewish refugee. They had the idea of using a purified mass of the uranium isotope ^{235}U, which had a cross section just determined, and which was much larger than that of ^{238}U or natural uranium (which is 99.3% the latter isotope). Assuming that the cross section for fast-neutron fission of ^{235}U was the same as for slow neutron fission, they determined that a pure ^{235}U bomb could have a critical mass of only 6 kg instead of tons, and that the resulting explosion would be tremendous. (The amount actually turned out to be 15 kg, although several times this amount was used in the actual uranium (Little Boy) bomb). In February 1940 they delivered the Frisch–Peierls memorandum. Ironically, they were still officially considered "enemy aliens" at the time. Glenn Seaborg, Joseph W. Kennedy, Arthur Wahl and Italian-Jewish refugee Emilio Segrè shortly thereafter discovered ^{239}Pu in the decay products of ^{239}U produced by bombarding ^{238}U with neutrons, and determined it to be a fissile material, like ^{235}U.

The possibility of isolating uranium-235 was technically daunting, because uranium-235 and uranium-238 are chemically identical, and vary in their mass by only the weight of three neutrons. However, if a sufficient quantity of uranium-235 could be isolated, it would allow for a fast neutron fission chain reaction. This would be extremely explosive, a true "atomic bomb." The discovery that plutonium-239 could be produced in a nuclear reactor pointed towards another approach to a fast neutron fission bomb. Both approaches were extremely novel and not yet well understood, and there was considerable scientific skepticism at the idea that they could be developed in a short amount of time.

On June 28, 1941, the Office of Scientific Research and Development was formed in the U.S. to mobilize scientific resources and apply the results of research to national defense. In September, Fermi assembled his first nuclear "pile" or reactor, in an attempt to create a slow neutron-induced chain reaction in uranium, but the experiment failed to achieve criticality, due to lack of proper materials, or not enough of the proper materials which were available.

Producing a fission chain reaction in natural uranium fuel was found to be far from trivial. Early nuclear reactors did not use isotopically enriched uranium, and in consequence they were required to use large quantities of highly purified graphite as neutron moderation materials. Use of ordinary water (as opposed to heavy water) in nuclear reactors requires enriched fuel — the partial separation and relative enrichment of the rare ^{235}U isotope from the far more common ^{238}U isotope. Typically, reactors also require inclusion of extremely chemically pure neutron moderator materials such as deuterium (in heavy water), helium, beryllium, or carbon, the latter usually as graphite. (The high purity for carbon is required because many chemical impurities such as the boron-10 component of natural boron, are very strong neutron absorbers and thus poison the chain reaction and end it prematurely.)

Production of such materials at industrial scale had to be solved for nuclear power generation and weapons production to be accomplished. Up to 1940, the total amount of uranium metal produced in the USA was not more than a few grams, and even this was of doubtful purity; of metallic beryllium not more than a few kilograms; and concentrated deuterium oxide (heavy water) not more than a few kilograms. Finally, carbon had never been produced in quantity with anything like the purity required of a moderator.

The problem of producing large amounts of high purity uranium was solved by Frank Spedding using the thermite or "Ames" process. Ames Laboratory was established in 1942 to produce the large amounts of natural (unenriched) uranium metal that would be necessary for the research to come. The critical nuclear chain-reaction success of the Chicago Pile-1 (December 2, 1942) which used unenriched (natural) uranium, like all of the atomic "piles" which produced the plutonium for the atomic bomb, was also due specifically to Szilard's realization that very pure graphite could be used for the moderator of even natural uranium "piles". In wartime Germany, failure to appreciate the qualities of very pure graphite led to reactor designs dependent on heavy water, which in turn was denied the Germans by Allied attacks in Norway, where heavy water was produced. These difficulties—among many others— prevented the Nazis from building a nuclear reactor capable of criticality during the war, although they never put as much effort as the United States into nuclear research, focusing on other technologies (see German nuclear energy project for more details).

7.2.3   Manhattan Project and beyond

See also: Manhattan Project

In the United States, an all-out effort for making atomic weapons was begun in late 1942. This work was taken over by the U.S. Army Corps of Engineers in 1943, and known as the Manhattan Engineer District. The top-secret Manhattan Project, as it was colloquially known, was led by General Leslie R. Groves. Among the project's dozens of sites were: Hanford Site in Washington state, which had the first industrial-scale nuclear reactors; Oak Ridge, Tennessee, which was primarily concerned with uranium enrichment; and Los Alamos, in New Mexico, which was the scientific hub for research on bomb development and design. Other sites, notably the Berkeley Radiation Laboratory and the Metallurgical Laboratory at the University of Chicago, played important contributing roles. Overall scientific direction of the project was managed by the physicist J. Robert Oppenheimer.

In July 1945, the first atomic bomb, dubbed "Trinity", was detonated in the New Mexico desert. It was fueled by plutonium created at Hanford. In August 1945, two more atomic bombs—"Little Boy", a uranium-235 bomb, and "Fat Man", a plutonium bomb—were used against the Japanese cities of Hiroshima and Nagasaki.

In the years after World War II, many countries were involved in the further development of nuclear fission for the purposes of nuclear reactors and nuclear weapons. The UK opened the first commercial nuclear power plant in 1956. In 2013, there are 437 reactors in 31 countries.

7.2.4   Natural fission chain-reactors on Earth

Criticality in nature is uncommon. At three ore deposits at Oklo in Gabon, sixteen sites (the so-called Oklo Fossil Reactors) have been discovered at which self-sustaining nuclear fission took place approximately 2 billion years ago. Unknown until 1972 (but postulated by Paul Kuroda in 1956[30]), when French physicist Francis Perrin discovered the Oklo Fossil Reactors, it was realized that nature had beaten humans to the punch. Large-scale natural uranium fission chain reactions, moderated by normal water, had occurred far in the past and would not be possible now. This ancient process was able to use normal water as a moderator only because 2 billion years before the present, natural uranium was richer in the shorter-lived fissile isotope ^{235}U (about 3%), than natural uranium available today (which is only 0.7%, and must be enriched to 3% to be usable in light-water reactors).

7.3   See also

- Hybrid fusion/fission

- Cold fission

- Nuclear propulsion

- Photofission

7.4 Notes

[1] M. G. Arora and M. Singh (1994). *Nuclear Chemistry*. Anmol Publications. p. 202. ISBN 81-261-1763-X.

[2] Gopal B. Saha (1 November 2010). *Fundamentals of Nuclear Pharmacy*. Springer. pp. 11–. ISBN 978-1-4419-5860-0.

[3] Петржак, Константин (1989). "Как было открыто спонтанное деление" [How spontaneous fission was discovered]. In Черникова, Вера. *Краткий Миг Торжества — О том, как делаются научные открытия* [*Brief Moment of Triumph — About making scientific discoveries*] (in Russian). Наука. pp. 108–112. ISBN 5-02-007779-8.

[4] S. Vermote, et al. (2008) "Comparative study of the ternary particle emission in 243-Cm (nth,f) and 244-Cm(SF)" in *Dynamical aspects of nuclear fission: proceedings of the 6th International Conference*. J. Kliman, M. G. Itkis, S. Gmuca (eds.). World Scientific Publishing Co. Pte. Ltd. Singapore.

[5] J. Byrne (2011) *Neutrons, Nuclei, and Matter*, Dover Publications, Mineola, NY, p. 259, ISBN 978-0-486-48238-5.

[6] Marion Brünglinghaus. "Nuclear fission". European Nuclear Society. Retrieved 2013-01-04.

[7] Hans A. Bethe (April 1950), "The Hydrogen Bomb", *Bulletin of the Atomic Scientists*, p. 99.

[8] These fission neutrons have a wide energy spectrum, with range from 0 to 14 MeV, with mean of 2 MeV and mode (statistics) of 0.75 Mev. See Byrne, op. cite.

[9] "Nuclear Fission and Fusion, and Nuclear Interactions". National Physical Laboratory. Retrieved 2013-01-04.

[10] L. Bonneau; P. Quentin. "Microscopic calculations of potential energy surfaces: fission and fusion properties" (PDF). Retrieved 2008-07-28.

[11] By R.D. Madan and Satya Prakash - *Modern Inorganic Chemistry*

[12] "Frequently Asked Questions #1". Radiation Effects Research Foundation. Retrieved September 18, 2007.

[13] E. Rutherford (1911). "The scattering of α and β particles by matter and the structure of the atom" (PDF). *Philosophical Magazine* **21** (4): 669–688. Bibcode:2012PMag...92..379R. doi:10.1080/14786435.2011.617037.

[14] "Cockcroft and Walton split lithium with high energy protons April 1932". Outreach.phy.cam.ac.uk. 1932-04-14. Retrieved 2013-01-04.

[15] "Sir Mark Oliphant (1901–2000)" (PDF). University of Adelaide. Retrieved 5 October 2013.

[16] Chadwick announced his initial findings in: J. Chadwick (1932). "Possible Existence of a Neutron" (PDF). *Nature* **129** (3252): 312. Bibcode:1932Natur.129Q.312C. doi:10.1038/129312a0. Subsequently he communicated his findings in more detail in: Chadwick, J. (1932). "The existence of a neutron". *Proceedings of the Royal Society A* **136** (830): 692–708. Bibcode:1932RSPSA.136..692C. doi:10.1098/rspa.1932.0112.; and Chadwick, J. (1933). "The Bakerian Lecture: The neutron". *Proceedings of the Royal Society A* **142** (846): 1–25. Bibcode:1933RSPSA.142....1C. doi:10.1098/rspa.1933.0152.

[17] E. Fermi, E. Amaldi, O. D'Agostino, F. Rasetti, and E. Segrè (1934) "Radioattività provocata da bombardamento di neutroni III," *La Ricerca Scientifica*, vol. 5, no. 1, pages 452–453.

[18] Ida Noddack (1934). "Über das Element 93". *Zeitschrift für Angewandte Chemie* **47** (37): 653. doi:10.1002/ange.19340473707.

[19] Tacke, Ida Eva. Astr.ua.edu. Retrieved on 2010-12-24.

[20] Bob Weintraub. *Lise Meitner (1878–1968): Protactinium, Fission, and Meitnerium*. Retrieved on June 8, 2009.

[21] O. Hahn and F. Strassmann (1939). "Über den Nachweis und das Verhalten der bei der Bestrahlung des Urans mittels Neutronen entstehenden Erdalkalimetalle ("On the detection and characteristics of the alkaline earth metals formed by irradiation of uranium with neutrons")". *Naturwissenschaften* **27** (1): 11–15. Bibcode:1939NW.....27...11H. doi:10.1007/BF01488241.. The authors were identified as being at the Kaiser-Wilhelm-Institut für Chemie, Berlin-Dahlem. Received 22 December 1938.

[22] L. Meitner and O. R. Frisch (1939). "Disintegration of Uranium by Neutrons: a New Type of Nuclear Reaction". *Nature* **143** (3615): 239. Bibcode:1939Natur.143..239M. doi:10.1038/143239a0.. The paper is dated 16 January 1939. Meitner is identified as being at the Physical Institute, Academy of Sciences, Stockholm. Frisch is identified as being at the Institute of Theoretical Physics, University of Copenhagen.

[23] O. R. Frisch (1939). "Physical Evidence for the Division of Heavy Nuclei under Neutron Bombardment". *Nature* **143** (3616): 276. Bibcode:1939Natur.143..276F. doi:10.1038/143276a0.

[24] "Physical Evidence for the Division of Heavy Nuclei under Neutron Bombardment". 17 January 1939. Archived from the original on 2008-01-08. The experiment for this letter to the editor was conducted on 13 January 1939; see Richard Rhodes (1986) *The Making of the Atomic Bomb*, Simon and Schuster. pp. 263 and 268, ISBN 0-671-44133-7.

[25] "The Nobel Prize in Chemistry 1944". Nobelprize.org. Retrieved 2008-10-06.

[26] Richard Rhodes. (1986) *The Making of the Atomic Bomb*, Simon and Schuster, p. 268, ISBN 0-671-44133-7.

[27] H. L. Anderson, E. T. Booth, J. R. Dunning, E. Fermi, G. N. Glasoe, and F. G. Slack (1939). "The Fission of Uranium". *Physical Review* **55** (5): 511. Bibcode:1939PhRv...55..511A. doi:10.1103/PhysRev.55.511.2.

[28] Richard Rhodes (1986). *The Making of the Atomic Bomb*, Simon and Schuster, pp. 267–270, ISBN 0-671-44133-7.

[29] H. Von Halban; F. Joliot and L. Kowarski (1939). "Number of Neutrons Liberated in the Nuclear Fission of Uranium". *Nature* **143** (3625): 680. Bibcode:1939Natur.143..680V. doi:10.1038/143680a0.

[30] P. K. Kuroda (1956). "On the Nuclear Physical Stability of the Uranium Minerals" (PDF). *The Journal of Chemical Physics* **25** (4): 781. Bibcode:1956JChPh..25..781K. doi:10.1063/1.1743058.

7.5   References

- *DOE Fundamentals Handbook: Nuclear Physics and Reactor Theory Volume 1* (PDF). U.S. Department of Energy. January 1993. Retrieved 2012-01-03.

- *DOE Fundamentals Handbook: Nuclear Physics and Reactor Theory Volume 2* (PDF). U.S. Department of Energy. January 1993. Retrieved 2012-01-03.

7.6   External links

- The Effects of Nuclear Weapons

- Annotated bibliography for nuclear fission from the Alsos Digital Library

- The Discovery of Nuclear Fission Historical account complete with audio and teacher's guides from the American Institute of Physics History Center

- atomicarchive.com Nuclear Fission Explained

- Nuclear Files.org What is Nuclear Fission?

- Nuclear Fission Animation

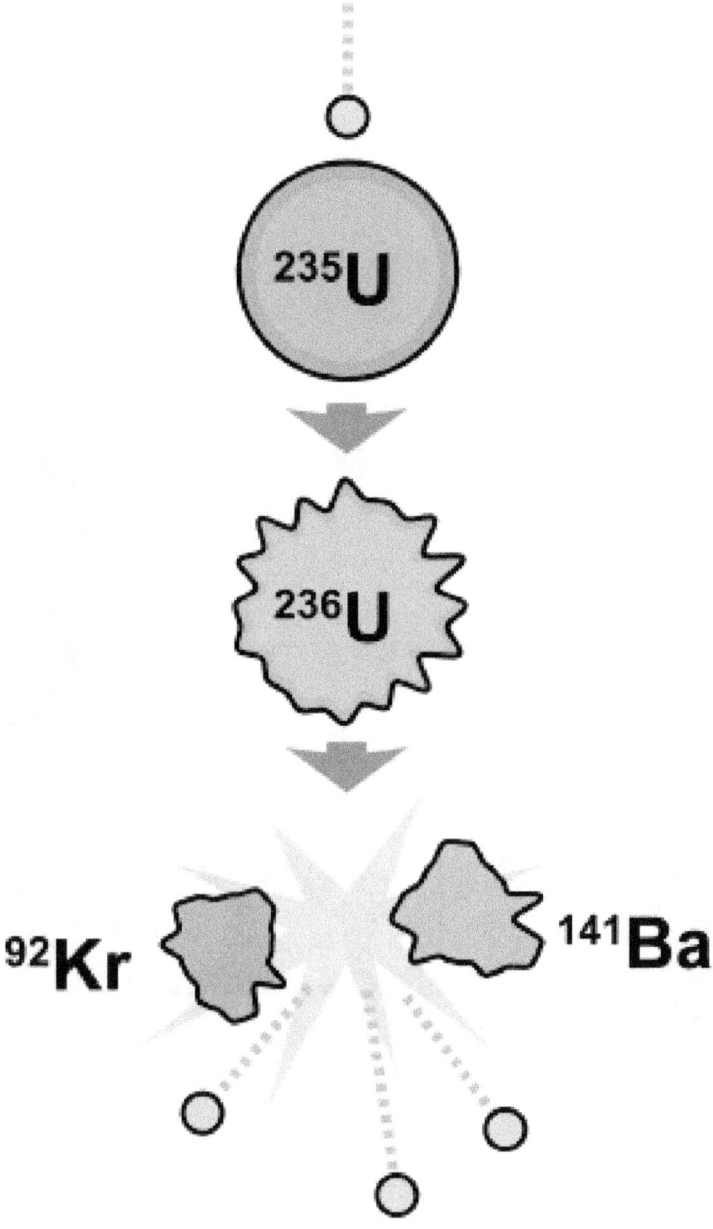

An induced fission reaction. A neutron is absorbed by a uranium-235 nucleus, turning it briefly into an excited uranium-236 nucleus, with the excitation energy provided by the kinetic energy of the neutron plus the forces that bind the neutron. The uranium-236, in turn, splits into fast-moving lighter elements (fission products) and releases three free neutrons. At the same time, one or more "prompt gamma rays" (not shown) are produced, as well.

Illustration From October 2002
Issue of "Popular Mechanics" (pg. 69)

The mushroom cloud produced by Tsar Bomba, currently the largest man-made nuclear device detonated in history, next to other mushroom clouds of various nuclear devices.

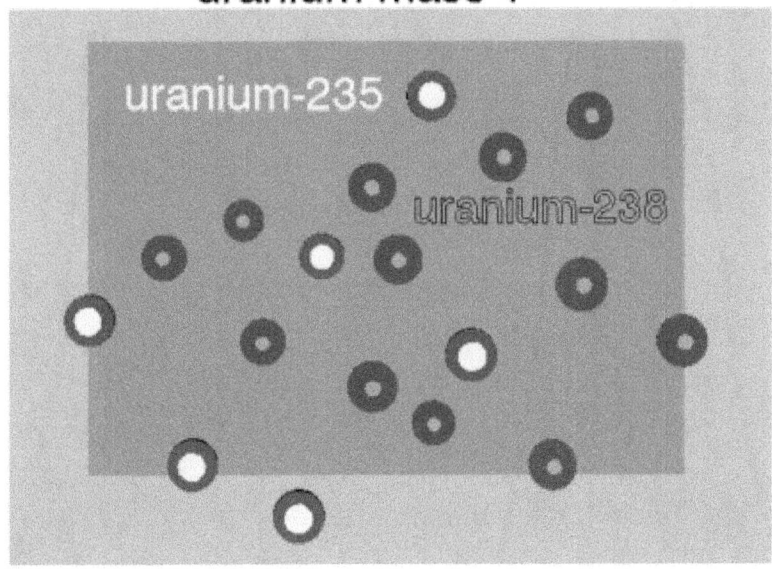

A visual representation of an induced nuclear fission event where a slow-moving neutron is absorbed by the nucleus of a uranium-235 atom, which fissions into two fast-moving lighter elements (fission products) and additional neutrons. Most of the energy released is in the form of the kinetic velocities of the fission products and the neutrons.

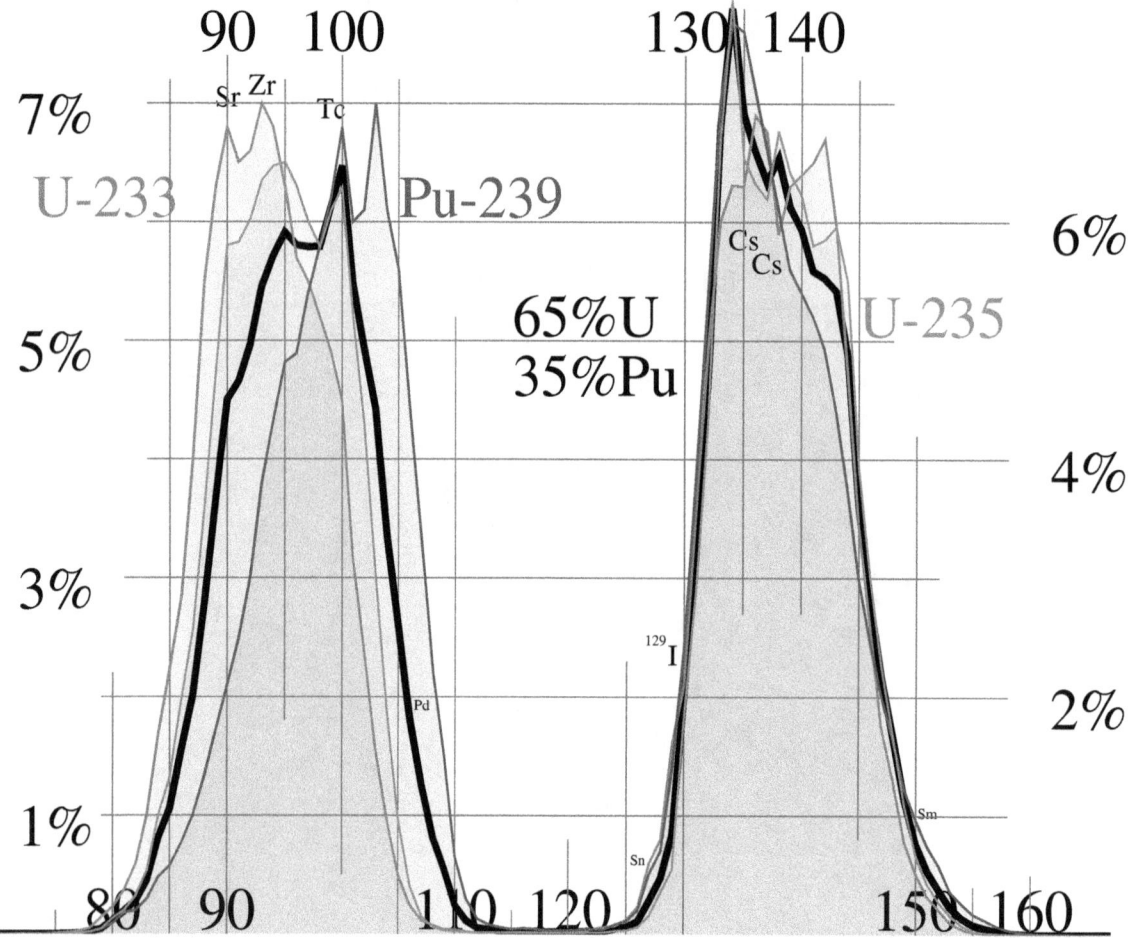

Fission product yields by mass for thermal neutron fission of U-235, Pu-239, a combination of the two typical of current nuclear power reactors, and U-233 used in the thorium cycle.

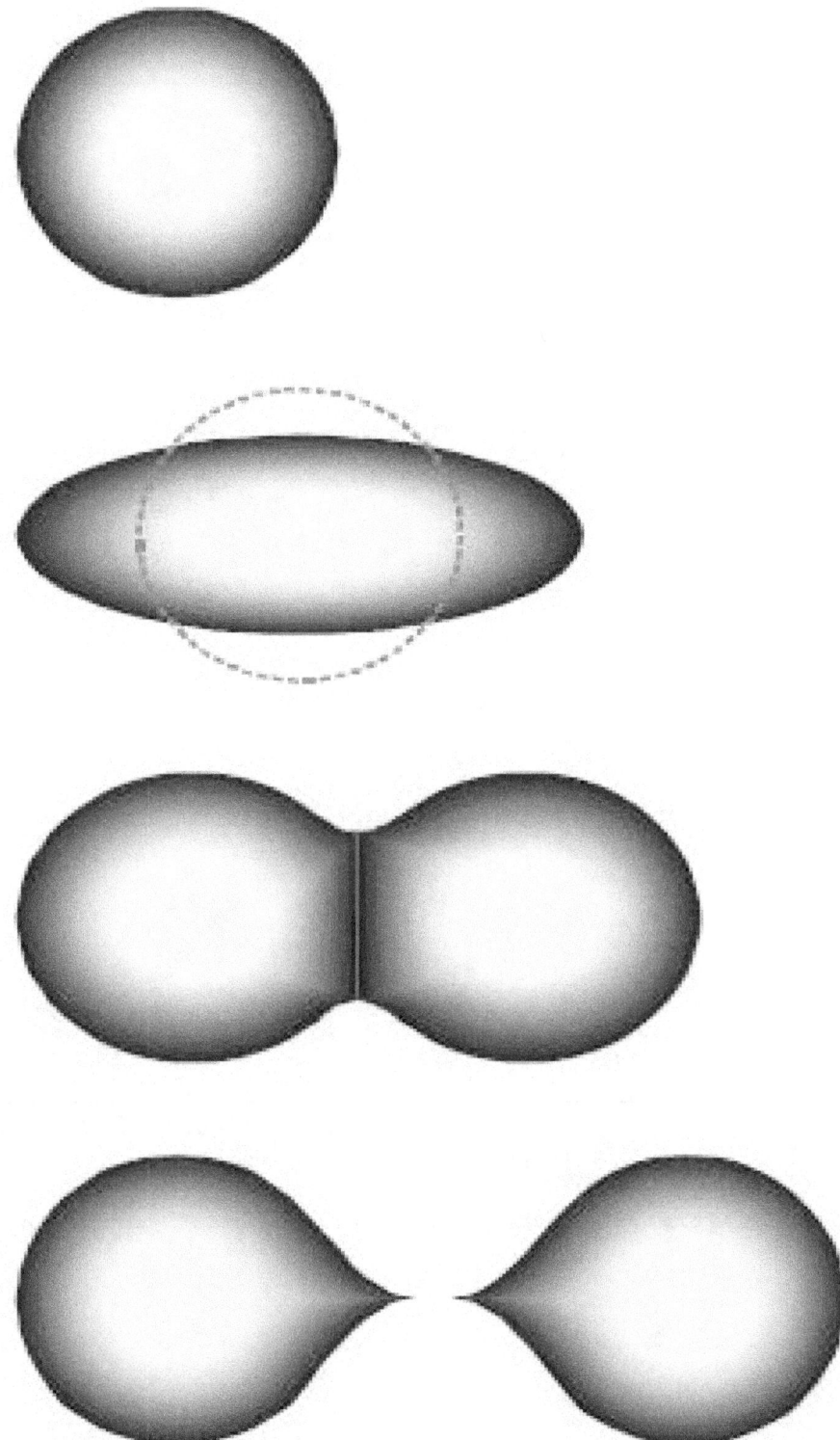

The stages of binary fission in a liquid drop model. Energy input deforms the nucleus into a fat "cigar" shape, then a "peanut" shape, followed by binary fission as the two lobes exceed the short-range nuclear force attraction distance, then are pushed apart and away by their electrical charge. In the liquid drop model, the two fission fragments are predicted to be the same size. The nuclear shell model allows for them to differ in size, as usually experimentally observed.

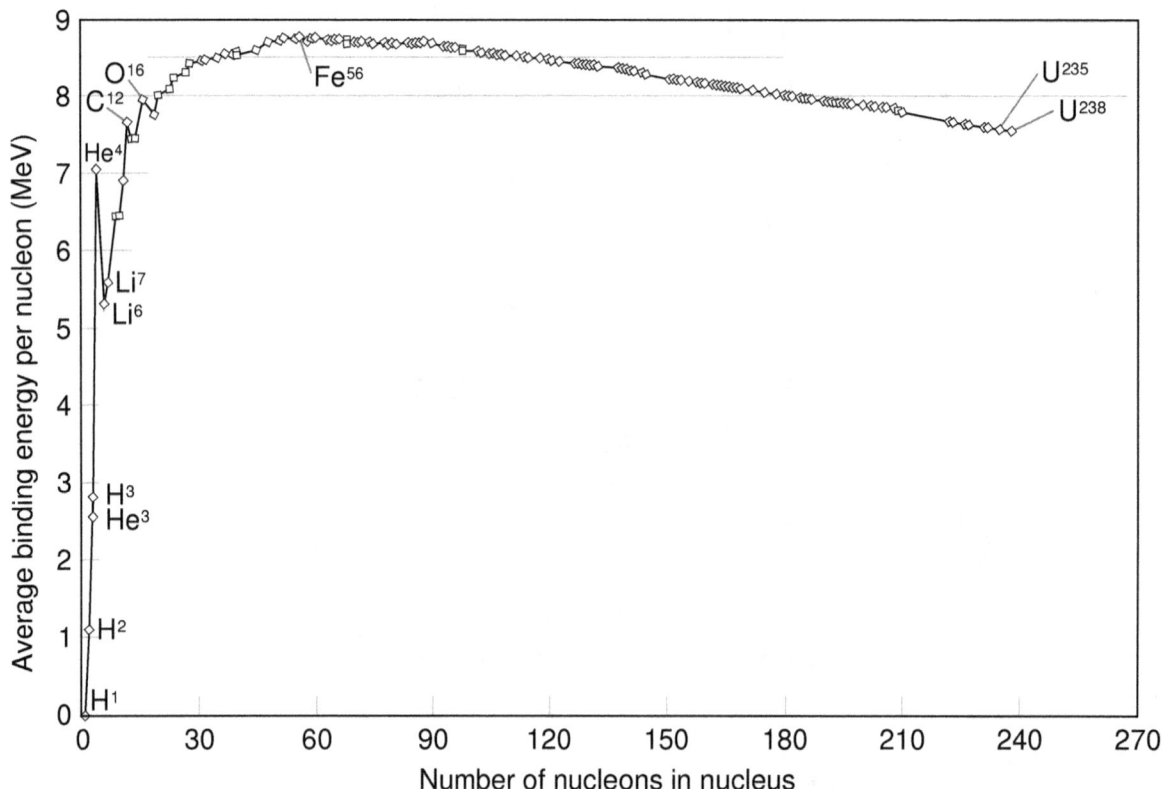

The "curve of binding energy": A graph of binding energy per nucleon of common isotopes.

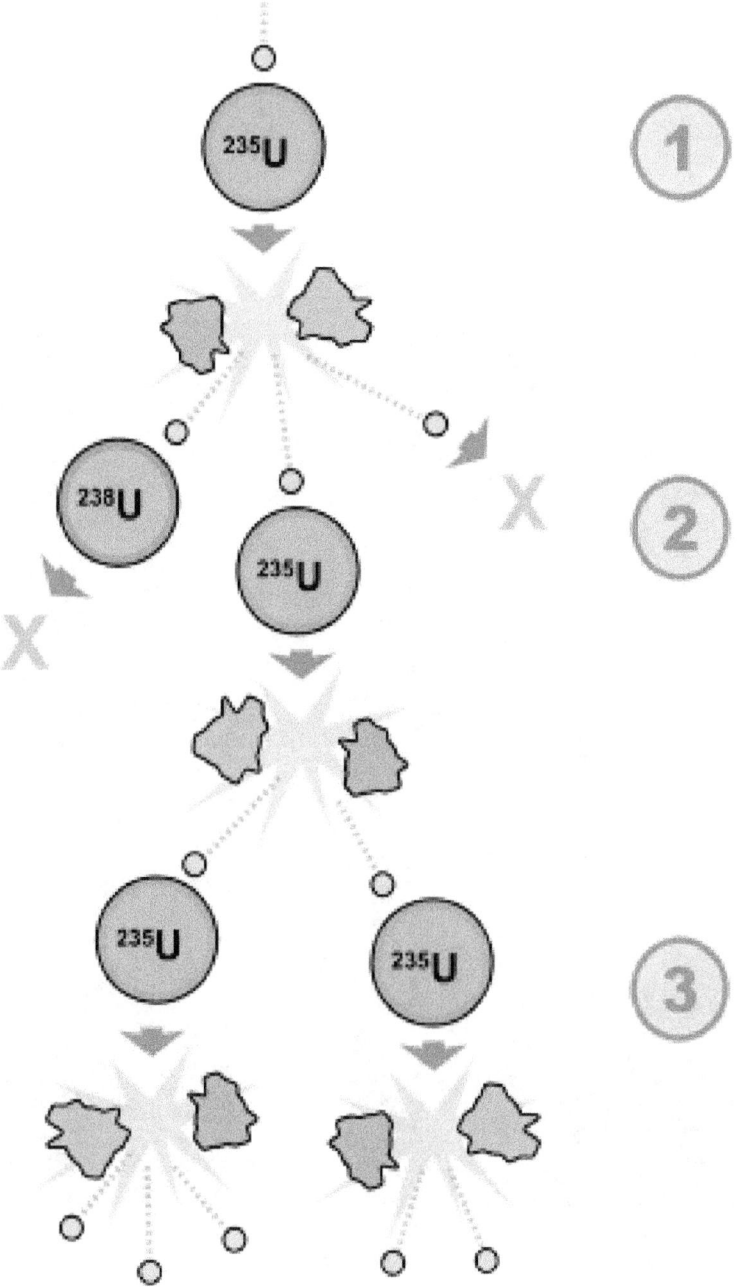

A schematic nuclear fission chain reaction. 1. A uranium-235 atom absorbs a neutron and fissions into two new atoms (fission fragments), releasing three new neutrons and some binding energy. 2. One of those neutrons is absorbed by an atom of uranium-238 and does not continue the reaction. Another neutron is simply lost and does not collide with anything, also not continuing the reaction. However, the one neutron does collide with an atom of uranium-235, which then fissions and releases two neutrons and some binding energy. 3. Both of those neutrons collide with uranium-235 atoms, each of which fissions and releases between one and three neutrons, which can then continue the reaction.

The cooling towers of the Philippsburg Nuclear Power Plant, in Germany.

The mushroom cloud of the atom bomb dropped on Nagasaki, Japan in 1945 rose some 18 kilometres (11 mi) above the bomb's hypocenter. The bomb killed at least 60,000 people.[12]

The experimental apparatus with which Otto Hahn and Fritz Strassmann discovered nuclear fission in 1938

German stamp honoring Otto Hahn and his discovery of nuclear fission (1979)

Drawing of the first artificial reactor, Chicago Pile-1.

Chapter 8

Nucleosynthesis

For the song by Vangelis, see Albedo 0.39.

Nucleosynthesis is the process that creates new atomic nuclei from pre-existing nucleons, primarily protons and neutrons. The first nuclei were formed about three minutes after the Big Bang, through the process called Big Bang nucleosynthesis. It was then that hydrogen and helium formed to become the content of the first stars, and this primeval process is responsible for the present hydrogen/helium ratio of the cosmos.

With the formation of stars, heavier nuclei were created from hydrogen and helium by stellar nucleosynthesis, a process that continues today. Some of these elements, particularly those lighter than iron, continue to be delivered to the interstellar medium when low mass stars eject their outer envelope before they collapse to form white dwarfs. The remains of their ejected mass form the planetary nebulae observable throughout our galaxy.

Supernova nucleosynthesis within exploding stars by fusing carbon and oxygen is responsible for the abundances of elements between magnesium (atomic number 12) and nickel (atomic number 28).[1] Supernova nucleosynthesis is also thought to be responsible for the creation of rarer elements heavier than iron and nickel, in the last few seconds of a type II supernova event. The synthesis of these heavier elements absorbs energy (endothermic) as they are created, from the energy produced during the supernova explosion. Some of those elements are created from the absorption of multiple neutrons (the R process) in the period of a few seconds during the explosion. The elements formed in supernovas include the heaviest elements known, such as the long-lived elements uranium and thorium.

Cosmic ray spallation, caused when cosmic rays impact the interstellar medium and fragment larger atomic species, is a significant source of the lighter nuclei, particularly ^{3}He, ^{9}Be and ^{10,11}B, that are not created by stellar nucleosynthesis.

In addition to the fusion processes responsible for the growing abundances of elements in the universe, a few minor natural processes continue to produce very small numbers of new nuclides on Earth. These nuclides contribute little to their abundances, but may account for the presence of specific new nuclei. These nuclides are produced via radiogenesis (decay) of long-lived, heavy, primordial radionuclides such as uranium and thorium. Cosmic ray bombardment of elements on Earth also contribute to the presence of rare, short-lived atomic species called cosmogenic nuclides.

8.1 Timeline

It is thought that the primordial nucleons themselves were formed from the quark–gluon plasma during the Big Bang as it cooled below two trillion degrees. A few minutes afterward, starting with only protons and neutrons, nuclei up to lithium and beryllium (both with mass number 7) were formed, but the abundances of other elements dropped sharply with growing atomic mass. Some boron may have been formed at this time, but the process stopped before significant carbon could be formed, as this element requires a far higher product of helium density and time than were present in the short nucleosynthesis period of the Big Bang. That fusion process essentially shut down at about 20 minutes, due to drops in temperature and density as the universe continued to expand. This first process, Big Bang nucleosynthesis, was

Periodic table showing the cosmogenic origin of each element. Elements from carbon up to sulfur may be made in small stars by the alpha process. Elements beyond iron are made in large stars with slow neutron capture (s-process), followed by expulsion to space in gas ejections (see planetary nebulae). Elements heavier than iron may be made in supernovae after the r-process, involving a dense burst of neutrons and rapid capture by the element.

the first type of nucleogenesis to occur in the universe.

The subsequent nucleosynthesis of the heavier elements requires the extreme temperatures and pressures found within stars and supernovas. These processes began as hydrogen and helium from the Big Bang collapsed into the first stars at 500 million years. Star formation has occurred continuously in the galaxy since that time. The elements found on Earth, the so-called primordial elements, were created prior to Earth's formation by stellar nucleosynthesis and by supernova nucleosynthesis. They range in atomic numbers from $Z=6$ (carbon) to $Z=94$ (plutonium). Synthesis of these elements occurred either by nuclear fusion (including both rapid and slow multiple neutron capture) or to a lesser degree by nuclear fission followed by beta decay.

A star gains heavier elements by combining its lighter nuclei, hydrogen, deuterium, beryllium, lithium, and boron, which were found in the initial composition of the interstellar medium and hence the star. Interstellar gas therefore contains declining abundances of these light elements, which are present only by virtue of their nucleosynthesis during the Big Bang. Larger quantities of these lighter elements in the present universe are therefore thought to have been restored through billions of years of cosmic ray (mostly high-energy proton) mediated breakup of heavier elements in interstellar gas and dust. The fragments of these cosmic-ray collisions include the light elements Li, Be and B.

8.2   History of nucleosynthesis theory

The first ideas on nucleosynthesis were simply that the chemical elements were created at the beginning of the universe, but no rational physical scenario for this could be identified. Gradually it became clear that hydrogen and helium are much more abundant than any of the other elements. All the rest constitute less than 2% of the mass of the Solar System, and of other star systems as well. At the same time it was clear that oxygen and carbon were the next two most common elements, and also that there was a general trend toward high abundance of the light elements, especially those composed of whole numbers of helium-4 nuclei.

Arthur Stanley Eddington first suggested in 1920, that stars obtain their energy by fusing hydrogen into helium and raised the possibility that the heavier elements may also form in stars.[2][3] This idea was not generally accepted, as the nuclear mechanism was not understood. In the years immediately before World War II, Hans Bethe first elucidated those nuclear mechanisms by which hydrogen is fused into helium.

Fred Hoyle's original work on nucleosynthesis of heavier elements in stars, occurred just after World War II.[4] His work explained the production of all heavier elements, starting from hydrogen. Hoyle proposed that hydrogen is continuously created in the universe from vacuum and energy, without need for universal beginning.

Hoyle's work explained how the abundances of the elements increased with time as the galaxy aged. Subsequently, Hoyle's picture was expanded during the 1960s by contributions from William A. Fowler, Alastair G. W. Cameron, and Donald D. Clayton, followed by many others. In the seminal 1957 review paper by E. M. Burbidge, G. R. Burbidge, Fowler and Hoyle (see Ref. list) is a well-known summary of the state of the field in 1957. That paper defined new processes for the transformation of one heavy nucleus into others within stars, processes that could be documented by astronomers.

The Big Bang itself had been proposed in 1931, long before this period, by Georges Lemaître, a Belgian physicist, who suggested that the evident expansion of the Universe in time required that the Universe, if contracted backwards in time, would continue to do so until it could contract no further. This would bring all the mass of the Universe to a single point, a "primeval atom", to a state before which time and space did not exist. Hoyle later gave Lemaître's model the derisive term of Big Bang, not realizing that Lemaître's model was needed to explain the existence of deuterium and nuclides between helium and carbon, as well as the fundamentally high amount of helium present, not only in stars but also in interstellar space. As it happened, both Lemaître and Hoyle's models of nucleosynthesis would be needed to explain the elemental abundances in the universe.

The goal of the theory of nucleosynthesis is to explain the vastly differing abundances of the chemical elements and their several isotopes from the perspective of natural processes. The primary stimulus to the development of this theory was the shape of a plot of the abundances versus the atomic number of the elements. Those abundances, when plotted on a graph as a function of atomic number, have a jagged sawtooth structure that varies by factors up to ten million. A very influential stimulus to nucleosynthesis research was an abundance table created by Hans Suess and Harold Urey that was based on the unfractionated abundances of the non-volatile elements found within unevolved meteorites.[5] Such a graph of the abundances is displayed on a logarithmic scale below, where the dramatically jagged structure is visually suppressed by the many powers of ten spanned in the vertical scale of this graph. See *Handbook of Isotopes in the Cosmos* for more data and discussion of abundances of the isotopes.[6]

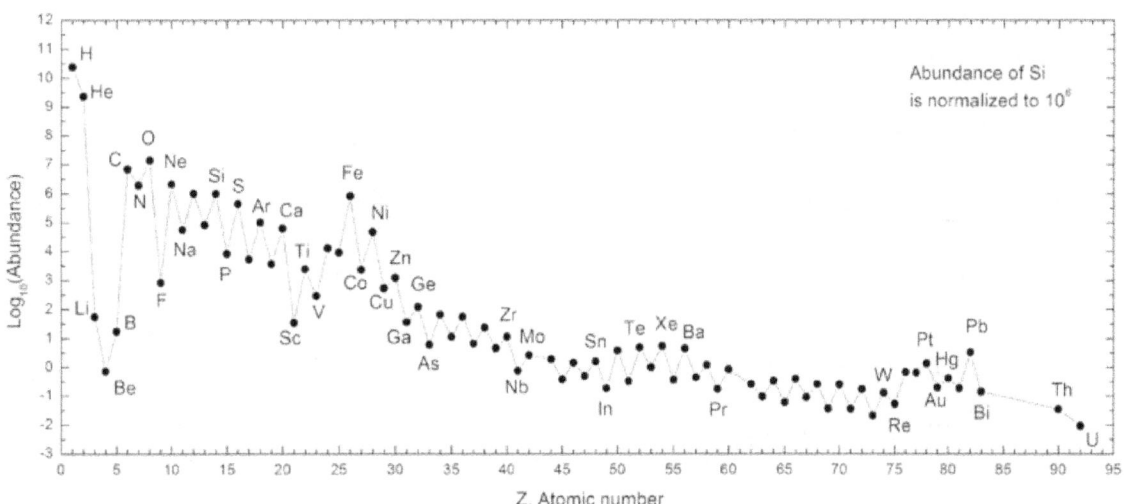

Abundances of the chemical elements in the Solar System. Hydrogen and helium are most common, residuals within the paradigm of the Big Bang.[7] The next three elements (Li, Be, B) are rare because they are poorly synthesized in the Big Bang and also in stars. The two general trends in the remaining stellar-produced elements are: (1) an alternation of abundance of elements according to whether they have even or odd atomic numbers, and (2) a general decrease in abundance, as elements become heavier. Within this trend is a peak at abundances of iron and nickel, which is especially visible on a logarithmic graph spanning fewer powers of ten, say between logA=2 (A=100) and logA=6 (A=1,000,000).

8.3   Processes

There are a number of astrophysical processes which are believed to be responsible for nucleosynthesis. The majority of these occur in shells within stars, and the chain of those nuclear fusion processes are known as hydrogen burning (via the proton-proton chain or the CNO cycle), helium burning, carbon burning, neon burning, oxygen burning and silicon burning. These processes are able to create elements up to and including iron and nickel. This is the region of nucleosynthesis within which the isotopes with the highest binding energy per nucleon are created. Heavier elements can be assembled within stars by a neutron capture process known as the s-process or in explosive environments, such as supernovae, by a number of other processes. Some of those others include the r-process, which involves rapid neutron captures, the rp-process, and the p-process (sometimes known as the gamma process), which results in the photodisintegration of existing nuclei.

8.4   The major types of nucleosynthesis

8.4.1   Big Bang nucleosynthesis

Main article: Big Bang nucleosynthesis

Big Bang nucleosynthesis occurred within the first three minutes of the beginning of the universe and is responsible for much of the abundance of ^{1}H (protium), ^{2}H (D, deuterium), ^{3}He (helium-3), and ^{4}He (helium-4). Although ^{4}He continues to be produced by stellar fusion and alpha decays and trace amounts of ^{1}H continue to be produced by spallation and certain types of radioactive decay, most of the mass of the isotopes in the universe are thought to have been produced in the Big Bang. The nuclei of these elements, along with some ^{7}Li and ^{7}Be are considered to have been formed between 100 and 300 seconds after the Big Bang when the primordial quark–gluon plasma froze out to form protons and neutrons. Because of the very short period in which nucleosynthesis occurred before it was stopped by expansion and cooling (about 20 minutes), no elements heavier than beryllium (or possibly boron) could be formed. Elements formed during this time were in the plasma state, and did not cool to the state of neutral atoms until much later.

$$n^0 \longrightarrow p^+ + e^- + \bar{\nu}_e \qquad\qquad p^+ + n^0 \longrightarrow {}^2_1D + \gamma$$

$$^2_1D + p^+ \longrightarrow {}^3_2He + \gamma \qquad\qquad {}^2_1D + {}^2_1D \longrightarrow {}^3_2He + n^0$$

$$^2_1D + {}^2_1D \longrightarrow {}^3_1T + p^+ \qquad\qquad {}^3_1T + {}^2_1D \longrightarrow {}^4_2He + n^0$$

$$^3_1T + {}^4_2He \longrightarrow {}^7_3Li + \gamma \qquad\qquad {}^3_2He + n^0 \longrightarrow {}^3_1T + p^+$$

$$^3_2He + {}^2_1D \longrightarrow {}^4_2He + p^+ \qquad\qquad {}^3_2He + {}^4_2He \longrightarrow {}^7_4Be + \gamma$$

$$^7_3Li + p^+ \longrightarrow {}^4_2He + {}^4_2He \qquad\qquad {}^7_4Be + n^0 \longrightarrow {}^7_3Li + p^+$$

Chief nuclear reactions responsible for the relative abundances of light atomic nuclei observed throughout the universe.

8.4.2   Stellar nucleosynthesis

Main articles: Stellar nucleosynthesis, Proton-proton chain, Triple-alpha process, CNO cycle, s-process, p-process and photodisintegration

Stellar nucleosynthesis is the nuclear process by which new nuclei are produced. It occurs in stars during stellar evolution. It is responsible for the galactic abundances of elements from carbon to iron. Stars are thermonuclear furnaces in which H and He are fused into heavier nuclei by increasingly high temperatures as the composition of the core evolves.[8] Of particular importance is carbon, because its formation from He is a bottleneck in the entire process. Carbon is produced by the triple-alpha process in all stars. Carbon is also the main element that causes the release of free neutrons within stars, giving rise to the s-process, in which the slow absorption of neutrons converts iron into elements heavier than iron and nickel.[9]

The products of stellar nucleosynthesis are generally dispersed into the interstellar gas through mass loss episodes and the stellar winds of low mass stars. The mass loss events can be witnessed today in the planetary nebulae phase of low-mass star evolution, and the explosive ending of stars, called supernovae, of those with more than eight times the mass of the Sun.

The first direct proof that nucleosynthesis occurs in stars was the astronomical observation that interstellar gas has become enriched with heavy elements as time passed. As a result, stars that were born from it late in the galaxy, formed with much higher initial heavy element abundances than those that had formed earlier. The detection of technetium in the atmosphere of a red giant star in 1952,[10] by spectroscopy, provided the first evidence of nuclear activity within stars. Because technetium is radioactive, with a half-life much less than the age of the star, its abundance must reflect its recent creation within that star. Equally convincing evidence of the stellar origin of heavy elements, is the large overabundances of specific stable elements found in stellar atmospheres of asymptotic giant branch stars. Observation of barium abundances some 20-50 times greater than found in unevolved stars is evidence of the operation of the s-process within such stars. Many modern proofs of stellar nucleosynthesis are provided by the isotopic compositions of stardust, solid grains that have condensed from the gases of individual stars and which have been extracted from meteorites. Stardust is one component of cosmic dust, and is frequently called presolar grains. The measured isotopic compositions in stardust grains demonstrate many aspects of nucleosynthesis within the stars from which the grains condensed during the star's late-life mass-loss episodes.[11]

8.4.3 Explosive nucleosynthesis

Main articles: r-process, rp-process and Supernova nucleosynthesis

Supernova nucleosynthesis occurs in the energetic environment in supernovae, in which the elements between silicon and nickel are synthesized in quasiequilibrium[12] established during fast fusion that attaches by reciprocating balanced nuclear reactions to ^{28}Si. Quasiequilibrium can be thought of as *almost equilibrium* except for a high abundance of the ^{28}Si nuclei in the feverishly burning mix. This concept[13] was the most important discovery in nucleosynthesis theory of the intermediate-mass elements since Hoyle's 1954 paper because it provided an overarching understanding of the abundant and chemically important elements between silicon (A=28) and nickel (A=60). It replaced the incorrect although much cited alpha process of the B2FH paper, which inadvertently obscured Hoyle's better 1954 theory.[14] Further nucleosynthesis processes can occur, in particular the r-process (rapid process) described by the B2FH paper and first calculated by Seeger, Fowler and Clayton,[15] in which the most neutron-rich isotopes of elements heavier than nickel are produced by rapid absorption of free neutrons. The creation of free neutrons by electron capture during the rapid compression of the supernova core along with assembly of some neutron-rich seed nuclei makes the r-process a *primary process*, and one that can occur even in a star of pure H and He. This is in contrast to the B2FH designation of the process as a *secondary process*. This promising scenario, though generally supported by supernova experts, has yet to achieve a totally satisfactory calculation of r-process abundances. The primary r-process has been confirmed by astronomers who have observed old stars born when galactic metallicity was still small, that nonetheless contain their complement of r-process nuclei; thereby demonstrating that the metallicity is a product of an internal process. The r-process is responsible for our natural cohort of radioactive elements, such as uranium and thorium, as well as the most neutron-rich isotopes of each heavy element.

The rp-process (rapid proton) involves the rapid absorption of free protons as well as neutrons, but its role and its existence are less certain.

Explosive nucleosynthesis occurs too rapidly for radioactive decay to decrease the number of neutrons, so that many abundant isotopes with equal and even numbers of protons and neutrons are synthesized by the silicon quasiequilibrium process.[16] During this process, the burning of oxygen and silicon fuses nuclei that themselves have equal numbers of protons and neutrons to produce nuclides which consist of whole numbers of helium nuclei, up to 15 (representing ^{60}Ni). Such multiple-alpha-particle nuclides are totally stable up to ^{40}Ca (made of 10 helium nuclei), but heavier nuclei with equal and even numbers of protons and neutrons are tightly bound but unstable. The quasiequilibrium produces radioactive isobars ^{44}Ti, ^{48}Cr, ^{52}Fe, and ^{56}Ni, which (except ^{44}Ti) are created in abundance but decay after the explosion and leave the most stable isotope of the corresponding element at the same atomic weight. The most abundant and extant isotopes of elements produced in this way are ^{48}Ti, ^{52}Cr, and ^{56}Fe. These decays are accompanied by the emission of gamma-rays (radiation from the nucleus), whose spectroscopic lines can be used to identify the isotope created by the decay. The detection of these emission lines were an important early product of gamma-ray astronomy.[17]

The most convincing proof of explosive nucleosynthesis in supernovae occurred in 1987 when those gamma-ray lines were detected emerging from supernova 1987A. Gamma ray lines identifying ^{56}Co and ^{57}Co nuclei, whose radioactive half-lives limit their age to about a year, proved that they were created by their radioactive cobalt parents. This nuclear astronomy observation was predicted in 1969[18] as a way to confirm explosive nucleosynthesis of the elements, and that prediction played an important role in the planning for NASA's Compton Gamma-Ray Observatory.

Other proofs of explosive nucleosynthesis are found within the stardust grains that condensed within the interiors of supernovae as they expanded and cooled. Stardust grains are one component of cosmic dust. In particular, radioactive ^{44}Ti was measured to be very abundant within supernova stardust grains at the time they condensed during the supernova expansion.[19] This confirmed a 1975 prediction of the identification of supernova stardust (SUNOCONs), which became part of the pantheon of presolar grains. Other unusual isotopic ratios within these grains reveal many specific aspects of explosive nucleosynthesis.

8.4.4   Cosmic ray spallation

Main article: Cosmic ray spallation

Cosmic ray spallation process reduces the atomic weight of interstellar matter by the impact with cosmic rays, to produce some of the lightest elements present in the universe (though not a significant amount of deuterium). Most notably spallation is believed to be responsible for the generation of almost all of ^{3}He and the elements lithium, beryllium, and boron, although some 7Li and 7Be are thought to have been produced in the Big Bang. The spallation process results from the impact of cosmic rays (mostly fast protons) against the interstellar medium. These impacts fragment carbon, nitrogen, and oxygen nuclei present. The process results in the light elements beryllium, boron, and lithium in cosmos at much greater abundances than they are within solar atmospheres. The light elements ^{1}H and ^{4}He nuclei are not a product of spallation and are represented in the cosmos with approximately primordial abundance.

Beryllium and boron are not significantly produced by stellar fusion processes, due to the instability of any ^{8}Be formed from two ^{4}He nuclei.

8.5   Empirical evidence

Theories of nucleosynthesis are tested by calculating isotope abundances and comparing those results with observed results. Isotope abundances are typically calculated from the transition rates between isotopes in a network. Often these calculations can be simplified as a few key reactions control the rate of other reactions.

8.6   Minor mechanisms and processes

Very small amounts of certain nuclides are produced on Earth by artificial means. Those are our primary source, for example, of technetium. However, some nuclides are also produced by a number of natural means that have continued after primordial elements were in place. These often act to produce new elements in ways that can be used to date rocks or to trace the source of geological processes. Although these processes do not produce the nuclides in abundance, they are assumed to be the entire source of the existing natural supply of those nuclides.

These mechanisms include:

- Radioactive decay may lead to radiogenic daughter nuclides. The nuclear decay of many long-lived primordial isotopes, especially uranium-235, uranium-238, and thorium-232 produce many intermediate daughter nuclides, before they too finally decay to isotopes of lead. The Earth's natural supply of elements like radon and polonium is via this mechanism. The atmosphere's supply of argon-40 is due mostly to the radioactive decay of potassium-40 in the time since the formation of the Earth. Little of the atmospheric argon is primordial. Helium-4 is produced by alpha-decay, and the helium trapped in Earth's crust is also mostly non-primordial. In other types of radioactive decay, such as cluster decay, larger species of nuclei are ejected (for example, neon-20), and these eventually become newly formed stable atoms.

- Radioactive decay may lead to spontaneous fission. This is not cluster decay, as the fission products may be split among nearly any type of atom. Thorium-232, uranium-235, and uranium-238 are primordial isotopes that undergo spontaneous fission. Natural technetium and promethium are produced in this manner.
- Nuclear reactions. Naturally-occurring nuclear reactions powered by radioactive decay give rise to so-called nucleogenicnuclides. This process happens when an energetic particle from a radioactive decay, often an alpha particle, reactswith a nucleus of another atom to change the nucleus into another nuclide. This process may also cause the pro-duction of further subatomic particles, such as neutrons. Neutrons can also be produced in spontaneous fission andby neutron emission. These neutrons can then go on to produce other nuclides via neutron-induced fission, or byneutron capture. For example, some stable isotopes such as neon-21 and neon-22 are produced by several routesof nucleogenic synthesis, and thus only part of their abundance is primordial.
- Nuclear reactions due to cosmic rays. By convention, these reaction-products are not termed "nucleogenic" nuclides, but rather cosmogenic nuclides. Cosmic rays continue to produce new elements on Earth by the same cosmogenic processes discussed above that produce primordial beryllium and boron. One important example is carbon-14, produced from nitrogen-14 in the atmosphere by cosmic rays. Iodine-129 is another example.

In addition to artificial processes, it is postulated that neutron star collision is the main source of elements heavier than iron.[20]

8.7 See also

- Stellar evolution

- Supernova nucleosynthesis

- Cosmic dust

- Metallicity

8.8 References

[1] Donald D. Clayton, *Handbook of isotopes in the cosmos*, Cambridge University Press (Cambridge 2003)

[2] A.S. Eddington, The Internal Constitution of the Stars, *The Observatory*, **43**, 341 (1920) http://adsabs.harvard.edu/abs/1920Obs. ...43..341E

[3] A.S. Eddington, The Internal Constitution of the Stars, *Nature*, **106**, 106 (1920) http://adsabs.harvard.edu/abs/1920Natur.106. ..14E

[4] Actually, before the war ended, he learned abut the problem of spherical implosion of plutonium in the Manhattan project. He saw an analogy between the plutonium fission reaction and the newly discovered supernovae, and he was able to show that exploding super novae produced all of the elements in the same proportion as existed on Earth. He felt that he had accidentally fallen into a subject that would make his career. Autobiography William A. Fowler

[5] H.E. Suess and H.C. Urey, Abundances of the elements, *Revs. Mod. Phys.*, **28**, 53 (1957)

[6] Donald D. Clayton, *Handbook of isotopes in the cosmos*, Cambridge University Press (Cambridge U.K. 2003)

[7] Massimo S. Stiavelli. From First Light to Reionization. John Wiley & Sons, Apr 22, 2009. Pg 8.

[8] Donald D. Clayton, *Principles of Stellar Evolution and Nucleosynthesis*, McGraw-Hill (New York 1968) Chapter 5; reissued by University of Chicago Press (Chicago 1883)

[9] D.D. Clayton, W.A. Fowler, T. Hull and B. Zimmerman, Neutron capture chains in heavy element synthesis, *Ann. Phys.*, **12**, 331-408 (1961); Donald D. Clayton, *Principles of Stellar Evolution and Nucleosynthesis*, McGraw-Hill (New York 1968) Chapter 7

[10] S. Paul W. Merrill (1952). "Spectroscopic Observations of Stars of Class S". *The Astrophysical Journal* **116**: 21. doi:10.1086/145589.

[11] Donald D. Clayton and L. R. Nittler (2004). "Astrophysics with Presolar Stardust". *Annual Review of Astronomy and Astrophysics* **42** (1): 39–78. Bibcode:2004ARA&A..42...39C. doi:10.1146/annurev.astro.42.053102.134022.

[12] D. Bodansky, Donald D. Clayton, and W. A. Fowler, (1968) Nuclear quasi-equilibrium during silicon burning, *Astrophys. J. Suppl.* No. 148, **16**, 299-371

[13] See also Chapter 7 of Donald D. Clayton, *Principles of Stellar Evolution and Nucleosynthesis*, McGraw-Hill, New York (1968)

[14] Donald D. Clayton, Hoyle's Equation, *Science*, **318**, 1876-77 (2007)

[15] P.A.Seeger, W. A. Fowler, and Donald D. Clayton, Nucleosynthesis of heavy elements by neutron capture, *Astrophys. J. Suppl*, **11**, 121-66, (1965)

[16] D. Bodansky, Donald D. Clayton, and W. A. Fowler, (1968) Nuclear quasi-equilibrium during silicon burning, *Astrophys. J. Suppl.* No. 148, **16** 299-371

[17] Donald D. Clayton, Stirling A. Colgate and G. J. Fishman (1969) Gamma ray lines from young supernova remnants, *Astrophys. J..* **155** 175

[18] D. D. Clayton; S.A. Colgate; G.J. Fishman (1969). "Gamma ray lines from young supernova remnants". *The Astrophysical Journal* **155**: 75–82. Bibcode:1969ApJ...155...75C. doi:10.1086/149849.

[19] D. D. Clayton; L. R.Nittler (2004). "Astrophysics with Presolar stardust". *Annual Reviews of Astronomy and Astrophysics* **42** (1): 39–78. Bibcode:2004ARA&A..42...39C. doi:10.1146/annurev.astro.42.053102.134022.

[20] Stromberg, Joseph. "All the Gold in the Universe Could Come From the Collisions of Neutron Stars". *Smithsonian*. Retrieved 27 April 2014.

8.9   Further reading

- E. M. Burbidge, G. R. Burbidge, W. A. Fowler, F. Hoyle, *Synthesis of the Elements in Stars*, Rev. Mod. Phys. 29 (1957) 547 (article at the Physical Review Online Archive (subscription required)).

- M. Meneguzzi, J. Audouze, H. Reeves, « The production of the elements Li, Be, B by galactic cosmic rays in space and its relation with stellar observations », Astronomy and Astrophysics, vol. 15, 1971, p. 337–359

- F. Hoyle, Monthly Notices Roy. Astron. Soc. 106, 366 (1946)

- F. Hoyle, Astrophys. J. Suppl. 1, 121 (1954)

- D. D. Clayton, "Principles of Stellar Evolution and Nucleosynthesis", McGraw-Hill, 1968; University of Chicago Press, 1983, ISBN 0-226-10952-6

- C. E. Rolfs, W. S. Rodney, *Cauldrons in the Cosmos*, Univ. of Chicago Press, 1988, ISBN 0-226-72457-3.

- D. D. Clayton, "Handbook of Isotopes in the Cosmos", Cambridge University Press, 2003, ISBN 0-521-82381-1.

- C. Iliadis, "Nuclear Physics of Stars", Wiley-VCH, 2007, ISBN 978-3-527-40602-9

Chapter 9

Ionization energy

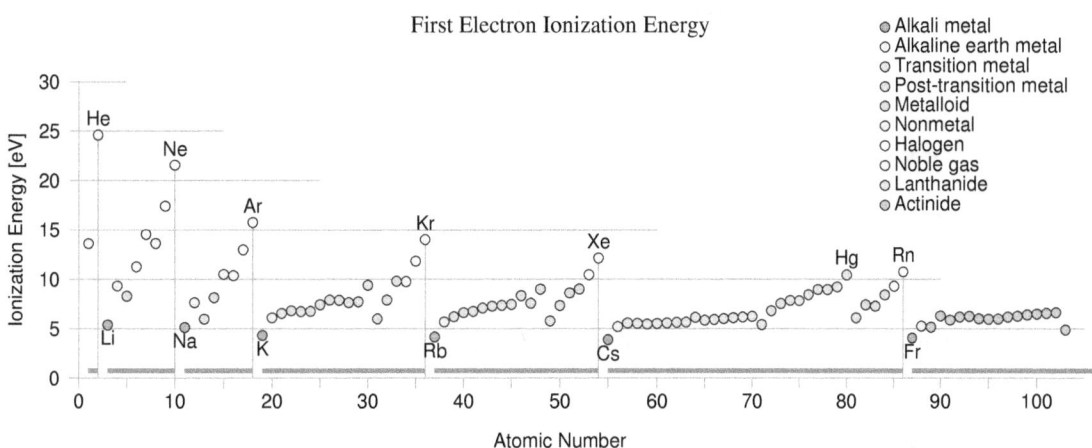

Periodic trends for ionization energy (IE) vs. proton number: note that within each of the seven periods the IE (colored circles) of an element begins at a minimum for the first column of the Periodic table (the alkali metals), and progresses to a maximum for the last column (the noble gases) which are indicated by vertical lines and labelled with a noble gas element symbol, and which also serve as lines dividing the 7 periods. Note that the maximum ionization energy for each row diminishes as one progresses from row 1 to row 7 in a given column, due to the increasing distance of the outer electron shell from the nucleus as inner shells are added.

The **ionization energy** (**IE**) is qualitatively defined as the amount of energy required to remove the most loosely bound electron of an isolated gaseous atom to form a cation. It is quantitatively expressed in symbols as:

$$X + energy \rightarrow X^+ + e^-$$

where X is any atom or molecule capable of being ionized, X^+ is that atom or molecule with an electron removed, and e^- is the removed electron. This is an endothermic process.

Comparison of IE's of atoms in the Periodic table reveals two patterns: 1)IE's generally increase as one moves from left to right within a given row (period); and 2) IE's decrease as one moves from row (period) 1 to period 7 down any given column. The latter is due to the outer electron shell being progressively further away from the nucleus with the addition of one inner shell per row as one moves down the column.

The units for ionization energy are different in physics and chemistry. In physics, the unit is the amount of energy required to remove a single electron from a single atom or molecule:expressed as an electron volt. In chemistry, the units are the amount of energy it takes for all the atoms in a mole of substance to lose one electron each: molar ionization energy or enthalpy, expressed as kilojoules per mole (kJ/mol) or kilocalories per mole (kcal/mol). [1]

The nth ionization energy refers to the amount of energy required to remove an electron from the species with a charge of $(n-1)$. For example, the first three ionization energies are defined as follows:

1st ionization energy

$$X \rightarrow X^+ + e^-$$

2nd ionization energy

$$X^+ \rightarrow X^{2+} + e^-$$

3rd ionization energy

$$X^{2+} \rightarrow X^{3+} + e^-$$

The term **ionization potential** is an older name for ionization energy,[2] because the oldest method of measuring ionization energies was based on ionizing a sample and accelerating the electron removed using an electrostatic potential. However this term is now considered obsolete.[3] Some factors affecting the ionization potential include:

1. Nuclear charge.

2. Number of electron shells.

3. Screening effect.

4. Type of orbital ionized.

5. Occupancy of the ionized orbital: completely filled or half filled.

9.1   Values and trends

Main articles: Molar ionization energies of the elements and Ionization energies of the elements (data page)

Generally the $(n+1)$th ionization energy is larger than the nth ionization energy. When the next ionization energy involves removing an electron from the same electron shell, the increase in ionization energy is primarily due to the increased net charge of the ion from which the electron is being removed. Electrons removed from more highly charged ions of a particular element experience greater forces of electrostatic attraction; thus, their removal requires more energy. In addition, when the next ionization energy involves removing an electron from a lower electron shell, the greatly decreased distance between the nucleus and the electron also increases both the electrostatic force and the distance over which that force must be overcome to remove the electron. Both of these factors further increase the ionization energy.

Some values for elements of the third period are given in the following table:

Large jumps in the successive molar ionization energies occur when passing noble gas configurations. For example, as can be seen in the table above, the first two molar ionization energies of magnesium (stripping the two 3s electrons from a magnesium atom) are much smaller than the third, which requires stripping off a 2p electron from the neon configuration of Mg^{2+}. That electron is much closer to the nucleus than the previous 3s electron.

Ionization energy is also a periodic trend within the periodic table organization. Moving left to right within a period or upward within a group, the first ionization energy generally increases with a few discrepancies (aluminum and sulphur). As the nuclear charge of the nucleus increases across the period, the atomic radius decreases and the electron cloud becomes closer towards the nucleus.

Ionization energy increases from left to right in a period and decreases from top to bottom in a group.

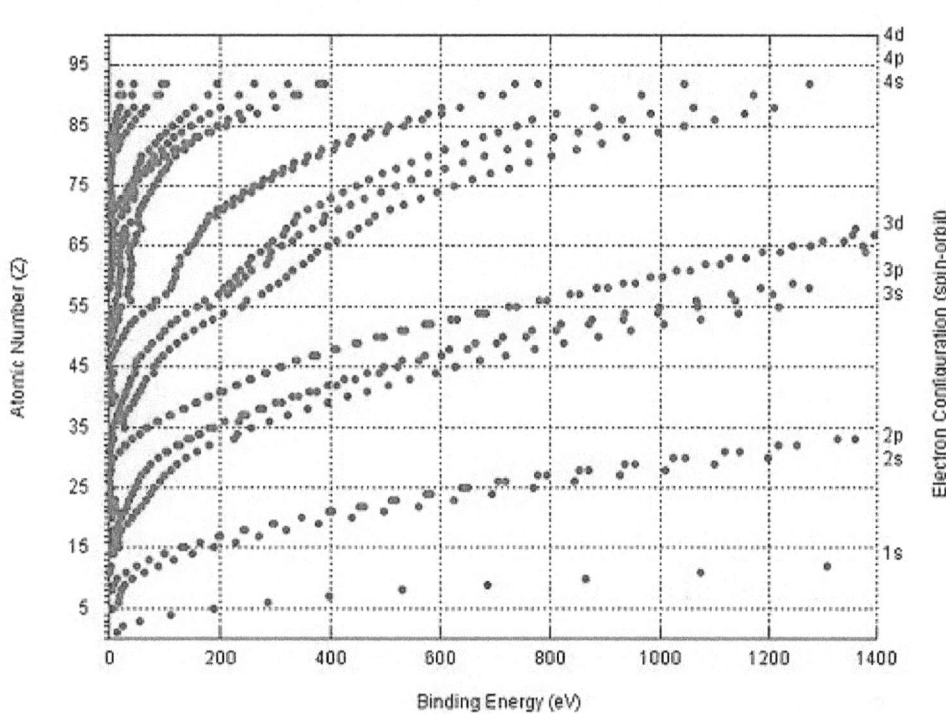

9.2 Electrostatic explanation

Atomic ionization energy can be predicted by an analysis using electrostatic potential and the Bohr model of the atom, as follows (note that the derivation uses Gaussian units).

Consider an electron of charge $-e$ and an atomic nucleus with charge $+Ze$, where Z is the number of protons in the nucleus. According to the Bohr model, if the electron were to approach and bond with the atom, it would come to rest at a certain radius a. The electrostatic potential V at distance a from the ionic nucleus, referenced to a point infinitely far away, is:

$V = \frac{Ze}{a}$

Since the electron is negatively charged, it is drawn inwards by this positive electrostatic potential. The energy required for the electron to "climb out" and leave the atom is:

$E = eV = \frac{Ze^2}{a}$

This analysis is incomplete, as it leaves the distance a as an unknown variable. It can be made more rigorous by assigning to each electron of every chemical element a characteristic distance, chosen so that this relation agrees with experimental data.

It is possible to expand this model considerably by taking a semi-classical approach, in which momentum is quantized. This approach works very well for the hydrogen atom, which only has one electron. The magnitude of the angular momentum for a circular orbit is:

$L = |\mathbf{r} \times \mathbf{p}| = rmv = n\hbar$

The total energy of the atom is the sum of the kinetic and potential energies, that is:

$$E = T + U = \frac{p^2}{2m_e} - \frac{Ze^2}{r} = \frac{m_e v^2}{2} - \frac{Ze^2}{r}$$

Velocity can be eliminated from the kinetic energy term by setting the Coulomb attraction equal to the centripetal force, giving:

$$T = \frac{Ze^2}{2r}$$

Solving the angular momentum for v and substituting this into the expression for kinetic energy, we have:

$$\frac{n^2 \hbar^2}{r m_e} = Ze^2$$

This establishes the dependence of the radius on n. That is:

$$r(n) = \frac{n^2 \hbar^2}{Z m_e e^2}$$

Now the energy can be found in terms of Z, e, and r. Using the new value for the kinetic energy in the total energy equation above, it is found that:

$$E = -\frac{Ze^2}{2r}$$

At its smallest value, n is equal to 1 and r is the Bohr radius a_0 which equals to $\frac{\hbar^2}{me^2}$. Now, the equation for the energy can be established in terms of the Bohr radius. Doing so gives the result:

$$E = -\frac{1}{n^2} \frac{Z^2 e^2}{2a_0} = -\frac{Z^2 \, 13.6eV}{n^2}$$

9.3   Quantum-mechanical explanation

According to the more complete theory of quantum mechanics, the location of an electron is best described as a probability distribution within an electron cloud, i.e. atomic orbital. The energy can be calculated by integrating over this cloud. The cloud's underlying mathematical representation is the wavefunction which is built from Slater determinants consisting of molecular spin orbitals. These are related by Pauli's exclusion principle to the antisymmetrized products of the atomic or molecular orbitals.

In general, calculating the nth ionization energy requires calculating the energies of $Z - n + 1$ and $Z - n$ electron systems. Calculating these energies exactly is not possible except for the simplest systems (i.e. hydrogen), primarily because of difficulties in integrating the electron correlation terms. Therefore, approximation methods are routinely employed, with different methods varying in complexity (computational time) and in accuracy compared to empirical data. This has become a well-studied problem and is routinely done in computational chemistry. At the lowest level of approximation, the ionization energy is provided by Koopmans' theorem.

9.4   Vertical and adiabatic ionization energy in molecules

Ionization of molecules often leads to changes in molecular geometry, and two types of (first) ionization energy are defined – *adiabatic* and *vertical*.[4]

Adiabatic ionization energy: The adiabatic ionization energy of a molecule is the *minimum* amount of energy required to remove an electron from a neutral molecule, i.e. the difference between the energy of the vibrational ground state of the neutral species (v" = 0 level) and that of the positive ion (v' = 0). The specific equilibrium geometry of each species does not affect this value.

Vertical ionization energy: Due to the possible changes in molecular geometry that may result from ionization, additional transitions may exist between the vibrational ground state of the neutral species and vibrational excited states of the positive ion. In other words, ionization is accompanied by vibrational excitation. The intensity of such transitions are explained by the Franck–Condon principle, which predicts that the most probable and intense transition corresponds to the vibrational excited state of the positive ion that has the same geometry as the neutral molecule. This transition is referred to as the "vertical" ionization energy since it is represented by a completely vertical line on a potential energy diagram (see Figure).

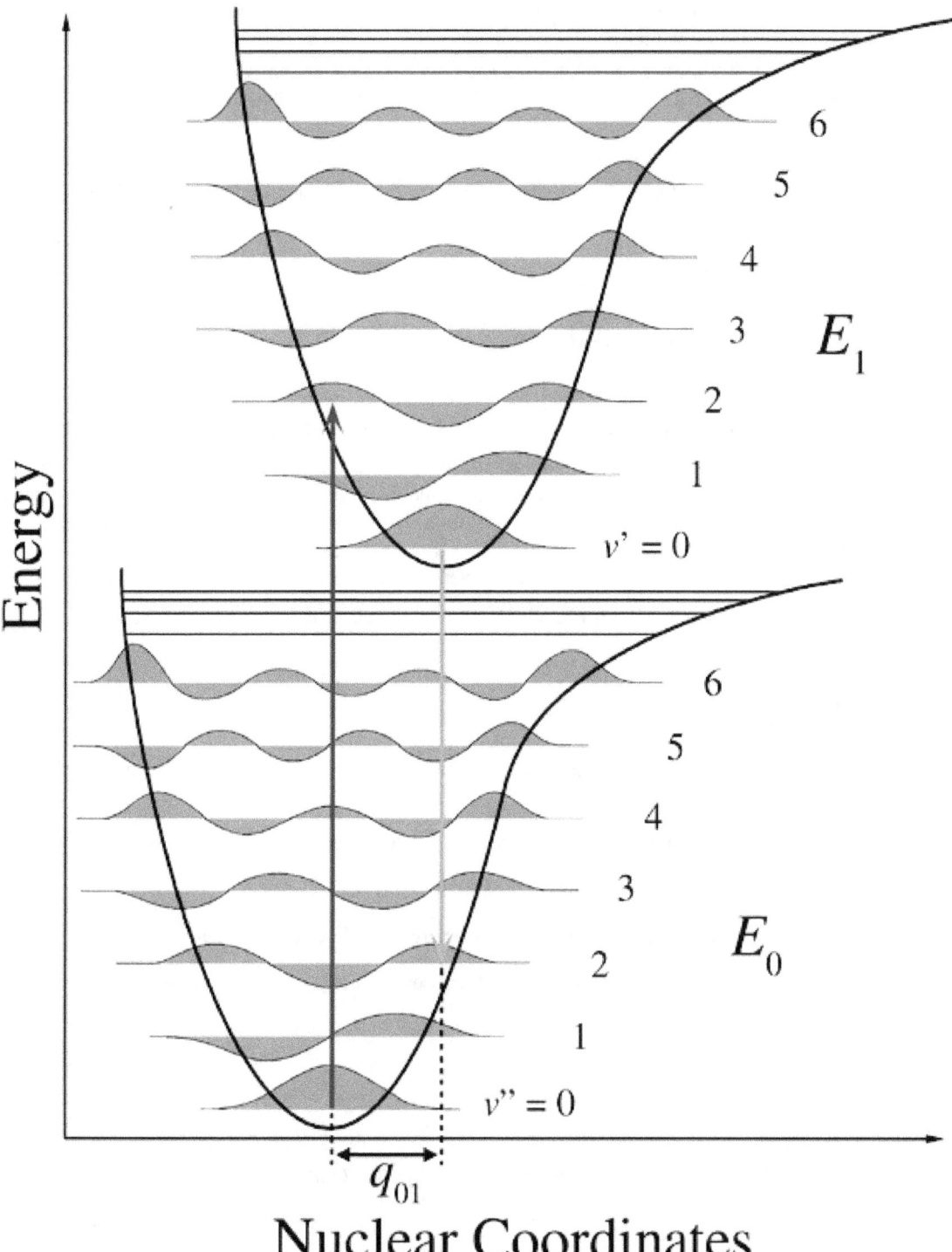

Figure 1. *Franck–Condon principle energy diagram. For ionization of a diatomic molecule the only nuclear coordinate is the bond length. The lower curve is the potential energy curve of the neutral molecule, and the upper curve is for the positive ion with a longer bond length. The blue arrow is vertical ionization, here from the ground state of the molecule to the v=2 level of the ion.*

For a diatomic molecule, the geometry is defined by the length of a single bond. The removal of an electron from a bonding molecular orbital weakens the bond and increases the bond length. In Figure 1, the lower potential energy curve is for the neutral molecule and the upper surface is for the positive ion. Both curves plot the potential energy as a function of bond length. The horizontal lines correspond to vibrational levels with their associated vibrational wave functions. Since the ion has a weaker bond, it will have a longer bond length. This effect is represented by shifting the minimum of the potential energy curve to the right of the neutral species. The adiabatic ionization is the diagonal transition to the vibrational ground state of the ion. Vertical ionization involves vibrational excitation of the ionic state and therefore requires greater energy.

In many circumstances, the adiabatic ionization energy is often a more interesting physical quantity since it describes the difference in energy between the two potential energy surfaces. However, due to experimental limitations, the adiabatic ionization energy is often difficult to determine, whereas the vertical detachment energy is easily identifiable and measurable.

9.5 Analogs of ionization energy to other systems

While the term ionization energy is largely used only for gas-phase atomic or molecular species, there are a number of analogous quantities that consider the amount of energy required to remove an electron from other physical systems.

Electron binding energy: A generic term for the ionization energy that can be used for species with any charge state. For example, the electron binding energy for the chloride ion is the minimum amount of energy required to remove an electron from the chlorine atom when it has a charge of -1. In this particular example, the electron binding energy has the same magnitude as the electron affinity for the neutral chlorine atom. In another example, the electron binding energy refers the minimum amount of energy required to remove an electron from the dicarboxylate dianion $^-O_2C(CH_2)_8CO_2^-$.

Work function: The minimum amount of energy required to remove an electron from a solid surface.

9.6 See also

- Electron affinity — a closely related concept describing the energy released by *adding* an electron to a neutral atom or molecule.

- Work function is the energy required to strip an electron from a solid to just outside its surface.

- Electronegativity is a number that shares some similarities with ionization energy.

- Koopmans' theorem, regarding the predicted ionization energies in Hartree–Fock theory.

- Di-tungsten tetra(hpp) has the lowest recorded ionization energy for a stable chemical compound.

9.7 References

[1] http://chemwiki.ucdavis.edu/Inorganic_Chemistry/Descriptive_Chemistry/Periodic_Table_of_the_Elements/Ionization_Energy

[2] F. Albert Cotton and Geoffrey Wilkinson, *Advanced Inorganic Chemistry* (5th ed., John Wiley 1988) p.1381 ISBN 0-471-84997-9

[3] IUPAC Gold Book

[4] The difference between a vertical ionization energy and adiabatic ionization energy.

Chapter 10

Strong interaction

In particle physics, the **strong interaction** is the mechanism responsible for the strong nuclear force (also called the **strong force**, **nuclear strong force** or **colour force**), one of the four fundamental interactions of nature, the others being electromagnetism, the weak interaction and gravitation. Effective only at a distance of a femtometer, it is approximately 100 times stronger than electromagnetism, a million times stronger than the weak force interaction and 10^{38} times stronger than gravitation at that range.[1] It ensures the stability of ordinary matter, as it confines the quark elementary particles into hadron particles, such as the proton and neutron, the largest components of the mass of ordinary matter. Furthermore, most of the mass-energy of a common proton or neutron is in the form of the strong force field energy; the individual quarks provide only about 1% of the mass-energy of a proton.

The strong interaction is observable in two areas: on a larger scale (about 1 to 3 femtometers (fm)), it is the force that binds protons and neutrons (nucleons) together to form the nucleus of an atom. On the smaller scale (less than about 0.8 fm, the radius of a nucleon), it is the force (carried by gluons) that holds quarks together to form protons, neutrons, and other hadron particles. The strong force inherently has so high a strength that the energy of an object bound by the strong force (a hadron) is high enough to produce new massive particles. Thus, if hadrons are struck by high-energy particles, they give rise to new hadrons instead of emitting freely moving radiation (gluons). This property of the strong force is called colour confinement, and it prevents the free "emission" of the strong force: instead, in practice, jets of massive particles are observed.

In the context of binding protons and neutrons together to form atomic nuclei, the strong interaction is called the nuclear force (or *residual strong force*). In this case, it is the residuum of the strong interaction between the quarks that make up the protons and neutrons. As such, the residual strong interaction obeys a quite different distance-dependent behavior between nucleons, from when it is acting to bind quarks within nucleons. The binding energy that is partly released on the breakup of a nucleus is related to the residual strong force and is harnessed in nuclear power and fission-type nuclear weapons.[2][3]

The strong interaction is thought to be mediated by massless particles called gluons, that are exchanged between quarks, antiquarks, and other gluons. Gluons, in turn, are thought to interact with quarks and gluons as all carry a type of charge called colour charge. Colour charge is analogous to electromagnetic charge, but it comes in three types rather than one (+/- red, +/- green, +/- blue) that results in a different type of force, with different rules of behavior. These rules are detailed in the theory of quantum chromodynamics (QCD), which is the theory of quark-gluon interactions.

Just after the Big Bang, and during the electroweak epoch, the electroweak force separated from the strong force. Although it is expected that a Grand Unified Theory exists to describe this, no such theory has been successfully formulated, and the unification remains an unsolved problem in physics.

10.1 History

Before the 1970s, physicists were uncertain about the binding mechanism of the atomic nucleus. It was known that the nucleus was composed of protons and neutrons and that protons possessed positive electric charge, while neutrons

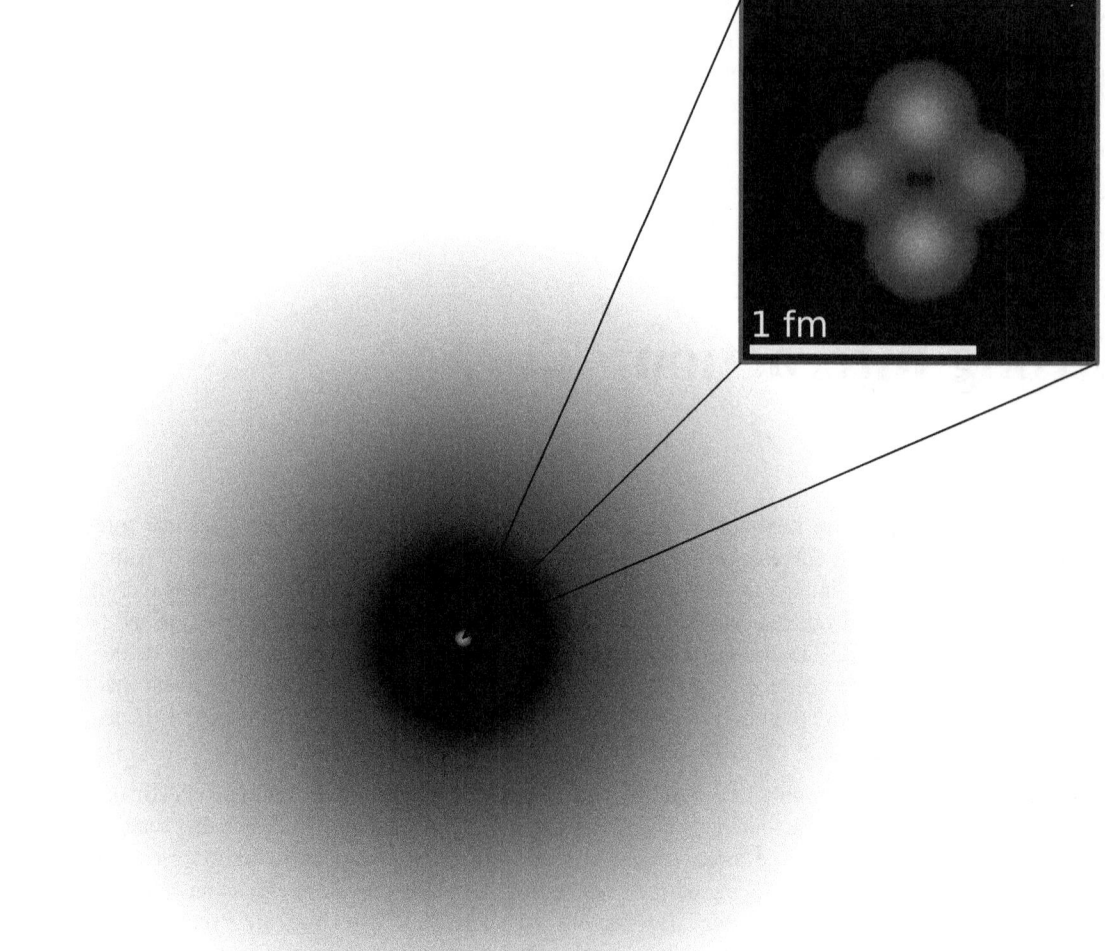

$$1 \text{ Å} = 100,000 \text{ fm}$$

The nucleus of a helium atom. The two protons have the same charge, but still stay together due to the residual nuclear force

were electrically neutral. However, these facts seemed to contradict one another. By physical understanding at that time, positive charges would repel one another and the nucleus should therefore fly apart. However, this was never observed. New physics was needed to explain this phenomenon.

A stronger attractive force was postulated to explain how the atomic nucleus was bound together despite the protons' mutual electromagnetic repulsion. This hypothesized force was called the *strong force*, which was believed to be a fundamental force that acted on the protons and neutrons that make up the nucleus.

It was later discovered that protons and neutrons were not fundamental particles, but were made up of constituent particles called quarks. The strong attraction between nucleons was the side-effect of a more fundamental force that bound the quarks together in the protons and neutrons. The theory of quantum chromodynamics explains that quarks carry what is called a colour charge, although it has no relation to visible colour.[4] Quarks with unlike colour charge attract one another as a result of the **strong interaction**, which is mediated by particles called gluons.

10.2 Details

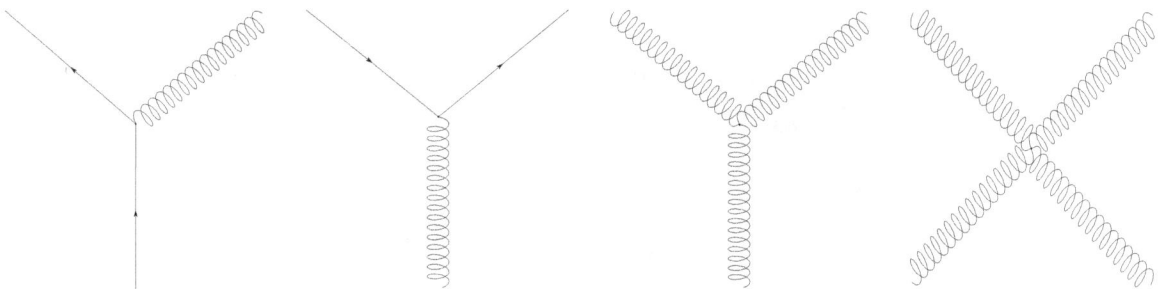

The fundamental couplings of the strong interaction, from left to right: gluon radiation, gluon splitting and gluon self-coupling.

The word *strong* is used since the strong interaction is the "strongest" of the four fundamental forces; its strength is around 10^2 times that of the electromagnetic force, some 10^6 times as great as that of the weak force, and about 10^{39} times that of gravitation, at a distance of a femtometer or less.

10.2.1 Behaviour of the strong force

The contemporary understanding of strong force is described by quantum chromodynamics (QCD), a part of the standard model of particle physics. Mathematically, QCD is a non-Abelian gauge theory based on a local (gauge) symmetry group called SU(3).

Quarks and gluons are the only fundamental particles that carry non-vanishing colour charge, and hence participate in strong interactions. The strong force itself acts directly only on elementary quark and gluon particles.

All quarks and gluons in QCD interact with each other through the strong force. The strength of interaction is parametrized by the strong coupling constant. This strength is modified by the gauge colour charge of the particle, a group theoretical property.

The strong force acts between quarks. Unlike all other forces (electromagnetic, weak, and gravitational), the strong force does not diminish in strength with increasing distance. After a limiting distance (about the size of a hadron) has been reached, it remains at a strength of about 10,000 newtons, no matter how much farther the distance between the quarks.[5] In QCD, this phenomenon is called colour confinement; it implies that only hadrons, not individual free quarks, can be observed. The explanation is that the amount of work done against a force of 10,000 newtons (about the weight of a one-metric ton mass on the surface of the Earth) is enough to create particle-antiparticle pairs within a very short distance of an interaction. In simple terms, the very energy applied to pull two quarks apart will create a pair of new quarks that will pair up with the original ones. The failure of all experiments that have searched for free quarks is considered to be evidence for this phenomenon.

The elementary quark and gluon particles affected are unobservable directly, but they instead emerge as jets of newly created hadrons, whenever energy is deposited into a quark-quark bond, as when a quark in a proton is struck by a very fast quark (in an impacting proton) during a particle accelerator experiment. However, quark–gluon plasmas have been observed.

Every quark in the universe does not attract every other quark in the above distance independent manner, since colour-confinement implies that the strong force acts without distance-diminishment only between pairs of single quarks, and that in collections of bound quarks (i.e., hadrons), the net colour-charge of the quarks cancels out, as seen from far away. Collections of quarks (hadrons) therefore appear (nearly) without colour-charge, and the strong force is therefore nearly absent between these hadrons (i.e., between baryons or mesons). However, the cancellation is not quite perfect. A small residual force remains (described below) known as the **residual strong force**. This residual force *does* diminish rapidly with distance, and is thus very short-range (effectively a few femtometers). It manifests as a force between the "colourless" hadrons, and is therefore sometimes known as the **strong nuclear force** or simply nuclear force.

10.2.2   Residual strong force

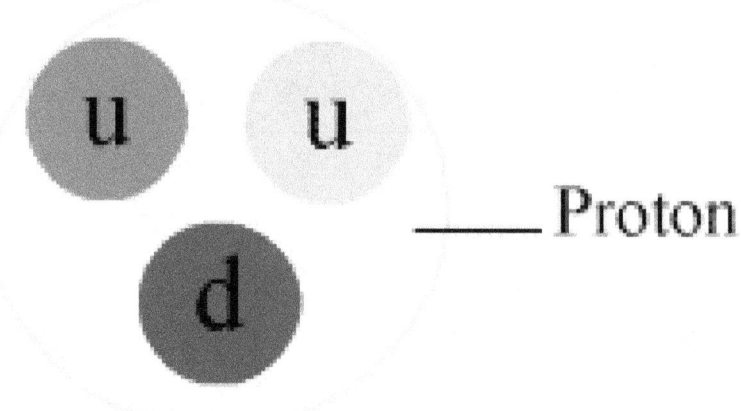

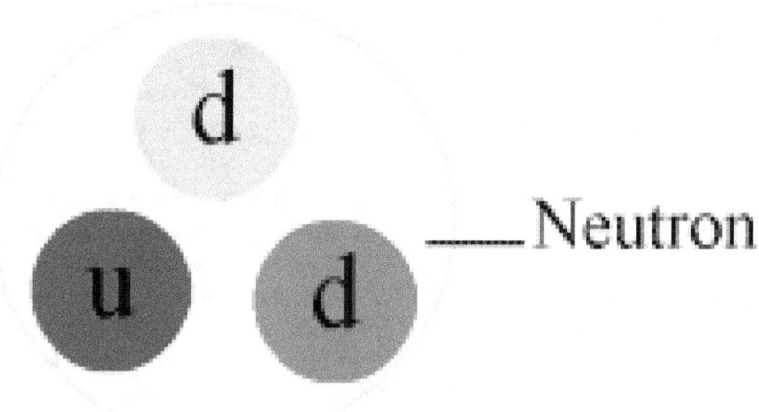

An animation of the nuclear force (or residual strong force) interaction between a proton and a neutron. The small coloured double circles are gluons, which can be seen binding the proton and neutron together. These gluons also hold the quark-antiquark combination called the pion together, and thus help transmit a residual part of the strong force even between colourless hadrons. Anticolours are shown as per this diagram. For a larger version, click here

The residual effect of the strong force is called the nuclear force. The nuclear force acts between hadrons, such as mesons or the nucleons in atomic nuclei. This "residual strong force", acting indirectly, transmits gluons that form part of the virtual pi and rho mesons, which, in turn, transmit the nuclear force between nucleons.

The residual strong force is thus a minor residuum of the strong force that binds quarks together into protons and neutrons.

This same force is much weaker *between* neutrons and protons, because it is mostly neutralized *within* them, in the same way that electromagnetic forces between neutral atoms (van der Waals forces) are much weaker than the electromagnetic forces that hold the atoms internally together.[6]

Unlike the strong force itself, the nuclear force, or residual strong force, *does* diminish in strength, and in fact diminishes rapidly with distance. The decrease is approximately as a negative exponential power of distance, though there is no simple expression known for this; see Yukawa potential. This fact, together with the less-rapid decrease of the disruptive electromagnetic force between protons with distance, causes the instability of larger atomic nuclei, such as all those with atomic numbers larger than 82 (the element lead).

10.3 See also

- Nuclear binding energy

- Colour charge

- Coupling constant

- Nuclear physics

- QCD matter

- Quantum field theory and Gauge theory

- Standard model of particle physics and Standard Model (mathematical formulation)

- Weak interaction, electromagnetism and gravity

- Intermolecular force

- Vortex

- Yukawa interaction

10.4 References

[1] Relative strength of interaction varies with distance. See for instance Matt Strassler's essay, "The strength of the known forces".

[2] on Binding energy: see Binding Energy, Mass Defect, Furry Elephant physics educational site, retr 2012 7 1

[3] on Binding energy: see Chapter 4 NUCLEAR PROCESSES, THE STRONG FORCE, M. Ragheb 1/27/2012, University of Illinois

[4] Feynman, R. P. (1985). *QED: The Strange Theory of Light and Matter*. Princeton University Press. p. 136. ISBN 0-691-08388-6. The idiot physicists, unable to come up with any wonderful Greek words anymore, call this type of polarization by the unfortunate name of 'colour,' which has nothing to do with colour in the normal sense.

[5] Fritzsch, op. cite, p. 164. The author states that the force between differently coloured quarks remains constant at any distance after they travel only a tiny distance from each other, and is equal to that need to raise one ton, which is 1000 kg x 9.8 m/s^2 = ~10,000 N.

[6] Fritzsch, H. (1983). *Quarks: The Stuff of Matter*. Basic Books. pp. 167–168. ISBN 978-0-465-06781-7.

10.5   Further reading

- Christman, J. R. (2001). "MISN-0-280: *The Strong Interaction*" (PDF). *Project PHYSNET*. External link in |work= (help)

- Griffiths, David (1987). *Introduction to Elementary Particles*. John Wiley & Sons. ISBN 0-471-60386-4.

- Halzen, F.; Martin, A. D. (1984). *Quarks and Leptons: An Introductory Course in Modern Particle Physics*. John Wiley & Sons. ISBN 0-471-88741-2.

- Kane, G. L. (1987). *Modern Elementary Particle Physics*. Perseus Books. ISBN 0-201-11749-5.

- Morris, R. (2003). *The Last Sorcerers: The Path from Alchemy to the Periodic Table*. Joseph Henry Press. ISBN 0-309-50593-3.

10.6   External links

- Strong force at *Encyclopædia Britannica*

Chapter 11

Mass–energy equivalence

"E=MC2" and "E=mc2" redirect here. For other uses, see E=MC2 (disambiguation).
In physics, **mass–energy equivalence** is the concept that mass and energy are the same thing, so that every mass has an

The four-metre tall sculpture of Einstein's 1905 formula E=mc² at the 2006 Walk of Ideas, Berlin, Germany.

energy equivalent and vice versa. This relationship is expressed using the formula

$$E = mc^2$$

where E is the energy of a physical system, m is the mass of the system, and c is the speed of light in a vacuum (about 3×10^8 m/s). In words, energy equals mass multiplied by the speed of light squared. Because the speed of light is a very large number in everyday units, the formula implies that any small amount of matter contains a very large amount of energy. Some of this energy may be released as heat and light by nuclear transformations. This also serves to convert units of mass to units of energy, no matter what system of measurement units used.

Mass–energy equivalence arose originally from special relativity as a paradox described by Henri Poincaré.[1] It was proposed by Albert Einstein in 1905, in the paper "Does the inertia of a body depend upon its energy-content?", one of his Annus Mirabilis ("Miraculous Year") Papers.[2] Einstein was the first to propose that the equivalence of mass and energy is a general principle and a consequence of the symmetries of space and time.

A consequence of the mass–energy equivalence is that if a body is stationary, it still has some internal or intrinsic energy, called its rest energy. Rest mass and rest energy are equivalent and remain proportional to one another. When the body is in motion (relative to an observer), its total energy is greater than its rest energy. The rest mass (or rest energy) remains

$$c^2 = 89{,}875{,}517{,}873{,}681{,}800 \ m^2/s^2$$

$E = mc^2$ *explicated.*

an important quantity in this case because it remains the same regardless of this motion, even for the extreme speeds or gravity considered in special and general relativity; thus it is also called the invariant mass.

11.1   Nomenclature

The formula was initially written in many different notations, and its interpretation and justification was further developed in several steps.[3][4]

- In "*Does the inertia of a body depend upon its energy content?*" (1905), Einstein used V to mean the speed of light in a vacuum and L to mean the energy lost by a body in the form of radiation.[2] Consequently, the equation $E = mc^2$ was not originally written as a formula but as a sentence in German saying that *if a body gives off the energy L in the form of radiation, its mass diminishes by* L/V^2. A remark placed above it informed that the equation was approximated by neglecting "magnitudes of fourth and higher orders" of a series expansion.[5]

- In May 1907, Einstein explained that the expression for energy ε of a moving mass point assumes the simplest form, when its expression for the state of rest is chosen to be $\varepsilon_0 = \mu V^2$ (where μ is the mass), which is in agreement with the "principle of the equivalence of mass and energy". In addition, Einstein used the formula $\mu = E_0/V^2$, with E_0 being the energy of a system of mass points, in order to describe the energy and mass increase of that system when the velocity of the differently moving mass points is increased.[6]

- In June 1907, Max Planck rewrote Einstein's mass–energy relationship as $M = (E_0 + pV_0)/c^2$, where p is the pressure and V the volume, in order to express the relation between mass, its "latent energy", and thermodynamic energy within the body.[7] Subsequently in October 1907, this was rewritten as $M_0 = E_0/c^2$ and given a quantum interpretation by Johannes Stark, who assumed its validity and correctness (*Gültigkeit*).[8]

- In December 1907, Einstein expressed the equivalence in the form $M = \mu + E_0/c^2$ and concluded: *A mass μ is equivalent, as regards inertia, to a quantity of energy μc^2.* [...] *It appears far more natural to consider every inertial mass as a store of energy.*[9][10]

- In 1909, Gilbert N. Lewis and Richard C. Tolman used two variations of the formula: $m = E/c^2$ and $m_0 = E_0/c^2$, with E being the energy of a moving body, E_0 its rest energy, m the relativistic mass, and m_0 the invariant mass.[11]

The same relations in different notation were used by Hendrik Lorentz in 1913 (published 1914), though he placed the energy on the left-hand side: $\varepsilon = Mc^2$ and $\varepsilon_0 = mc^2$, with ε being the total energy (rest energy plus kinetic energy) of a moving material point, ε_0 its rest energy, M the relativistic mass, and m the invariant (or rest) mass.[12]

- In 1911, Max von Laue gave a more comprehensive proof of $M_0 = E_0/c^2$ from the stress–energy tensor,[13] which was later (1918) generalized by Felix Klein.[14]

- Einstein returned to the topic once again after World War II and this time he wrote $E = mc^2$ in the title of his article[15] intended as an explanation for a general reader by analogy.[16]

11.2 Conservation of mass and energy

Main articles: Conservation of energy and Conservation of mass

Mass and energy can be seen as two names (and two measurement units) for the same underlying, conserved physical quantity.[17] Thus, the laws of conservation of energy and conservation of (total) mass are equivalent and both hold true.[18] Einstein elaborated in a 1946 essay that "the principle of the conservation of mass [...] proved inadequate in the face of the special theory of relativity. It was therefore merged with the energy [conservation] principle—just as, about 60 years before, the principle of the conservation of mechanical energy had been combined with the principle of the conservation of heat [thermal energy]. We might say that the principle of the conservation of energy, having previously swallowed up that of the conservation of heat, now proceeded to swallow that of the conservation of mass—and holds the field alone."[19]

If the conservation of mass law is interpreted as conservation of *rest* mass, it does not hold true in special relativity. The *rest* energy (equivalently, rest mass) of a particle can be converted, not "to energy" (it already *is* energy (mass)), but rather to *other* forms of energy (mass) which require motion, such as kinetic energy, thermal energy, or radiant energy; similarly, kinetic or radiant energy can be converted to other kinds of particles which have rest energy (rest mass). In the transformation process, neither the total amount of mass nor the total amount of energy changes, since both are properties which are connected to each other via a simple constant.[20][21] This view requires that if either energy or (total) mass disappears from a system, it will always be found that both have simply moved off to another place, where they may both be measured as an increase of both energy and mass corresponding to the loss in the first system.

11.2.1 Fast-moving objects and systems of objects

When an object is pushed in the direction of motion, it gains momentum and energy, but when the object is already traveling near the speed of light, it cannot move much faster, no matter how much energy it absorbs. Its momentum and energy continue to increase without bounds, whereas its speed approaches a constant value—the speed of light. This implies that in relativity the momentum of an object cannot be a constant times the velocity, nor can the kinetic energy be a constant times the square of the velocity.

A property called the relativistic mass is defined as the ratio of the momentum of an object to its velocity.[22] Relativistic mass depends on the motion of the object, so that different observers in relative motion see different values for it. If the object is moving slowly, the relativistic mass is nearly equal to the rest mass and both are nearly equal to the usual Newtonian mass. If the object is moving quickly, the relativistic mass is greater than the rest mass by an amount equal to the mass associated with the kinetic energy of the object. As the object approaches the speed of light, the relativistic mass grows infinitely, because the kinetic energy grows infinitely and this energy is associated with mass.

The relativistic mass is always equal to the total energy (rest energy plus kinetic energy) divided by c^2.[23] Because the relativistic mass is exactly proportional to the energy, relativistic mass and relativistic energy are nearly synonyms; the only difference between them is the units. If length and time are measured in natural units, the speed of light is equal to 1, and even this difference disappears. Then mass and energy have the same units and are always equal, so it is redundant to speak about relativistic mass, because it is just another name for the energy. This is why physicists usually reserve the useful short word "mass" to mean rest mass, or invariant mass, and not relativistic mass.

The relativistic mass of a moving object is larger than the relativistic mass of an object that is not moving, because a moving object has extra kinetic energy. The *rest mass* of an object is defined as the mass of an object when it is at rest, so that the rest mass is always the same, independent of the motion of the observer: it is the same in all inertial frames.

For things and systems made up of many parts, like an atomic nucleus, planet, or star, the relativistic mass is the sum of the relativistic masses (or energies) of the parts, because energies are additive in isolated systems. This is not true in systems which are open, however, if energy is subtracted. For example, if a system is *bound* by attractive forces, and the energy gained due to the forces of attraction in excess of the work done is removed from the system, then mass will be lost with this removed energy. For example, the mass of an atomic nucleus is less than the total mass of the protons and neutrons that make it up, but this is only true after this energy from binding has been removed in the form of a gamma ray (which in this system, carries away the mass of the energy of binding). This mass decrease is also equivalent to the energy required to break up the nucleus into individual protons and neutrons (in this case, work and mass would need to be supplied). Similarly, the mass of the solar system is slightly less than the sum of the individual masses of the sun and planets.

For a system of particles going off in different directions, the invariant mass of the system is the analog of the rest mass, and is the same for all observers, even those in relative motion. It is defined as the total energy (divided by c^2) in the center of mass frame (where by definition, the system total momentum is zero). A simple example of an object with moving parts but zero total momentum is a container of gas. In this case, the mass of the container is given by its total energy (including the kinetic energy of the gas molecules), since the system total energy and invariant mass are the same in any reference frame where the momentum is zero, and such a reference frame is also the only frame in which the object can be weighed. In a similar way, the theory of special relativity posits that the thermal energy in all objects (including solids) contributes to their total masses and weights, even though this energy is present as the kinetic and potential energies of the atoms in the object, and it (in a similar way to the gas) is not seen in the rest masses of the atoms that make up the object.

In a similar manner, even photons (light quanta), if trapped in a container space (as a photon gas or thermal radiation), would contribute a mass associated with their energy to the container. Such an extra mass, in theory, could be weighed in the same way as any other type of rest mass. This is true in special relativity theory, even though individually photons have no rest mass. The property that trapped energy *in any form* adds weighable mass to systems that have no net momentum is one of the characteristic and notable consequences of relativity. It has no counterpart in classical Newtonian physics, in which radiation, light, heat, and kinetic energy never exhibit weighable mass under any circumstances.

Just as the relativistic mass of an isolated system is conserved through time, so also is its invariant mass. It is this property which allows the conservation of all types of mass in systems, and also conservation of all types of mass in reactions where matter is destroyed (annihilated), leaving behind the energy that was associated with it (which is now in non-material form, rather than material form). Matter may appear and disappear in various reactions, but mass and energy are both unchanged in this process.

11.3   Applicability of the strict mass–energy equivalence formula, $E = mc^2$

As is noted above, two different definitions of mass have been used in special relativity, and also two different definitions of energy. The simple equation $E = mc^2$ is not generally applicable to all these types of mass and energy, except in the special case that the total additive momentum is zero for the system under consideration. In such a case, which is always guaranteed when observing the system from either its center of mass frame or its center of momentum frame, $E = mc^2$ is always true for any type of mass and energy that are chosen. Thus, for example, in the center of mass frame, the total energy of an object or system is equal to its rest mass times c^2, a useful equality. This is the relationship used for the container of gas in the previous example. It is *not* true in other reference frames where the center of mass is in motion. In these systems or for such an object, its total energy will depend on both its rest (or invariant) mass, and also its (total) momentum.[24]

In inertial reference frames other than the rest frame or center of mass frame, the equation $E = mc^2$ remains true if the energy is the relativistic energy *and* the mass is the relativistic mass. It is also correct if the energy is the rest or invariant energy (also the minimum energy), *and* the mass is the rest mass, or the invariant mass. However, connection of the **total or relativistic energy** (E_r) with the **rest or invariant mass** (m_0) requires consideration of the system total momentum, in systems and reference frames where the total momentum has a non-zero value. The formula then required to connect the two different kinds of mass and energy, is the extended version of Einstein's equation, called the relativistic energy–momentum relation:[25]

$$E_r^2 - |\vec{p}|^2 c^2 = m_0^2 c^4$$
$$E_r^2 - (pc)^2 = (m_0 c^2)^2$$

or

$$E_r = \sqrt{(m_0 c^2)^2 + (pc)^2}$$

Here the $(pc)^2$ term represents the square of the Euclidean norm (total vector length) of the various momentum vectors in the system, which reduces to the square of the simple momentum magnitude, if only a single particle is considered. This equation reduces to $E = mc^2$ when the momentum term is zero. For photons where $m_0 = 0$, the equation reduces to $Er = pc$.

11.4 Meanings of the strict mass–energy equivalence formula, $E = mc^2$

Mass–energy equivalence states that any object has a certain energy, even when it is stationary. In Newtonian mechanics, a motionless body has no kinetic energy, and it may or may not have other amounts of internal stored energy, like chemical energy or thermal energy, in addition to any potential energy it may have from its position in a field of force. In Newtonian mechanics, all of these energies are much smaller than the mass of the object times the speed of light squared.

In relativity, all of the energy that moves along with an object (that is, all the energy which is present in the object's rest frame) contributes to the total mass of the body, which measures how much it resists acceleration. Each potential and kinetic energy makes a proportional contribution to the mass. As noted above, even if a box of ideal mirrors "contains" light, then the individually massless photons still contribute to the total mass of the box, by the amount of their energy divided by c².[26]

In relativity, removing energy is removing mass, and for an observer in the center of mass frame, the formula $m = E/c^2$ indicates how much mass is lost when energy is removed. In a nuclear reaction, the mass of the atoms that come out is less than the mass of the atoms that go in, and the difference in mass shows up as heat and light which has the same relativistic mass as the difference (and also the same invariant mass in the center of mass frame of the system). In this case, the E in the formula is the energy released and removed, and the mass m is how much the mass decreases. In the same way, when any sort of energy is added to an isolated system, the increase in the mass is equal to the added energy divided by c^2. For example, when water is heated it gains about 1.11×10^{-17} kg of mass for every joule of heat added to the water.

An object moves with different speed in different frames, depending on the motion of the observer, so the kinetic energy in both Newtonian mechanics and relativity is *frame dependent*. This means that the amount of relativistic energy, and therefore the amount of relativistic mass, that an object is measured to have depends on the observer. The *rest mass* is defined as the mass that an object has when it is not moving (or when an inertial frame is chosen such that it is not moving). The term also applies to the invariant mass of systems when the system as a whole is not "moving" (has no net momentum). The rest and invariant masses are the smallest possible value of the mass of the object or system. They also are conserved quantities, so long as the system is isolated. Because of the way they are calculated, the effects of moving observers are subtracted, so these quantities do not change with the motion of the observer.

The rest mass is almost never additive: the rest mass of an object is not the sum of the rest masses of its parts. The rest mass of an object is the total energy of all the parts, including kinetic energy, as measured by an observer that sees the center of the mass of the object to be standing still. The rest mass adds up only if the parts are standing still and do not attract or repel, so that they do not have any extra kinetic or potential energy. The other possibility is that they have a positive kinetic energy and a negative potential energy that exactly cancels.

The mass–energy equivalence formula was displayed on Taipei 101 during the event of the World Year of Physics 2005.

11.4.1   Binding energy and the "mass defect"

Main article: binding energy

Whenever any type of energy is removed from a system, the mass associated with the energy is also removed, and the system therefore loses mass. This mass defect in the system may be simply calculated as $\Delta m = \Delta E/c^2$, and this was the form of the equation historically first presented by Einstein in 1905. However, use of this formula in such circumstances has led to the false idea that mass has been "converted" to energy. This may be particularly the case when the energy (and mass) removed from the system is associated with the *binding energy* of the system. In such cases, the binding energy is observed as a "mass defect" or deficit in the new system.

The fact that the released energy is not easily weighed in many such cases, may cause its mass to be neglected as though it no longer existed. This circumstance has encouraged the false idea of conversion of *mass* to energy, rather than the correct idea that the binding energy of such systems is relatively large, and exhibits a measurable mass, which is removed when the binding energy is removed. This energy is often released in the form of light and heat, which is too quickly and widely dispersed to be easily weighed, though it does carry mass.

The difference between the rest mass of a bound system and of the unbound parts is the binding energy of the system, if this energy has been removed after binding. For example, a water molecule weighs a little less than two free hydrogen atoms and an oxygen atom; the minuscule mass difference is the energy that is needed to split the molecule into three individual atoms (divided by c^2), and which was given off as heat when the molecule formed (this heat had mass). Likewise, a stick of dynamite in theory weighs a little bit more than the fragments after the explosion, but this is true only so long as the fragments are cooled and the heat removed. In this case the mass difference is the energy/heat that is released when the dynamite explodes, and when this heat escapes, the mass associated with it escapes, only to be deposited in the surroundings which absorb the heat (so that total mass is conserved).

Such a change in mass may only happen when the system is open, and the energy and mass escapes. Thus, if a stick of dynamite is blown up in a hermetically sealed chamber, the mass of the chamber and fragments, the heat, sound, and light would still be equal to the original mass of the chamber and dynamite. If sitting on a scale, the weight and mass would not change. This would in theory also happen even with a nuclear bomb, if it could be kept in an ideal box of infinite strength, which did not rupture or pass radiation.[21] Thus, a 21.5 kiloton (9×10^{13} joule) nuclear bomb produces about one gram of heat and electromagnetic radiation, but the mass of this energy would not be detectable in an exploded bomb in an ideal box sitting on a scale; instead, the contents of the box would be heated to millions of degrees without changing total mass and weight. If then, however, a transparent window (passing only electromagnetic radiation) were opened in such an ideal box after the explosion, and a beam of X-rays and other lower-energy light allowed to escape the box, it would eventually be found to weigh one gram less than it had before the explosion. This weight loss and mass loss would happen as the box was cooled by this process, to room temperature. However, any surrounding mass which had absorbed the X-rays (and other "heat") would *gain* this gram of mass from the resulting heating, so the mass "loss" would represent merely its relocation. Thus, no mass (or, in the case of a nuclear bomb, no matter) would be "converted" to energy in such a process. Mass and energy, as always, would both be separately conserved.

11.4.2   Massless particles

Massless particles have zero rest mass. Their relativistic mass is simply their relativistic energy, divided by c^2, or $m_{rel} = E/c^2$.[27][28] The energy for photons is $E = hf$, where h is Planck's constant and f is the photon frequency. This frequency and thus the relativistic energy are frame-dependent.

If an observer runs away from a photon in the direction it travels from a source, having it catch up with the observer, then when the photon catches up it will be seen as having less energy than it had at the source. The faster the observer is traveling with regard to the source when the photon catches up, the less energy the photon will have. As an observer approaches the speed of light with regard to the source, the photon looks redder and redder, by relativistic Doppler effect (the Doppler shift is the relativistic formula), and the energy of a very long-wavelength photon approaches zero. This is why a photon is *massless*; this means that the rest mass of a photon is zero.

11.4.3   Massless particles contribute rest mass and invariant mass to systems

Two photons moving in different directions cannot both be made to have arbitrarily small total energy by changing frames, or by moving toward or away from them. The reason is that in a two-photon system, the energy of one photon is decreased by chasing after it, but the energy of the other will increase with the same shift in observer motion. Two photons not moving in the same direction will exhibit an inertial frame where the combined energy is smallest, but not zero. This is called the center of mass frame or the center of momentum frame; these terms are almost synonyms (the center of mass frame is the special case of a center of momentum frame where the center of mass is put at the origin). The most that chasing a pair of photons can accomplish to decrease their energy is to put the observer in a frame where the photons have equal energy and are moving directly away from each other. In this frame, the observer is now moving in the same direction and speed as the center of mass of the two photons. The total momentum of the photons is now zero, since their momenta are equal and opposite. In this frame the two photons, as a system, have a mass equal to their total energy divided by c^2. This mass is called the invariant mass of the pair of photons together. It is the smallest mass and energy the system may be seen to have, by any observer. It is only the invariant mass of a two-photon system that can be used to make a single particle with the same rest mass.

If the photons are formed by the collision of a particle and an antiparticle, the invariant mass is the same as the total energy of the particle and antiparticle (their rest energy plus the kinetic energy), in the center of mass frame, where they will automatically be moving in equal and opposite directions (since they have equal momentum in this frame). If the photons are formed by the disintegration of a *single* particle with a well-defined rest mass, like the neutral pion, the invariant mass of the photons is equal to rest mass of the pion. In this case, the center of mass frame for the pion is just the frame where the pion is at rest, and the center of mass does not change after it disintegrates into two photons. After the two photons are formed, their center of mass is still moving the same way the pion did, and their total energy in this frame adds up to the mass energy of the pion. Thus, by calculating the invariant mass of pairs of photons in a particle detector, pairs can be identified that were probably produced by pion disintegration.

A similar calculation illustrates that the invariant mass of systems is conserved, even when massive particles (particles with rest mass) within the system are converted to massless particles (such as photons). In such cases, the photons contribute invariant mass to the system, even though they individually have no invariant mass or rest mass. Thus, an electron and positron (each of which has rest mass) may undergo annihilation with each other to produce two photons, each of which is massless (has no rest mass). However, in such circumstances, no system mass is lost. Instead, the system of both photons moving away from each other has an invariant mass, which acts like a rest mass for any system in which the photons are trapped, or that can be weighed. Thus, not only the quantity of relativistic mass, but also the quantity of invariant mass does not change in transformations between "matter" (electrons and positrons) and energy (photons).

11.4.4   Relation to gravity

In physics, there are two distinct concepts of mass: the gravitational mass and the inertial mass. The gravitational mass is the quantity that determines the strength of the gravitational field generated by an object, as well as the gravitational force acting on the object when it is immersed in a gravitational field produced by other bodies. The inertial mass, on the other hand, quantifies how much an object accelerates if a given force is applied to it. The mass–energy equivalence in special relativity refers to the inertial mass. However, already in the context of Newton gravity, the Weak Equivalence Principle is postulated: the gravitational and the inertial mass of every object are the same. Thus, the mass–energy equivalence, combined with the Weak Equivalence Principle, results in the prediction that all forms of energy contribute to the gravitational field generated by an object. This observation is one of the pillars of the general theory of relativity.

The above prediction, that all forms of energy interact gravitationally, has been subject to experimental tests. The first observation testing this prediction was made in 1919.[29] During a solar eclipse, Arthur Eddington observed that the light from stars passing close to the Sun was bent. The effect is due to the gravitational attraction of light by the Sun. The observation confirmed that the energy carried by light indeed is equivalent to a gravitational mass. Another seminal experiment, the Pound–Rebka experiment, was performed in 1960.[30] In this test a beam of light was emitted from the top of a tower and detected at the bottom. The frequency of the light detected was higher than the light emitted. This result confirms that the energy of photons increases when they fall in the gravitational field of the Earth. The energy, and therefore the gravitational mass, of photons is proportional to their frequency as stated by the Planck's relation.

11.5 Application to nuclear physics

Task Force One, *the world's first nuclear-powered task force.* Enterprise, Long Beach *and* Bainbridge *in formation in the Mediterranean, 18 June 1964.* Enterprise *crew members are spelling out Einstein's mass–energy equivalence formula* E = mc² *on the flight deck.*

Max Planck pointed out that the mass–energy equivalence formula implied that bound systems would have a mass less than the sum of their constituents, once the binding energy had been allowed to escape. However, Planck was thinking about chemical reactions, where the binding energy is too small to measure. Einstein suggested that radioactive materials such as radium would provide a test of the theory, but even though a large amount of energy is released per atom in radium, due to the half-life of the substance (1602 years), only a small fraction of radium atoms decay over an experimentally measurable period of time.

Once the nucleus was discovered, experimenters realized that the very high binding energies of the atomic nuclei should allow calculation of their binding energies, simply from mass differences. But it was not until the discovery of the neutron in 1932, and the measurement of the neutron mass, that this calculation could actually be performed (see nuclear binding energy for example calculation). A little while later, the first transmutation reactions (such as[31] the Cockcroft–Walton experiment: $^7\text{Li} + p \rightarrow 2\ ^4\text{He}$) verified Einstein's formula to an accuracy of ±0.5%. In 2005, Rainville et al. published a direct test of the energy-equivalence of mass lost in the binding energy of a neutron to atoms of particular isotopes of silicon and sulfur, by comparing the mass lost to the energy of the emitted gamma ray associated with the neutron capture. The binding mass-loss agreed with the gamma ray energy to a precision of ±0.00004%, the most accurate test of $E = mc^2$ to date.[32]

The mass–energy equivalence formula was used in the understanding of nuclear fission reactions, and implies the great amount of energy that can be released by a nuclear fission chain reaction, used in both nuclear weapons and nuclear power. By measuring the mass of different atomic nuclei and subtracting from that number the total mass of the protons and neutrons as they would weigh separately, one gets the exact binding energy available in an atomic nucleus. This is used to calculate the energy released in any nuclear reaction, as the difference in the total mass of the nuclei that enter and exit the reaction.

11.6   Practical examples

Einstein used the CGS system of units (centimeters, grams, seconds, dynes, and ergs), but the formula is independent of the system of units. In natural units, the numerical value of the speed of light is set to equal 1, and the formula expresses an equality of numerical values: $E = m$. In the SI system (expressing the ratio E / m in joules per kilogram using the value of c in meters per second):

$$E / m = c^2 = (299{,}792{,}458 \text{ m/s})^2 = 89{,}875{,}517{,}873{,}681{,}764 \text{ J/kg} \ (\approx 9.0 \times 10^{16} \text{ joules per kilogram}).$$

So the energy equivalent of one gram (1/1000 of a kilogram) of mass is equivalent to:

89.9 terajoules

25.0 million kilowatt-hours (≈ 25 GW·h)

21.5 billion kilocalories (≈ 21 Tcal)[33]

85.2 billion BTUs[33]

or to the energy released by combustion of the following:

21.5 kilotons of TNT-equivalent energy (≈ 21 kt)[33]

568,000 US gallons of automotive gasoline

Any time energy is generated, the process can be evaluated from an $E = mc^2$ perspective. For instance, the "Gadget"-style bomb used in the Trinity test and the bombing of Nagasaki had an explosive yield equivalent to 21 kt of TNT. About 1 kg of the approximately 6.15 kg of plutonium in each of these bombs fissioned into lighter elements totaling almost exactly one gram less, after cooling. The electromagnetic radiation and kinetic energy (thermal and blast energy) released in this explosion carried the missing one gram of mass.[34] This occurs because nuclear binding energy is released whenever elements with more than 62 nucleons fission.

Another example is hydroelectric generation. The electrical energy produced by Grand Coulee Dam's turbines every 3.7 hours represents one gram of mass. This mass passes to the electrical devices (such as lights in cities) which are powered by the generators, where it appears as a gram of heat and light.[35] Turbine designers look at their equations in terms of pressure, torque, and RPM. However, Einstein's equations show that all energy has mass, and thus the electrical energy produced by a dam's generators, and the heat and light which result from it, all retain their mass, which is equivalent to the energy. The potential energy—and equivalent mass—represented by the waters of the Columbia River as it descends to the Pacific Ocean would be converted to heat due to viscous friction and the turbulence of white water rapids and waterfalls were it not for the dam and its generators. This heat would remain as mass on site at the water, were it not for the equipment which converted some of this potential and kinetic energy into electrical energy, which can be moved from place to place (taking mass with it).

Whenever energy is added to a system, the system gains mass:

- A spring's mass increases whenever it is put into compression or tension. Its added mass arises from the added potential energy stored within it, which is bound in the stretched chemical (electron) bonds linking the atoms within the spring.

- Raising the temperature of an object (increasing its heat energy) increases its mass. For example, consider the world's primary mass standard for the kilogram, made of platinum/iridium. If its temperature is allowed to change by 1 °C, its mass will change by 1.5 picograms (1 pg = 1×10^{-12} g).[36]

- A spinning ball will weigh more than a ball that is not spinning. Its increase of mass is exactly the equivalent of the mass of energy of rotation, which is itself the sum of the kinetic energies of all the moving parts of the ball. For example, the Earth itself is more massive due to its daily rotation, than it would be with no rotation. This rotational energy (2.14×10^{29} J) represents 2.38 billion metric tons of added mass.[37]

Note that no net mass or energy is really created or lost in any of these examples and scenarios. Mass/energy simply moves from one place to another. These are some examples of the *transfer* of energy and mass in accordance with the *principle of mass–energy conservation*.

11.7 Efficiency

Although mass cannot be converted to energy,[21] in some reactions matter particles (which contain a form of rest energy) can be destroyed and converted to other types of energy which are more usable and obvious as forms of energy, such as light and energy of motion (heat, etc.). However, the total amount of energy and mass does not change in such a transformation. Even when particles are not destroyed, a certain fraction of the ill-defined "matter" in ordinary objects can be destroyed, and its associated energy liberated and made available as the more dramatic energies of light and heat, even though no identifiable real particles are destroyed, and even though (again) the total energy is unchanged (as also the total mass). Such conversions between types of energy (resting to active energy) happen in nuclear weapons, in which the protons and neutrons in atomic nuclei lose a small fraction of their average mass, but this mass loss is not due to the destruction of any protons or neutrons (or even, in general, lighter particles like electrons). Also the mass is not destroyed, but simply removed from the system. in the form of heat and light from the reaction.

In nuclear reactions, typically only a small fraction of the total mass–energy of the bomb is converted into the mass–energy of heat, light, radiation and motion, which are "active" forms which can be used. When an atom fissions, it loses only about 0.1% of its mass (which escapes from the system and does not disappear), and additionally, in a bomb or reactor not all the atoms can fission. In a modern fission-based atomic bomb, the efficiency is only about 40%, so only 40% of the fissionable atoms actually fission, and only about 0.03% of the fissile core mass appears as energy in the end. In nuclear fusion, more of the mass is released as usable energy, roughly 0.3%. But in a fusion bomb, the bomb mass is partly casing and non-reacting components, so that in practicality, again (coincidentally) no more than about 0.03% of the total mass of the entire weapon is released as usable energy (which, again, retains the "missing" mass). See nuclear weapon yield for practical details of this ratio in modern nuclear weapons.

In theory, it should be possible to destroy matter and convert all of the rest-energy associated with matter into heat and light (which would of course have the same mass), but none of the theoretically known methods are practical. One way to convert all the energy within matter into usable energy is to annihilate matter with antimatter. But antimatter is rare in our universe, and must be made first. Due to inefficient mechanisms of production, making antimatter always requires far more usable energy than would be released when it was annihilated.

Since most of the mass of ordinary objects resides in protons and neutrons, in order to convert all of the energy of ordinary matter into a more useful type of energy, the protons and neutrons must be converted to lighter particles, or else particles with no rest-mass at all. In the Standard Model of particle physics, the number of protons plus neutrons is nearly exactly conserved. Still, Gerard 't Hooft showed that there is a process which will convert protons and neutrons to antielectrons and neutrinos.[38] This is the weak SU(2) instanton proposed by Belavin Polyakov Schwarz and Tyupkin.[39] This process, can in principle destroy matter and convert all the energy of matter into neutrinos and usable energy, but it is normally extraordinarily slow. Later it became clear that this process will happen at a fast rate at very high temperatures,[40] since then instanton-like configurations will be copiously produced from thermal fluctuations. The temperature required is so high that it would only have been reached shortly after the big bang.

Many extensions of the standard model contain magnetic monopoles, and in some models of grand unification, these monopoles catalyze proton decay, a process known as the Callan–Rubakov effect.[41] This process would be an efficient mass–energy conversion at ordinary temperatures, but it requires making monopoles and anti-monopoles first. The energy required to produce monopoles is believed to be enormous, but magnetic charge is conserved, so that the lightest monopole is stable. All these properties are deduced in theoretical models—magnetic monopoles have never been observed, nor have they been produced in any experiment so far.

A third known method of total matter–energy "conversion" (which again in practice only means conversion of one type of energy into a different type of energy), is using gravity, specifically black holes. Stephen Hawking theorized[42] that black holes radiate thermally with no regard to how they are formed. So it is theoretically possible to throw matter into a black hole and use the emitted heat to generate power. According to the theory of Hawking radiation, however, the black hole used will radiate at a higher rate the smaller it is, producing usable powers at only small black hole masses, where usable may for example be something greater than the local background radiation. It is also worth noting that the ambient irradiated power would change with the mass of the black hole, increasing as the mass of the black hole decreases, or decreasing as the mass increases, at a rate where power is proportional to the inverse square of the mass. In a "practical" scenario, mass and energy could be dumped into the black hole to regulate this growth, or keep its size, and thus power output, near constant. This could result from the fact that mass and energy are lost from the hole with its thermal radiation.

11.8 Background

11.8.1 Mass–velocity relationship

In developing special relativity, Einstein found that the kinetic energy of a moving body is

$$E_k = m_0(\gamma - 1)c^2 = \frac{m_0 c^2}{\sqrt{1 - \frac{v^2}{c^2}}} - m_0 c^2,$$

with v the velocity, m_0 the rest mass, and γ the Lorentz factor.

He included the second term on the right to make sure that for small velocities the energy would be the same as in classical mechanics, thus satisfying the correspondence principle:

$$E_k = \frac{1}{2} m_0 v^2 + \cdots$$

Without this second term, there would be an additional contribution in the energy when the particle is not moving.

Einstein found that the total momentum of a moving particle is:

$$P = \frac{m_0 v}{\sqrt{1 - \frac{v^2}{c^2}}}.$$

and it is this quantity which is conserved in collisions. The ratio of the momentum to the velocity is the relativistic mass, m.

$$m = \frac{m_0}{\sqrt{1 - \frac{v^2}{c^2}}}$$

And the relativistic mass and the relativistic kinetic energy are related by the formula:

$$E_k = mc^2 - m_0 c^2.$$

Einstein wanted to omit the unnatural second term on the right-hand side, whose only purpose is to make the energy at rest zero, and to declare that the particle has a total energy which obeys:

$$E = mc^2$$

which is a sum of the rest energy m_0c^2 and the kinetic energy. This total energy is mathematically more elegant, and fits better with the momentum in relativity. But to come to this conclusion, Einstein needed to think carefully about collisions. This expression for the energy implied that matter at rest has a huge amount of energy, and it is not clear whether this energy is physically real, or just a mathematical artifact with no physical meaning.

In a collision process where all the rest-masses are the same at the beginning as at the end, either expression for the energy is conserved. The two expressions only differ by a constant which is the same at the beginning and at the end of the collision. Still, by analyzing the situation where particles are thrown off a heavy central particle, it is easy to see that the inertia of the central particle is reduced by the total energy emitted. This allowed Einstein to conclude that the inertia of a heavy particle is increased or diminished according to the energy it absorbs or emits.

11.8.2 Relativistic mass

Main article: Mass in special relativity

After Einstein first made his proposal, it became clear that the word mass can have two different meanings. Some denote the *relativistic mass* with an explicit index:

$$m_{\text{rel}} = \frac{m_0}{\sqrt{1 - \frac{v^2}{c^2}}}.$$

This mass is the ratio of momentum to velocity, and it is also the relativistic energy divided by c^2 (it is not Lorentz-invariant, in contrast to m_0). The equation $E = m_{\text{rel}}c^2$ holds for moving objects. When the velocity is small, the relativistic mass and the rest mass are almost exactly the same.

- $E = mc^2$ either means $E = m_0c^2$ for an object at rest, or $E = m_{\text{rel}}c^2$ when the object is moving.

Also Einstein (following Hendrik Lorentz and Max Abraham) used velocity- and direction-dependent mass concepts (longitudinal and transverse mass) in his 1905 electrodynamics paper and in another paper in 1906.[43][44] However, in his first paper on $E = mc^2$ (1905), he treated m as what would now be called the *rest mass*.[2] Some claim that (in later years) he did not like the idea of "relativistic mass".[45] When modern physicists say "mass", they are usually talking about rest mass, since if they meant "relativistic mass", they would just say "energy".

Considerable debate has ensued over the use of the concept "relativistic mass" and the connection of "mass" in relativity to "mass" in Newtonian dynamics. For example, one view is that only rest mass is a viable concept and is a property of the particle; while relativistic mass is a conglomeration of particle properties and properties of spacetime. A perspective that avoids this debate, due to Kjell Vøyenli, is that the Newtonian concept of mass as a particle property and the relativistic concept of mass have to be viewed as embedded in their own theories and as having no precise connection.[46][47]

11.8.3 Low speed expansion

We can rewrite the expression $E = \gamma m_0 c^2$ as a Taylor series:

$$E = m_0c^2 \left[1 + \frac{1}{2} \left(\frac{v}{c} \right)^2 + \frac{3}{8} \left(\frac{v}{c} \right)^4 + \frac{5}{16} \left(\frac{v}{c} \right)^6 + \dots \right].$$

For speeds much smaller than the speed of light, higher-order terms in this expression get smaller and smaller because v/c is small. For low speeds we can ignore all but the first two terms:

$$E \approx m_0c^2 + \frac{1}{2}m_0v^2.$$

The total energy is a sum of the rest energy and the Newtonian kinetic energy.

The classical energy equation ignores both the m_0c^2 part, and the high-speed corrections. This is appropriate, because all the high-order corrections are small. Since only *changes* in energy affect the behavior of objects, whether we include the m_0c^2 part makes no difference, since it is constant. For the same reason, it is possible to subtract the rest energy from the total energy in relativity. By considering the emission of energy in different frames, Einstein could show that the rest energy has a real physical meaning.

The higher-order terms are extra correction to Newtonian mechanics which become important at higher speeds. The Newtonian equation is only a low-speed approximation, but an extraordinarily good one. All of the calculations used in putting astronauts on the moon, for example, could have been done using Newton's equations without any of the higher-order corrections. The total mass energy equivalence should also include the rotational and vibrational kinetic energies as well as the linear kinetic energy at low speeds.

11.9   History

While Einstein was the first to have correctly deduced the mass–energy equivalence formula, he was not the first to have related energy with mass. But nearly all previous authors thought that the energy which contributes to mass comes only from electromagnetic fields.[48][49][50][51]

11.9.1   Newton: matter and light

In 1717 Isaac Newton speculated that light particles and matter particles were inter-convertible in "Query 30" of the *Opticks*, where he asks:

> Are not the gross bodies and light convertible into one another, and may not bodies receive much of their activity from the particles of light which enter their composition?

11.9.2   Swedenborg: matter composed of "pure and total motion"

In 1734 the Swedish scientist and theologian Emanuel Swedenborg in his *Principia* theorized that all matter is ultimately composed of dimensionless points of "pure and total motion." He described this motion as being without force, direction or speed, but having the potential for force, direction and speed everywhere within it.[52][53]

11.9.3   Electromagnetic mass

Main article: Electromagnetic mass

There were many attempts in the 19th and the beginning of the 20th century—like those of J. J. Thomson (1881), Oliver Heaviside (1888), and George Frederick Charles Searle (1897), Wilhelm Wien (1900), Max Abraham (1902), Hendrik Antoon Lorentz (1904) — to understand how the mass of a charged object depends on the electrostatic field.[48][49] This concept was called electromagnetic mass, and was considered as being dependent on velocity and direction as well. Lorentz (1904) gave the following expressions for longitudinal and transverse electromagnetic mass:

$$m_L = \frac{m_0}{\left(\sqrt{1-\frac{v^2}{c^2}}\right)^3}, \quad m_T = \frac{m_0}{\sqrt{1-\frac{v^2}{c^2}}}$$

where

$$m_0 = \frac{4}{3}\frac{E_{em}}{c^2} .$$

11.9.4   Radiation pressure and inertia

Main article: Electromagnetic mass § Inertia of energy and radiation paradoxes

Another way of deriving some sort of electromagnetic mass was based on the concept of radiation pressure. In 1900, Henri Poincaré associated electromagnetic radiation energy with a "fictitious fluid" having momentum and mass[1]

$$m_{em} = E_{em}/c^2 .$$

By that, Poincaré tried to save the center of mass theorem in Lorentz's theory, though his treatment led to radiation paradoxes.[51]

Friedrich Hasenöhrl showed in 1904, that electromagnetic cavity radiation contributes the "apparent mass"

$$m_0 = \frac{4}{3}\frac{E_{em}}{c^2}$$

to the cavity's mass. He argued that this implies mass dependence on temperature as well.[54]

11.9.5   Einstein: mass–energy equivalence

Albert Einstein did not formulate exactly the formula $E = mc^2$ in his 1905 *Annus Mirabilis* paper "Does the Inertia of an object Depend Upon Its Energy Content?";[2] rather, the paper states that if a body gives off the energy L in the form of radiation, its mass diminishes by L/c^2. (Here, "radiation" means electromagnetic radiation, or light, and mass means the ordinary Newtonian mass of a slow-moving object.) This formulation relates only a change Δm in mass to a change L in energy without requiring the absolute relationship.

Objects with zero mass presumably have zero energy, so the extension that all mass is proportional to energy is obvious from this result. In 1905, even the hypothesis that changes in energy are accompanied by changes in mass was untested. Not until the discovery of the first type of antimatter (the positron in 1932) was it found that all of the mass of pairs of resting particles could be converted to radiation.

The first derivation by Einstein (1905)

Already in his relativity paper "On the electrodynamics of moving bodies", Einstein derived the correct expression for the kinetic energy of particles:

$$E_k = mc^2 \left(\frac{1}{\sqrt{1 - \frac{v^2}{c^2}}} - 1 \right)$$

Now the question remained open as to which formulation applies to bodies at rest. This was tackled by Einstein in his paper "Does the inertia of a body depend upon its energy content?". Einstein used a body emitting two light pulses in opposite directions, having energies of E_0 before and E_1 after the emission as seen in its rest frame. As seen from a moving frame, this becomes H_0 and H_1. Einstein obtained:

$$(H_0 - E_0) - (H_1 - E_1) = E \left(\frac{1}{\sqrt{1 - \frac{v^2}{c^2}}} - 1 \right)$$

then he argued that $H - E$ can only differ from the kinetic energy K by an additive constant, which gives

$$K_0 - K_1 = E \left(\frac{1}{\sqrt{1 - \frac{v^2}{c^2}}} - 1 \right)$$

Neglecting effects higher than third order in v/c after a Taylor series expansion of the right side of this gives:

$$K_0 - K_1 = \frac{E}{c^2} \frac{v^2}{2}.$$

Einstein concluded that the emission reduces the body's mass by E/c^2, and that the mass of a body is a measure of its energy content.

The correctness of Einstein's 1905 derivation of $E = mc^2$ was criticized by Max Planck (1907), who argued that it is only valid to first approximation. Another criticism was formulated by Herbert Ives (1952) and Max Jammer (1961), asserting that Einstein's derivation is based on begging the question.[3][55] On the other hand, John Stachel and Roberto Torretti (1982) argued that Ives' criticism was wrong, and that Einstein's derivation was correct.[56] Hans Ohanian (2008) agreed with Stachel/Torretti's criticism of Ives, though he argued that Einstein's derivation was wrong for other reasons.[57] For a recent review, see Hecht (2011).[4]

Alternative version

An alternative version of Einstein's thought experiment was proposed by Fritz Rohrlich (1990), who based his reasoning on the Doppler effect.[58] Like Einstein, he considered a body at rest with mass M. If the body is examined in a frame moving with nonrelativistic velocity v, it is no longer at rest and in the moving frame it has momentum $P = Mv$. Then he supposed the body emits two pulses of light to the left and to the right, each carrying an equal amount of energy E/2. In its rest frame, the object remains at rest after the emission since the two beams are equal in strength and carry opposite momentum.

But if the same process is considered in a frame moving with velocity v to the left, the pulse moving to the left will be redshifted while the pulse moving to the right will be blue shifted. The blue light carries more momentum than the red light, so that the momentum of the light in the moving frame is not balanced: the light is carrying some net momentum to the right.

The object has not changed its velocity before or after the emission. Yet in this frame it has lost some right-momentum to the light. The only way it could have lost momentum is by losing mass. This also solves Poincaré's radiation paradox, discussed above.

The velocity is small, so the right-moving light is blueshifted by an amount equal to the nonrelativistic Doppler shift factor $1 - v/c$. The momentum of the light is its energy divided by c, and it is increased by a factor of v/c. So the right-moving light is carrying an extra momentum ΔP given by:

$$\Delta P = \frac{v}{c} \frac{E}{2c}.$$

The left-moving light carries a little less momentum, by the same amount ΔP. So the total right-momentum in the light is twice ΔP. This is the right-momentum that the object lost.

$$2\Delta P = v\frac{E}{c^2}.$$

The momentum of the object in the moving frame after the emission is reduced to this amount:

$$P' = Mv - 2\Delta P = \left(M - \frac{E}{c^2}\right)v.$$

So the change in the object's mass is equal to the total energy lost divided by c^2. Since any emission of energy can be carried out by a two step process, where first the energy is emitted as light and then the light is converted to some other form of energy, any emission of energy is accompanied by a loss of mass. Similarly, by considering absorption, a gain in energy is accompanied by a gain in mass.

Relativistic center-of-mass theorem (1906)

Like Poincaré, Einstein concluded in 1906 that the inertia of electromagnetic energy is a necessary condition for the center-of-mass theorem to hold. On this occasion, Einstein referred to Poincaré's 1900 paper and wrote:[59]

> Although the merely formal considerations, which we will need for the proof, are already mostly contained in a work by H. Poincaré[2], for the sake of clarity I will not rely on that work.[60]

In Einstein's more physical, as opposed to formal or mathematical, point of view, there was no need for fictitious masses. He could avoid the *perpetuum mobile* problem, because on the basis of the mass–energy equivalence he could show that the transport of inertia which accompanies the emission and absorption of radiation solves the problem. Poincaré's rejection of the principle of action–reaction can be avoided through Einstein's $E = mc^2$, because mass conservation appears as a special case of the energy conservation law.

11.9.6 Others

During the nineteenth century there were several speculative attempts to show that mass and energy were proportional in various ether theories.[61] In 1873 Nikolay Umov pointed out a relation between mass and energy for ether in the form of $E = kmc^2$, where $0.5 \leq k \leq 1$.[62] The writings of Samuel Tolver Preston,[63][64] and a 1903 paper by Olinto De Pretto,[65][66] presented a mass–energy relation. De Pretto's paper received recent press coverage when Umberto Bartocci discovered that there were only three degrees of separation linking De Pretto to Einstein, leading Bartocci to conclude that Einstein was probably aware of De Pretto's work.[67]

Preston and De Pretto, following Le Sage, imagined that the universe was filled with an ether of tiny particles which are always moving at speed c. Each of these particles have a kinetic energy of mc^2 up to a small numerical factor. The nonrelativistic kinetic energy formula did not always include the traditional factor of 1/2, since Leibniz introduced kinetic energy without it, and the 1/2 is largely conventional in prerelativistic physics.[68] By assuming that every particle has a mass which is the sum of the masses of the ether particles, the authors would conclude that all matter contains an amount of kinetic energy either given by $E = mc^2$ or $2E = mc^2$ depending on the convention. A particle ether was usually considered unacceptably speculative science at the time,[69] and since these authors did not formulate relativity, their reasoning is completely different from that of Einstein, who used relativity to change frames.

Independently, Gustave Le Bon in 1905 speculated that atoms could release large amounts of latent energy, reasoning from an all-encompassing qualitative philosophy of physics.[70][71]

11.9.7 Radioactivity and nuclear energy

It was quickly noted after the discovery of radioactivity in 1897, that the total energy due to radioactive processes is about one *million times* greater than that involved in any known molecular change. However, it raised the question where this energy is coming from. After eliminating the idea of absorption and emission of some sort of Lesagian ether particles, the existence of a huge amount of latent energy, stored within matter, was proposed by Ernest Rutherford and Frederick Soddy in 1903. Rutherford also suggested that this internal energy is stored within normal matter as well. He went on to speculate in 1904:[72][73]

> If it were ever found possible to control at will the rate of disintegration of the radio-elements, an enormous amount of energy could be obtained from a small quantity of matter.

Einstein's equation is in no way an explanation of the large energies released in radioactive decay (this comes from the powerful nuclear forces involved; forces that were still unknown in 1905). In any case, the enormous energy released from radioactive decay (which had been measured by Rutherford) was much more easily measured than the (still small) change in the gross mass of materials, as a result. Einstein's equation, by theory, can give these energies by measuring mass differences before and after reactions, but in practice, these mass differences in 1905 were still too small to be measured in bulk. Prior to this, the ease of measuring radioactive decay energies with a calorimeter was thought possibly likely to allow measurement of changes in mass difference, as a check on Einstein's equation itself. Einstein mentions in his 1905 paper that mass–energy equivalence might perhaps be tested with radioactive decay, which releases enough energy (the quantitative amount known roughly by 1905) to possibly be "weighed," when missing from the system (having been given off as heat). However, radioactivity seemed to proceed at its own unalterable (and quite slow, for radioactives known then) pace, and even when simple nuclear reactions became possible using proton bombardment, the idea that these great amounts of usable energy could be liberated at will with any practicality, proved difficult to substantiate. Rutherford was reported in 1933 to have declared that this energy could not be exploited efficiently: "Anyone who expects a source of power from the transformation of the atom is talking moonshine."[74]

This situation changed dramatically in 1932 with the discovery of the neutron and its mass, allowing mass differences for single nuclides and their reactions to be calculated directly, and compared with the sum of masses for the particles that made up their composition. In 1933, the energy released from the reaction of lithium-7 plus protons giving rise to 2 alpha particles (as noted above by Rutherford), allowed Einstein's equation to be tested to an error of ±0.5%. However, scientists still did not see such reactions as a source of power.

After the very public demonstration of huge energies released from nuclear fission after the atomic bombings of Hiroshima and Nagasaki in 1945, the equation $E = mc^2$ became directly linked in the public eye with the power and peril of nuclear weapons. The equation was featured as early as page 2 of the Smyth Report, the official 1945 release by the US government on the development of the atomic bomb, and by 1946 the equation was linked closely enough with Einstein's work that the cover of *Time* magazine prominently featured a picture of Einstein next to an image of a mushroom cloud emblazoned with the equation.[75] Einstein himself had only a minor role in the Manhattan Project: he had cosigned a letter to the U.S. President in 1939 urging funding for research into atomic energy, warning that an atomic bomb was theoretically possible. The letter persuaded Roosevelt to devote a significant portion of the wartime budget to atomic research. Without a security clearance, Einstein's only scientific contribution was an analysis of an isotope separation method in theoretical terms. It was inconsequential, on account of Einstein not being given sufficient information (for security reasons) to fully work on the problem.[76]

While $E = mc^2$ is useful for understanding the amount of energy potentially released in a fission reaction, it was not strictly necessary to develop the weapon, once the fission process was known, and its energy measured at 200 MeV (which was directly possible, using a quantitative Geiger counter, at that time). As the physicist and Manhattan Project participant Robert Serber put it: "Somehow the popular notion took hold long ago that Einstein's theory of relativity, in particular his famous equation $E = mc^2$, plays some essential role in the theory of fission. Albert Einstein had a part in alerting the United States government to the possibility of building an atomic bomb, but his theory of relativity is not required in discussing fission. The theory of fission is what physicists call a non-relativistic theory, meaning that relativistic effects are too small to affect the dynamics of the fission process significantly."[77] However the association between $E = mc^2$ and nuclear energy has since stuck, and because of this association, and its simple expression of the ideas of Albert Einstein himself, it has become "the world's most famous equation".[78]

While Serber's view of the strict lack of need to use mass–energy equivalence in designing the atomic bomb is correct, it does not take into account the pivotal role which this relationship played in making the fundamental leap to the initial hypothesis that large atoms were energetically *allowed* to split into approximately equal parts (before this energy was in fact measured). In late 1938, while on the winter walk on which they solved the meaning of Hahn's experimental

The popular connection between Einstein, E = mc², and the atomic bomb was prominently indicated on the cover of Time *magazine in July 1946 by the writing of the equation on the mushroom cloud.*

results and introduced the idea that would be called atomic fission, Lise Meitner and Otto Robert Frisch made direct use of Einstein's equation to help them understand the quantitative energetics of the reaction which overcame the "surface tension-like" forces holding the nucleus together, and allowed the fission fragments to separate to a configuration from which their charges could force them into an energetic "fission". To do this, they made use of "packing fraction", or nuclear binding energy values for elements, which Meitner had memorized. These, together with use of $E = mc^2$ allowed them to realize on the spot that the basic fission process was energetically possible:

> ...We walked up and down in the snow, I on skis and she on foot. ...and gradually the idea took shape... explained by Bohr's idea that the nucleus is like a liquid drop; such a drop might elongate and divide itself... We knew there were strong forces that would resist, ..just as surface tension. But nuclei differed from ordinary drops. At this point we both sat down on a tree trunk and started to calculate on scraps of paper. ...the Uranium nucleus might indeed be a very wobbly, unstable drop, ready to divide itself... But, ...when the two drops separated they would be driven apart by electrical repulsion, about 200 MeV in all. Fortunately Lise Meitner remembered how to compute the masses of nuclei... and worked out that the two nuclei formed... would be lighter by about one-fifth the mass of a proton. Now whenever mass disappears energy is created, according to Einstein's formula E = mc^2, and... the mass was just equivalent to 200 MeV; it all fitted![79][80]

11.10   See also

- Energy density

- Index of energy articles

- Index of wave articles

- Outline of energy

11.11   References

[1] Poincaré, H. (1900), "La théorie de Lorentz et le principe de réaction", *Archives néerlandaises des sciences exactes et naturelles* **5**: 252–278. See also the English translation

[2] Einstein, A. (1905), "Ist die Trägheit eines Körpers von seinem Energieinhalt abhängig?", *Annalen der Physik* **18**: 639–643, doi:10.1002/andp.19053231314. See also the English translation.

[3] Jammer, Max (1997) [1961], *Concepts of Mass in Classical and Modern Physics*, New York: Dover, ISBN 0-486-29998-8

[4] Hecht, Eugene (2011), "How Einstein confirmed E0=mc2", *American Journal o f Physics* **79** (6): 591–600, doi:10.1119/1.3549223

[5] See the sentence on the last page (p. 641) of the original German edition, above the equation $K_0 - K_1 = L/V^2\ v^2/2$. See also the sentence above the last equation in the English translation, $K_0 - K_1 = (1/2)(L/c^2)v^2$, and the comment on the symbols used in *About this edition* that follows the translation.

[6] Einstein, Albert (1907), "Über die vom Relativitätsprinzip geforderte Trägheit der Energie" (PDF), *Annalen der Physik* **328** (7): 371–384, Bibcode:1907AnP...328..371E, doi:10.1002/andp.19073280713

[7] Planck, Max (1907), "Zur Dynamik bewegter Systeme", *Sitzungsberichte der Königlich-Preussischen Akademie der Wissenschaften, Berlin*, Erster Halbband (29): 542–570

   English Wikisource translation: On the Dynamics of Moving Systems

[8] Stark, J. (1907), "Elementarquantum der Energie, Modell der negativen und der positiven Elekrizität", *Physikalische Zeitschrift* **24** (8): 881

[9] Einstein, Albert (1908), "Über das Relativitätsprinzip und die aus demselben gezogenen Folgerungen" (PDF), *Jahrbuch der Radioaktivität und Elektronik* **4**: 411–462, Bibcode:1908JRE.....4..411E

[10] Schwartz, H. M. (1977), "Einstein's comprehensive 1907 essay on relativity, part II", *American Journal of Physics* **45** (9): 811–817, Bibcode:1977AmJPh..45..811S, doi:10.1119/1.11053

[11] Lewis, Gilbert N. & Tolman, Richard C. (1909), "The Principle of Relativity, and Non-Newtonian Mechanics", *Proceedings of the American Academy of Arts and Sciences* **44** (25): 709–726, doi:10.2307/20022495

[12] Lorentz, Hendrik Antoon (1914), *Das Relativitätsprinzip. Drei Vorlesungen gehalten in Teylers Stiftung zu Haarlem (1913)*, Leipzig and Berlin: B.G. Teubner

[13] Laue, Max von (1911), "Zur Dynamik der Relativitätstheorie", *Annalen der Physik* **340** (8): 524–542, Bibcode: doi:10.1002/andp.19113400808

   English Wikisource translation: On the Dynamics of the Theory of Relativity

[14] Klein, Felix (1918), "Über die Integralform der Erhaltungssätze und die Theorie der räumlich-geschlossenen Welt", *Göttinger Nachrichten*: 394–423

[15] A.Einstein $E = mc^2$: *the most urgent problem of our time* Science illustrated, vol. 1 no. 1, April issue, pp. 16–17, 1946 (item 417 in the "Bibliography"

[16] M.C.Shields *Bibliography of the Writings of Albert Einstein to May 1951* in Albert Einstein: Philosopher-Scientist by Paul Arthur Schilpp (Editor) Albert Einstein Philosopher – Scientist

[17] "Einstein was unequivocally against the traditional idea of conservation of mass. He had concluded that mass and energy were essentially one and the same; 'inert[ial] mass is simply latent energy.'[ref...]. He made his position known publicly time and again[ref...]...", Eugene Hecht, "Einstein on mass and energy." Am. J. Phys., Vol. 77, No. 9, September 2009, online.

[18] "There followed also the principle of the equivalence of mass and energy, with the laws of conservation of mass and energy becoming one and the same.", Albert Einstein, "Considerations Concerning the Fundaments of Theoretical Physics", Science, Washington, DC, vol. 91, no. 2369, May 24th, 1940 scanned image online

[19] page 14 (preview online) of Albert Einstein, *The Theory of Relativity (And Other Essays)*, Citadel Press, 1950.

[20] In F. Fernflores. The Equivalence of Mass and Energy. Stanford Encyclopedia of Philosophy.

[21] E. F. Taylor and J. A. Wheeler, *Spacetime Physics*, W.H. Freeman and Co., NY. 1992. ISBN 0-7167-2327-1, see pp. 248–9 for discussion of mass remaining constant after detonation of nuclear bombs, until heat is allowed to escape.

[22] Note that the relativistic mass, in contrast to the rest mass m_0, *is not a relativistic invariant, and that the velocity* $v = dx^{(4)}/dt$ *is not a Minkowski four-vector, in contrast to the quantity* $\tilde{v} = dx^{(4)}/d\tau$, *where* $d\tau = dt \cdot \sqrt{1 - (v^2/c^2)}$ *is the differential of the proper time. However, the energy–momentum four-vector* $p^{(4)} = m_0 \cdot dx^{(4)}/d\tau$ *is a genuine Minkowski four-vector, and the intrinsic origin of the square root in the definition of the relativistic mass is the distinction between* $d\tau$ *and* dt.

[23] Paul Allen Tipler, Ralph A. Llewellyn (January 2003), *Modern Physics*, W. H. Freeman and Company, pp. 87–88, ISBN 0-7167-4345-0

[24] Relativity DeMystified, D. McMahon, Mc Graw Hill (USA), 2006, ISBN 0-07-145545-0

[25] Dynamics and Relativity, J.R. Forshaw, A.G. Smith, Wiley, 2009, ISBN 978-0-470-01460-8

[26] Hans, H. S.; Puri, S. P. (2003), *Mechanics* (2 ed.), Tata McGraw-Hill, p. 433, ISBN 0-07-047360-9, Chapter 12 page 433

[27] Mould, Richard A. (2002), *Basic relativity* (2 ed.), Springer, p. 126, ISBN 0-387-95210-1, Chapter 5 page 126

[28] Chow, Tail L. (2006), *Introduction to electromagnetic theory: a modern perspective*, Jones & Bartlett Learning, p. 392, ISBN 0-7637-3827-1, Chapter 10 page 392

[29] Dyson, F.W.; Eddington, A.S. & Davidson, C.R. (1920), "A Determination of the Deflection of Light by the Sun's Gravitational Field, from Observations Made at the Solar eclipse of May 29, 1919", *Phil. Trans. Roy. Soc. A* **220** (571–581): 291–333, Bibcode:1920RSPTA.220..291D, doi:10.1098/rsta.1920.0009

[30] Pound, R. V.; Rebka Jr. G. A. (April 1, 1960), "Apparent weight of photons", *Physical Review Letters* **4** (7): 337–341, Bibcode:1960PhRvL...4..337P, doi:10.1103/PhysRevLett.4.337

[31] Cockcroft–Walton experiment

[32] Rainville, S. et al. World Year of Physics: A direct test of E = mc2. *Nature* 438, 1096–1097 (22 December 2005). Published online 21 December 2005.

[33] Conversions used: 1956 International (Steam) Table (IT) values where one calorie ≡ 4.1868 J and one BTU ≡ 1055.05585262 J. Weapons designers' conversion value of one gram TNT ≡ 1000 calories used.

[34] The 6.2 kg core comprised 0.8% gallium by weight. Also, about 20% of the Gadget's yield was due to fast fissioning in its natural uranium tamper. This resulted in 4.1 moles of Pu fissioning with 180 MeV per atom actually contributing prompt kinetic energy to the explosion. Note too that the term *"Gadget"-style* is used here instead of "Fat Man" because this general design of bomb was very rapidly upgraded to a more efficient one requiring only 5 kg of the Pu/gallium alloy.

[35] Assuming the dam is generating at its peak capacity of 6,809 MW.

[36] Assuming a 90/10 alloy of Pt/Ir by weight, a Cp of 25.9 for Pt and 25.1 for Ir, a Pt-dominated average Cp of 25.8, 5.134 moles of metal, and 132 $J \cdot K^{-1}$ for the prototype. A variation of ±1.5 picograms is of course, much smaller than the actual uncertainty in the mass of the international prototype, which is ±2 micrograms.

[37] InfraNet Lab (2008-12-07). Harnessing the Energy from the Earth's Rotation. Article on Earth rotation energy. Divided by c^2. InfraNet Lab, 7 December 2008. Retrieved from http://infranetlab.org/blog/harnessing-energy-earth%E2%80%99s-rotation

[38] G. 't Hooft, "Computation of the quantum effects due to a four-dimensional pseudoparticle", Physical Review D14:3432–3450 (1976).

[39] A. Belavin, A. M. Polyakov, A. Schwarz, Yu. Tyupkin, "Pseudoparticle Solutions to Yang Mills Equations", Physics Letters 59B:85 (1975).

[40] F. Klinkhammer, N. Manton, "A Saddle Point Solution in the Weinberg Salam Theory", Physical Review D 30:2212.

[41] Rubakov V. A. "Monopole Catalysis of Proton Decay", Reports on Progress in Physics 51:189–241 (1988).

[42] S.W. Hawking "Black Holes Explosions?" *Nature* 248:30 (1974).

[43] Einstein, A. (1905), "Zur Elektrodynamik bewegter Körper" (PDF), *Annalen der Physik* **17** (10): 891–921, Bibcode: doi:10.1002/andp.19053221004. English translation.

[44] Einstein, A. (1906), "Über eine Methode zur Bestimmung des Verhältnisses der transversalen und longitudinalen Masse des Elektrons" (PDF), *Annalen der Physik* **21** (13): 583–586, Bibcode:1906AnP...326..583E, doi:10.1002/andp.19063261310

[45] See e.g. Lev B.Okun, *The concept of Mass*, Physics Today **42** (6), June 1969, p. 31–36, http://www.physicstoday.org/vol-42/iss-6/vol42no6p31_36.pdf

[46] Max Jammer (1999), *Concepts of mass in contemporary physics and philosophy*, Princeton University Press, p. 51, ISBN 0-691-01017-X

[47] Eriksen, Erik; Vøyenli, Kjell (1976), "The classical and relativistic concepts of mass", *Foundations of Physics* (Springer) **6**: 115–124, Bibcode:1976FoPh....6..115E, doi:10.1007/BF00708670

[48] Jannsen, M., Mecklenburg, M. (2007), V. F. Hendricks; et al., eds., "From classical to relativistic mechanics: Electromagnetic models of the electron.", *Interactions: Mathematics, Physics and Philosophy* (Dordrecht: Springer): 65–134

[49] Whittaker, E.T. (1951–1953), 2. *Edition: A History of the theories of aether and electricity, vol. 1: The classical theories / vol. 2: The modern theories 1900–1926*, London: Nelson

[50] Miller, Arthur I. (1981), *Albert Einstein's special theory of relativity. Emergence (1905) and early interpretation (1905–1911)*, Reading: Addison–Wesley, ISBN 0-201-04679-2

[51] Darrigol, O. (2005), "The Genesis of the theory of relativity" (PDF), *Séminaire Poincaré* **1**: 1–22, doi:10.1007/3-7643-7436-5_1

[52] Swedenborg, Emanuel (1734), "De Simplici Mundi vel Puncto naturali", *Principia Rerum Naturalia* (in Latin), Leipzig, p. 32

[53] Swedenborg, Emanuel (1845), *The Principia; or The First Principles of Natural Things*, Translated by Augustus Clissold, London: W. Newbery, pp. 55–57

[54] Philip Ball (Aug 23, 2011). "Did Einstein discover E = mc²?". Physics World.

[55] Ives, Herbert E. (1952), "Derivation of the mass–energy relation", *Journal of the Optical Society of America* **42** (8): 540–543, doi:10.1364/JOSA.42.000540

[56] Stachel, John; Torretti, Roberto (1982), "Einstein's first derivation of mass–energy equivalence", *American Journal of Physics* **50** (8): 760–763, Bibcode:1982AmJPh..50..760S, doi:10.1119/1.12764

[57] Ohanian, Hans (2008), "Did Einstein prove E=mc2?", *Studies in History and Philosophy of Science Part B* **40** (2): 167–173, arXiv:0805.1400, doi:10.1016/j.shpsb.2009.03.002

[58] Rohrlich, Fritz (1990), "An elementary derivation of E=mc2", *American Journal o f Physics* **58** (4): 348–349, doi:10.1119/1.16168

[59] Einstein, A. (1906), "Das Prinzip von der Erhaltung der Schwerpunktsbewegung und die Trägheit der Energie" (PDF), *Annalen der Physik* **20** (8): 627–633, Bibcode:1906AnP...325..627E, doi:10.1002/andp.19063250814

[60] Einstein 1906: Trotzdem die einfachen formalen Betrachtungen, die zum Nachweis dieser Behauptung durchgeführt werden müssen, in der Hauptsache bereits in einer Arbeit von H. Poincaré enthalten sind[2], werde ich mich doch der Übersichtlichkeit halber nicht auf jene Arbeit stützen.

[61] Helge Kragh, "Fin-de-Siècle Physics: A World Picture in Flux" in *Quantum Generations: A History of Physics in the Twentieth Century* (Princeton, NJ: Princeton University Press, 1999).

[62] *Умов Н. А.* Избранные сочинения. М. — Л., 1950. (Russian)

[63] Preston, S. T., Physics of the Ether, E. & F. N. Spon, London, (1875).

[64] Bjerknes: S. Tolver Preston's Explosive Idea $E = mc^2$.

[65] MathPages: Who Invented Relativity?

[66] De Pretto, O. *Reale Instituto Veneto Di Scienze, Lettere Ed Arti*, LXIII, II, 439–500, reprinted in Bartocci.

[67] Umberto Bartocci, *Albert Einstein e Olinto De Pretto—La vera storia della formula più famosa del mondo*, editore Andromeda, Bologna, 1999.

[68] Prentiss, J.J. (August 2005), "Why is the energy of motion proportional to the square of the velocity?", *American Journal of Physics* **73** (8): 705, Bibcode:2005AmJPh..73..701P, doi:10.1119/1.1927550

[69] John Worrall, review of the book *Conceptions of Ether. Studies in the History of Ether Theories* by Cantor and Hodges, The British Journal of the Philosophy of Science vol 36, no 1, March 1985, p. 84. The article contrasts a particle ether with a wave-carrying ether, the latter *was* acceptable.

[70] Le Bon: The Evolution of Forces.

[71] Bizouard: Poincaré $E = mc^2$ l'équation de Poincaré, Einstein et Planck.

[72] Rutherford, Ernest (1904), *Radioactivity*, Cambridge: University Press, pp. 336–338

[73] Heisenberg, Werner (1958), *Physics And Philosophy: The Revolution In Modern Science*, New York: Harper & Brothers, pp. 118–119

[74] "We might in these processes obtain very much more energy than the proton supplied, but on the average we could not expect to obtain energy in this way. It was a very poor and inefficient way of producing energy, and anyone who looked for a source of power in the transformation of the atoms was talking moonshine. But the subject was scientifically interesting because it gave insight into the atoms." *The Times* archives, September 12, 1933, "The British association—breaking down the atom"

[75] Cover. *Time* magazine, July 1, 1946.

[76] Isaacson, *Einstein: His Life and Universe*.

[77] Robert Serber, *The Los Alamos Primer: The First Lectures on How to Build an Atomic Bomb* (University of California Press, 1992), page 7. Note that the quotation is taken from Serber's 1992 version, and is not in the original 1943 Los Alamos Primer of the same name.

[78] David Bodanis, $E = mc^2$: *A Biography of the World's Most Famous Equation* (New York: Walker, 2000).

[79] A quote from Frisch about the discovery day. Accessed April 4, 2009.

[80] Sime, Ruth (1996), *Lise Meitner: A Life in Physics*, California Studies in the History of Science **13**, Berkeley: University of California Press, pp. 236–237, ISBN 0-520-20860-9

- Lasky, Ronald C. (April 23, 2007), "What is the significance of E = mc^2? And what does it mean?", *Scientific American* (Scientific American)

11.12   External links

- A shortcut to $E=mc^2$ – An easy to understand, high-school level derivation of the $E=mc^2$ formula.

- Einstein on the Inertia of Energy – MathPages

- Mass and Energy – Conversations About Science with Theoretical Physicist Matt Strassler

- Ask an Astrophysicist | Energy–Matter Conversion, NASA, 1997

- The Equivalence of Mass and Energy – Entry in the Stanford Encyclopedia of Philosophy

- Gail Wilson (May 2014) Scientists discover how to turn light into matter after 80-year quest Imperial College

- Living Reviews in Relativity – An open access, peer-referred, solely online physics journal publishing invited reviews covering all areas of relativity research.

- Merrifield, Michael; Copeland, Ed; Bowley, Roger. "E=mc^2 – Mass–Energy Equivalence". *Sixty Symbols*. Brady Haran for the University of Nottingham.

- Einstein on mass and energy, Eugene Hecht, Am. J. Phys. 77, 799 (2009). For example, "Early on, Einstein embraced the idea of a speed-dependent mass but changed his mind in 1906 and thereafter carefully avoided that notion entirely. He shunned, and explicitly rejected, what later came to be known as 'relativistic mass'. ... He consistently related the rest energy of a system to its invariant inertial mass."

Chapter 12

Nuclear fusion

In nuclear physics, **nuclear fusion** is a nuclear reaction in which two or more atomic nuclei come very close and then collide at a very high speed and join to form a new nucleus. During this process, matter is not conserved because some of the matter of the fusing nuclei is converted to photons (energy). Fusion is the process that powers active or "main sequence" stars.

The fusion of two nuclei with lower masses than iron-56 (which, along with nickel-62, has the largest binding energy per nucleon) generally releases energy, while the fusion of nuclei heavier than iron *absorbs* energy. The opposite is true for the reverse process, nuclear fission. This means that fusion generally occurs for lighter elements only, and likewise, that fission normally occurs only for heavier elements. There are extreme astrophysical events that can lead to short periods of fusion with heavier nuclei. This is the process that gives rise to nucleosynthesis, the creation of the heavy elements during events such as supernova.

Following the discovery of quantum tunneling by Friedrich Hund, in 1929 Robert Atkinson and Fritz Houtermans used the measured masses of light elements to predict that large amounts of energy could be released by fusing small nuclei. Building upon the nuclear transmutation experiments by Ernest Rutherford, carried out several years earlier, the laboratory fusion of hydrogen isotopes was first accomplished by Mark Oliphant in 1932. During the remainder of that decade the steps of the main cycle of nuclear fusion in stars were worked out by Hans Bethe. Research into fusion for military purposes began in the early 1940s as part of the Manhattan Project. Fusion was accomplished in 1951 with the Greenhouse Item nuclear test. Nuclear fusion on a large scale in an explosion was first carried out on November 1, 1952, in the Ivy Mike hydrogen bomb test.

Research into developing controlled thermonuclear fusion for civil purposes also began in earnest in the 1950s, and it continues to this day. The present article is about the theory of fusion. For details of the quest for controlled fusion and its history, see the article Fusion power.

12.1 Process

The origin of the energy released in fusion of light elements is due to interplay of two opposing forces, the nuclear force which combines together protons and neutrons, and the Coulomb force which causes protons to repel each other. The protons are positively charged and repel each other but they nonetheless stick together, demonstrating the existence of another force referred to as nuclear attraction. This force, called the strong nuclear force, overcomes electric repulsion in a very close range. The effect of this force is not observed outside the nucleus, hence the force has a strong dependence on distance, making it a short-range force. The same force also pulls the nucleons together, or neutrons and protons together.[2] Because the nuclear force is stronger than the Coulomb force for atomic nuclei smaller than iron and nickel, building up these nuclei from lighter nuclei by **fusion** releases the extra energy from the net attraction of these particles. For larger nuclei, however, no energy is released, since the nuclear force is short-range and cannot continue to act across still larger atomic nuclei. Thus, energy is no longer released when such nuclei are made by fusion; instead, energy is absorbed in such processes.

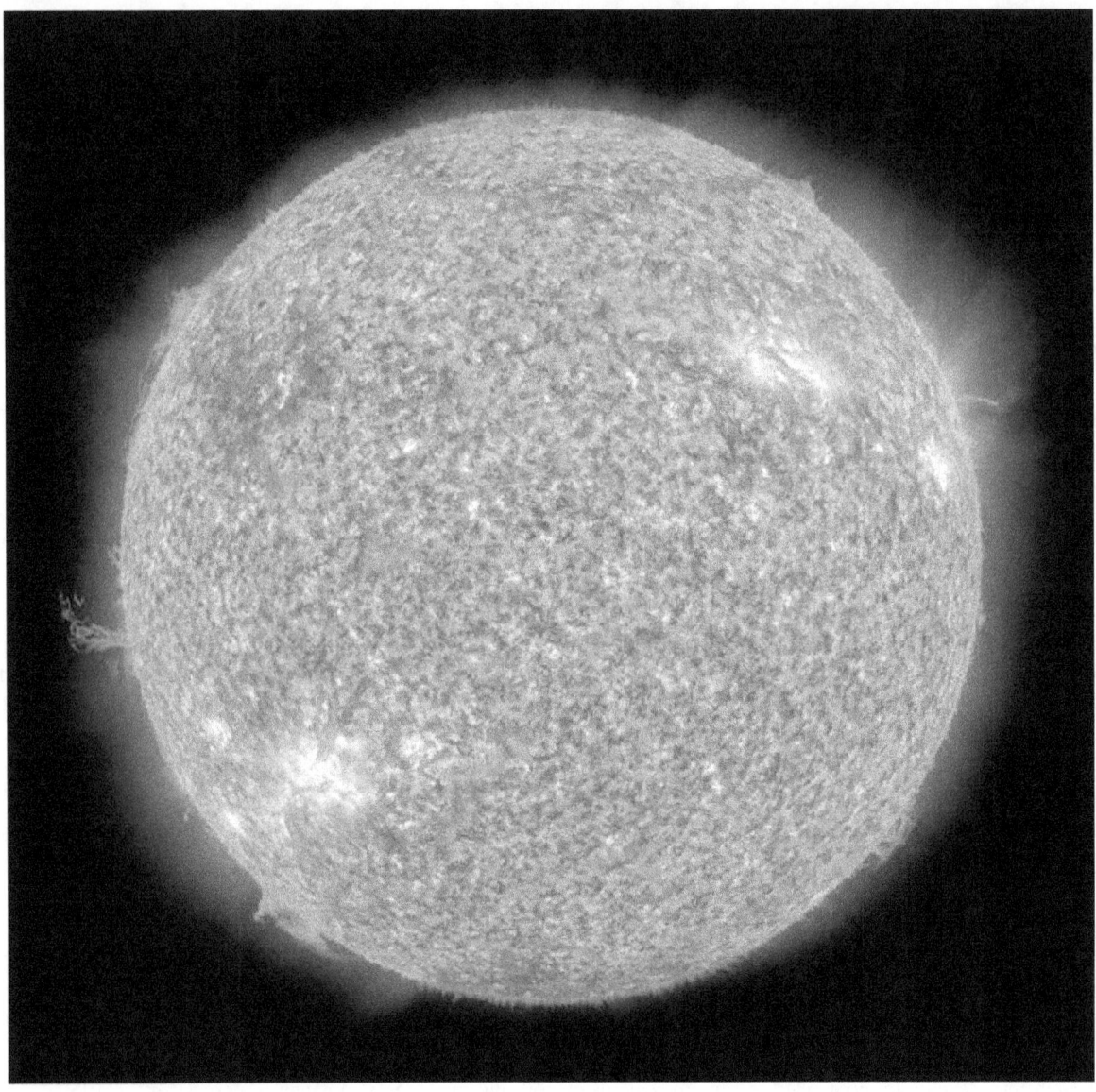

The Sun is a main-sequence star, and thus generates its energy by nuclear fusion of hydrogen nuclei into helium. In its core, the Sun fuses 620 million metric tons of hydrogen each second.

Fusion reactions of light elements power the stars and produce virtually all elements in a process called nucleosynthesis. The fusion of lighter elements in stars releases energy (and the mass that always accompanies it). For example, in the fusion of two hydrogen nuclei to form helium, 0.7% of the mass is carried away from the system in the form of kinetic energy or other forms of energy (such as electromagnetic radiation).[3]

Research into controlled fusion, with the aim of producing fusion power for the production of electricity, has been conducted for over 60 years. It has been accompanied by extreme scientific and technological difficulties, but has resulted in progress. At present, controlled fusion reactions have been unable to produce break-even (self-sustaining) controlled fusion reactions.[4] Workable designs for a reactor that theoretically will deliver ten times more fusion energy than the amount needed to heat up plasma to required temperatures are in development (see ITER). The ITER facility is expected to finish its construction phase in 2019. It will start commissioning the reactor that same year and initiate plasma experiments in 2020, but is not expected to begin full deuterium-tritium fusion until 2027.[5]

It takes considerable energy to force nuclei to fuse, even those of the lightest element, hydrogen. This is because all nuclei have a positive charge due to their protons, and as like charges repel, nuclei strongly resist being put close together.

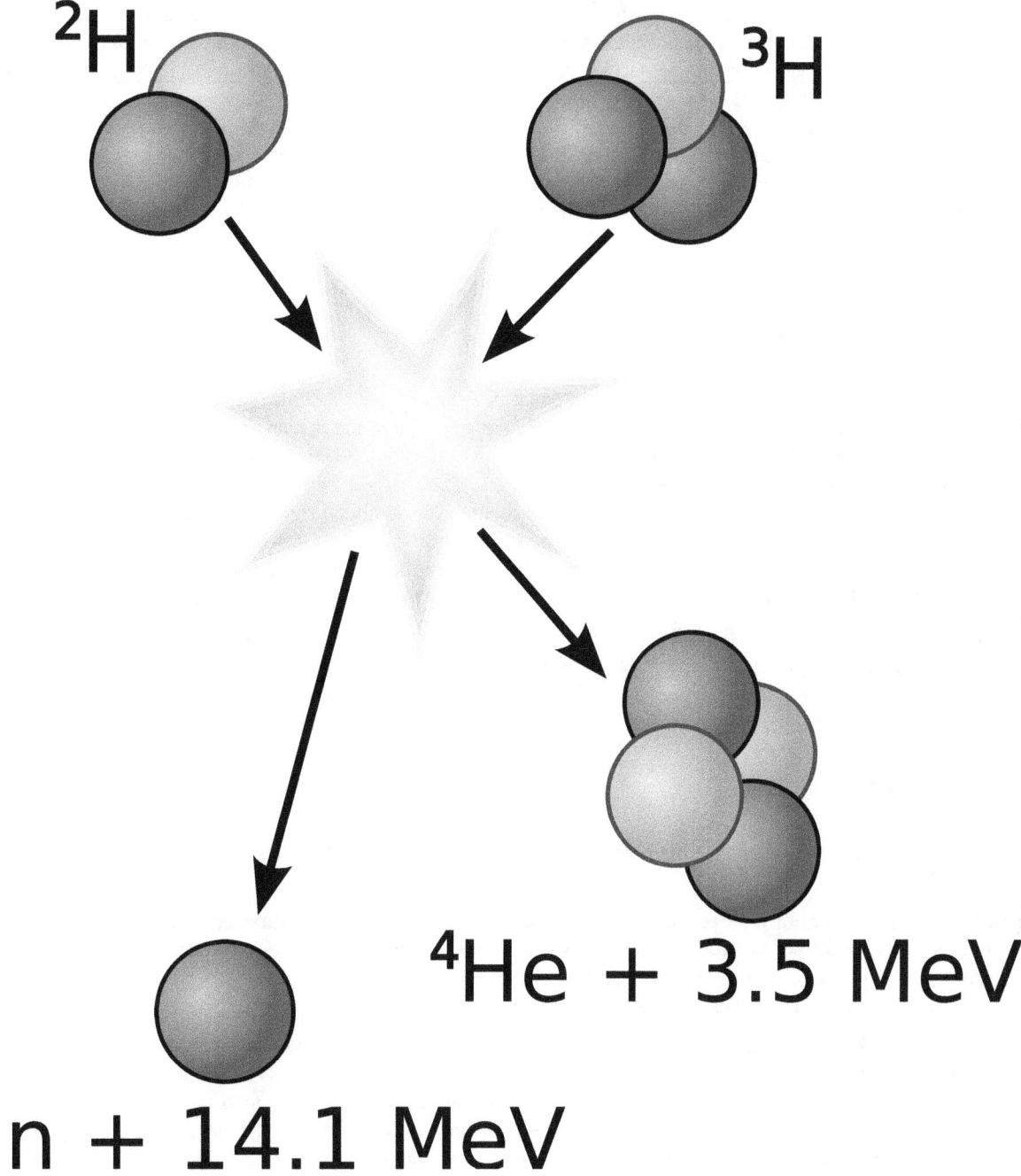

Fusion of deuterium with tritium creating helium-4, freeing a neutron, and releasing 17.59 MeV of energy, as an appropriate amount of mass changing forms to appear as the kinetic energy of the products, in agreement with kinetic $E = \Delta mc^2$, where Δm is the change in rest mass of particles.[1]

Accelerated to high speeds, they can overcome this electrostatic repulsion and be forced close enough for the attractive nuclear force to be sufficiently strong to achieve fusion. The fusion of lighter nuclei, which creates a heavier nucleus and often a free neutron or proton, generally releases more energy than it takes to force the nuclei together; this is an exothermic process that can produce self-sustaining reactions. The US National Ignition Facility, which uses laser-driven inertial confinement fusion, is thought to be capable of break-even fusion.

The first large-scale laser target experiments were performed in June 2009 and ignition experiments began in early 2011.[6][7]

Energy released in most nuclear reactions is much larger than in chemical reactions, because the binding energy that holds a nucleus together is far greater than the energy that holds electrons to a nucleus. For example, the ionization energy gained by adding an electron to a hydrogen nucleus is 13.6 eV—less than one-millionth of the 17.6 MeV released in the deuterium–tritium (D–T) reaction shown in the diagram to the right (one gram of matter would release 339 GJ of energy). Fusion reactions have an energy density many times greater than nuclear fission; the reactions produce far greater energy per unit of mass even though *individual* fission reactions are generally much more energetic than *individual* fusion ones, which are themselves millions of times more energetic than chemical reactions. Only direct conversion of mass into energy, such as that caused by the annihilatory collision of matter and antimatter, is more energetic per unit of mass than nuclear fusion.

12.2 Nuclear fusion in stars

The most important fusion process in nature is the one that powers stars. In the 20th century, it was realized that the energy released from nuclear fusion reactions accounted for the longevity of the Sun and other stars as a source of heat and light. The fusion of nuclei in a star, starting from its initial hydrogen and helium abundance, provides that energy and synthesizes new nuclei as a byproduct of that fusion process. The prime energy producer in the Sun is the fusion of hydrogen to form helium, which occurs at a solar-core temperature of 14 million kelvin. The net result is the fusion of four protons into one alpha particle, with the release of two positrons, two neutrinos (which changes two of the protons into neutrons), and energy. Different reaction chains are involved, depending on the mass of the star. For stars the size of the sun or smaller, the proton-proton chain dominates. In heavier stars, the CNO cycle is more important.

As a star uses up a substantial fraction of its hydrogen, it begins to synthesize heavier elements, as part of stellar nucleosynthesis. However the heaviest elements are synthesized by fusion that occurs as a more massive star undergoes a violent supernova at the end of its life, a process known as supernova nucleosynthesis.

12.3 Requirements

Details and supporting references on the material in this section can be found in textbooks on nuclear physics or nuclear fusion.[8]

A substantial energy barrier of electrostatic forces must be overcome before fusion can occur. At large distances, two naked nuclei repel one another because of the repulsive electrostatic force between their positively charged protons. If two nuclei can be brought close enough together, however, the electrostatic repulsion can be overcome by the attractive nuclear force, which is stronger at close distances.

When a nucleon such as a proton or neutron is added to a nucleus, the nuclear force attracts it to other nucleons, but primarily to its immediate neighbours due to the short range of the force. The nucleons in the interior of a nucleus have more neighboring nucleons than those on the surface. Since smaller nuclei have a larger surface area-to-volume ratio, the binding energy per nucleon due to the nuclear force generally increases with the size of the nucleus but approaches a limiting value corresponding to that of a nucleus with a diameter of about four nucleons. It is important to keep in mind that the above picture is a toy model because nucleons are quantum objects, and so, for example, since two neutrons in a nucleus are identical to each other, distinguishing one from the other, such as which one is in the interior and which is on the surface, is in fact meaningless, and the inclusion of quantum mechanics is necessary for proper calculations.

The electrostatic force, on the other hand, is an inverse-square force, so a proton added to a nucleus will feel an electrostatic repulsion from *all* the other protons in the nucleus. The electrostatic energy per nucleon due to the electrostatic force thus increases without limit as nuclei get larger.

The net result of these opposing forces is that the binding energy per nucleon generally increases with increasing size, up to the elements iron and nickel, and then decreases for heavier nuclei. Eventually, the binding energy becomes negative and very heavy nuclei (all with more than 208 nucleons, corresponding to a diameter of about 6 nucleons) are not stable. The four most tightly bound nuclei, in decreasing order of binding energy per nucleon, are 62Ni, 58Fe, 56Fe, and 60Ni.[9] Even though the nickel isotope, 62Ni, is more stable, the iron isotope 56Fe is an order of magnitude more common. This is due to the fact that there is no easy way for stars to create 62Ni through the alpha process.

An exception to this general trend is the helium-4 nucleus, whose binding energy is higher than that of lithium, the next heaviest element. This is because protons and neutrons are fermions, which according to the Pauli exclusion principle cannot exist in the same nucleus in exactly the same state. Each proton or neutron energy state in a nucleus can accommodate both a spin up particle and a spin down particle. Helium-4 has an anomalously large binding energy because its nucleus consists of two protons and two neutrons, so all four of its nucleons can be in the ground state. Any additional nucleons would have to go into higher energy states. Indeed, the helium-4 nucleus is so tightly bound that it is commonly treated as a single particle in nuclear physics, namely, the alpha particle.

The situation is similar if two nuclei are brought together. As they approach each other, all the protons in one nucleus repel all the protons in the other. Not until the two nuclei actually come in contact can the strong nuclear force take over. Consequently, even when the final energy state is lower, there is a large energy barrier that must first be overcome. It is called the Coulomb barrier.

The Coulomb barrier is smallest for isotopes of hydrogen, as their nuclei contain only a single positive charge. A diproton is not stable, so neutrons must also be involved, ideally in such a way that a helium nucleus, with its extremely tight binding, is one of the products.

Using deuterium-tritium fuel, the resulting energy barrier is about 0.1 MeV. In comparison, the energy needed to remove an electron from hydrogen is 13.6 eV, about 7500 times less energy. The (intermediate) result of the fusion is an unstable ^{5}He nucleus, which immediately ejects a neutron with 14.1 MeV. The recoil energy of the remaining ^{4}He nucleus is 3.5 MeV, so the total energy liberated is 17.6 MeV. This is many times more than what was needed to overcome the energy barrier.

The reaction **cross section** σ is a measure of the probability of a fusion reaction as a function of the relative velocity of the two reactant nuclei. If the reactants have a distribution of velocities, e.g. a thermal distribution, then it is useful to perform an average over the distributions of the product of cross section and velocity. This average is called the 'reactivity', denoted <σv>. The reaction rate (fusions per volume per time) is <σv> times the product of the reactant number densities:

$$f = n_1 n_2 \langle \sigma v \rangle.$$

If a species of nuclei is reacting with itself, such as the DD reaction, then the product $n_1 n_2$ must be replaced by $(1/2)n^2$.

$\langle \sigma v \rangle$ increases from virtually zero at room temperatures up to meaningful magnitudes at temperatures of 10–100 keV. At these temperatures, well above typical ionization energies (13.6 eV in the hydrogen case), the fusion reactants exist in a plasma state.

The significance of $\langle \sigma v \rangle$ as a function of temperature in a device with a particular energy confinement time is found by considering the Lawson criterion. This is an extremely challenging barrier to overcome on Earth, which explains why fusion research has taken many years to reach the current high state of technical prowess.[10]

12.4 Methods for achieving fusion

Main article: Fusion power

12.4.1 Thermonuclear fusion

Main article: Thermonuclear fusion

If the matter is sufficiently heated (hence being plasma), the fusion reaction may occur due to collisions with extreme thermal kinetic energies of the particles. In the form of thermonuclear weapons, thermonuclear fusion is the only fusion technique so far to yield undeniably large amounts of useful fusion energy. Usable amounts of thermonuclear fusion energy released in a controlled manner have yet to be achieved. In nature, this is what produces energy in stars through stellar nucleosynthesis.

12.4.2   Inertial confinement fusion

Main article: Inertial confinement fusion

Inertial confinement fusion (**ICF**) is a type of fusion energy research that attempts to initiate nuclear fusion reactions by heating and compressing a fuel target, typically in the form of a pellet that most often contains a mixture of deuterium and tritium.

12.4.3   Inertial electrostatic confinement

Main article: Inertial electrostatic confinement

Inertial electrostatic confinement is a set of devices that use an electric field to heat ions to fusion conditions. The most well known is the fusor. Starting in 1999, a number of amateurs have been able to do amateur fusion using these homemade devices.[11][12][13][14][15] Other IEC devices include: the Polywell, MIX POPS[16] and Marble concepts.[17]

12.4.4   Beam-beam or beam-target fusion

If the energy to initiate the reaction comes from accelerating one of the nuclei, the process is called *beam-target* fusion; if both nuclei are accelerated, it is *beam-beam* fusion.

Accelerator-based light-ion fusion is a technique using particle accelerators to achieve particle kinetic energies sufficient to induce light-ion fusion reactions. Accelerating light ions is relatively easy, and can be done in an efficient manner—all it takes is a vacuum tube, a pair of electrodes, and a high-voltage transformer; fusion can be observed with as little as 10 kV between electrodes. The key problem with accelerator-based fusion (and with cold targets in general) is that fusion cross sections are many orders of magnitude lower than Coulomb interaction cross sections. Therefore, the vast majority of ions end up expending their energy on bremsstrahlung and ionization of atoms of the target. Devices referred to as sealed-tube neutron generators are particularly relevant to this discussion. These small devices are miniature particle accelerators filled with deuterium and tritium gas in an arrangement that allows ions of these nuclei to be accelerated against hydride targets, also containing deuterium and tritium, where fusion takes place. Hundreds of neutron generators are produced annually for use in the petroleum industry where they are used in measurement equipment for locating and mapping oil reserves.

12.4.5   Muon-catalyzed fusion

Muon-catalyzed fusion is a well-established and reproducible fusion process that occurs at ordinary temperatures. It was studied in detail by Steven Jones in the early 1980s. Net energy production from this reaction cannot occur because of the high energy required to create muons, their short 2.2 µs half-life, and the high chance that a muon will bind to the new alpha particle and thus stop catalyzing fusion.[18]

12.4.6   Other principles

Some other confinement principles have been investigated.

Antimatter-initialized fusion uses small amounts of antimatter to trigger a tiny fusion explosion. This has been studied primarily in the context of making nuclear pulse propulsion, and pure fusion bombs feasible. This is not near becoming a practical power source, due to the cost of manufacturing antimatter alone.

Pyroelectric fusion was reported in April 2005 by a team at UCLA. The scientists used a pyroelectric crystal heated from −34 to 7 °C (−29 to 45 °F), combined with a tungsten needle to produce an electric field of about 25 gigavolts per meter to ionize and accelerate deuterium nuclei into an erbium deuteride target. At the estimated energy levels,[19] the D-D fusion reaction may occur, producing helium-3 and a 2.45 MeV neutron. Although it makes a useful neutron generator, the apparatus is not intended for power generation since it requires far more energy than it produces.[20][21][22][23]

Hybrid nuclear fusion-fission (hybrid nuclear power) is a proposed means of generating power by use of a combination of nuclear fusion and fission processes. The concept dates to the 1950s, and was briefly advocated by Hans Bethe during the 1970s, but largely remained unexplored until a revival of interest in 2009, due to the delays in the realization of pure fusion.[24] Project PACER, carried out at Los Alamos National Laboratory (LANL) in the mid-1970s, explored the possibility of a fusion power system that would involve exploding small hydrogen bombs (fusion bombs) inside an underground cavity. As an energy source, the system is the only fusion power system that could be demonstrated to work using existing technology. However it would also require a large, continuous supply of nuclear bombs, making the economics of such a system rather questionable.

12.5 Important reactions

12.5.1 Astrophysical reaction chains

At the temperatures and densities in stellar cores the rates of fusion reactions are notoriously slow. For example, at solar core temperature ($T \approx 15$ MK) and density (160 g/cm^3), the energy release rate is only 276 μW/cm^3—about a quarter of the volumetric rate at which a resting human body generates heat.[25] Thus, reproduction of stellar core conditions in a lab for nuclear fusion power production is completely impractical. Because nuclear reaction rates strongly depend on temperature ($\exp(-E/kT)$), achieving reasonable power levels in terrestrial fusion reactors requires 10–100 times higher temperatures (compared to stellar interiors): $T \approx 0.1$–1.0 GK.

12.5.2 Criteria and candidates for terrestrial reactions

Main article: Fusion power § Fuels

In artificial fusion, the primary fuel is not constrained to be protons and higher temperatures can be used, so reactions with larger cross-sections are chosen. Another concern is the production of neutrons, which activate the reactor structure radiologically, but also have the advantages of allowing volumetric extraction of the fusion energy and tritium breeding. Reactions that release no neutrons are referred to as *aneutronic*.

To be a useful energy source, a fusion reaction must satisfy several criteria. It must:

- **Be exothermic**: This limits the reactants to the low Z (number of protons) side of the curve of binding energy. It also makes helium 4He the most common product because of its extraordinarily tight binding, although 3He and 3H also show up.

- **Involve low Z nuclei**: This is because the electrostatic repulsion must be overcome before the nuclei are close enough to fuse.

- **Have two reactants**: At anything less than stellar densities, three body collisions are too improbable. In inertial confinement, both stellar densities and temperatures are exceeded to compensate for the shortcomings of the third parameter of the Lawson criterion, ICF's very short confinement time.

- **Have two or more products**: This allows simultaneous conservation of energy and momentum without relying on the electromagnetic force.

- **Conserve both protons and neutrons**: The cross sections for the weak interaction are too small.

Few reactions meet these criteria. The following are those with the largest cross sections:

For reactions with two products, the energy is divided between them in inverse proportion to their masses, as shown. In most reactions with three products, the distribution of energy varies. For reactions that can result in more than one set of products, the branching ratios are given.

Some reaction candidates can be eliminated at once.[26] The D-^{6}Li reaction has no advantage compared to p^+-$^{11}_5B$ because it is roughly as difficult to burn but produces substantially more neutrons through ^2_1D-^2_1D side reactions. There is also a p^+-^7_3Li reaction, but the cross section is far too low, except possibly when $T_i > 1$ MeV, but at such high temperatures an endothermic, direct neutron-producing reaction also becomes very significant. Finally there is also a p^+-^9_4Be reaction, which is not only difficult to burn, but ^9_4Be can be easily induced to split into two alpha particles and a neutron.

In addition to the fusion reactions, the following reactions with neutrons are important in order to "breed" tritium in "dry" fusion bombs and some proposed fusion reactors:

The latter of the two equations was unknown when the U.S. conducted the Castle Bravo fusion bomb test in 1954. Being just the second fusion bomb ever tested (and the first to use lithium), the designers of the Castle Bravo "Shrimp" had understood the usefulness of Lithium-6 in tritium production, but had failed to recognize that Lithium-7 fission would greatly increase the yield of the bomb. While Li-7 has a small neutron cross-section for low neutron energies, it has a higher cross section above 5 MeV.[27] Li-7 also undergoes a chain reaction due to its release of a neutron after fissioning. The 15 Mt yield was 150% greater than the predicted 6 Mt and caused casualties from the fallout generated.

To evaluate the usefulness of these reactions, in addition to the reactants, the products, and the energy released, one needs to know something about the cross section. Any given fusion device has a maximum plasma pressure it can sustain, and an economical device would always operate near this maximum. Given this pressure, the largest fusion output is obtained when the temperature is chosen so that $<\sigma v>/T^2$ is a maximum. This is also the temperature at which the value of the triple product $nT\tau$ required for ignition is a minimum, since that required value is inversely proportional to $<\sigma v>/T^2$ (see Lawson criterion). (A plasma is "ignited" if the fusion reactions produce enough power to maintain the temperature without external heating.) This optimum temperature and the value of $<\sigma v>/T^2$ at that temperature is given for a few of these reactions in the following table.

Note that many of the reactions form chains. For instance, a reactor fueled with ^3_1T and ^3_2He creates some ^2_1D, which is then possible to use in the ^2_1D-^3_2He reaction if the energies are "right". An elegant idea is to combine the reactions (8) and (9). The ^3_2He from reaction (8) can react with ^6_3Li in reaction (9) before completely thermalizing. This produces an energetic proton, which in turn undergoes reaction (8) before thermalizing. Detailed analysis shows that this idea would not work well, but it is a good example of a case where the usual assumption of a Maxwellian plasma is not appropriate.

12.5.3   Neutronicity, confinement requirement, and power density

Any of the reactions above can in principle be the basis of fusion power production. In addition to the temperature and cross section discussed above, we must consider the total energy of the fusion products E_{fus}, the energy of the charged fusion products E_{ch}, and the atomic number Z of the non-hydrogenic reactant.

Specification of the ^2_1D-^2_1D reaction entails some difficulties, though. To begin with, one must average over the two branches (2i) and (2ii). More difficult is to decide how to treat the ^3_1T and ^{3}

$^{}_{2}$He products. $^{3}_{1}$T burns so well in a deuterium plasma that it is almost impossible to extract from the plasma. The $^{2}_{1}$D-$^{3}_{2}$He reaction is optimized at a much higher temperature, so the burnup at the optimum $^{2}_{1}$D-$^{2}_{1}$D temperature may be low, so it seems reasonable to assume the $^{3}_{1}$T but not the $^{3}_{2}$He gets burned up and adds its energy to the net reaction. Thus the total reaction would be the sum of (2i), (2ii), and (1):

$$5\ ^{2}_{1}D \rightarrow\ ^{4}_{2}He + 2\, n^0 +\ ^{3}_{2}He + p^+,\quad E_{\text{fus}} = 4.03+17.6+3.27 = 24.9\text{ MeV},\ E_{\text{ch}} = 4.03+3.5+0.82 = 8.35\text{ MeV}.$$

We count the $^{2}_{1}$D-$^{2}_{1}$D fusion energy *per D-D reaction* (not per pair of deuterium atoms) as E_{fus} = (4.03 MeV + 17.6 MeV)×50% + (3.27 MeV)×50% = 12.5 MeV and the energy in charged particles as E_{ch} = (4.03 MeV + 3.5 MeV)×50% + (0.82 MeV)×50% = 4.2 MeV. (Note: if the tritium ion reacts with a deuteron while it still has a large kinetic energy, then the kinetic energy of the helium-4 produced may be quite different from 3.5 MeV, so this calculation of energy in charged particles is only approximate.)

Another unique aspect of the $^{2}_{1}$D-$^{2}_{1}$D reaction is that there is only one reactant, which must be taken into account when calculating the reaction rate.

With this choice, we tabulate parameters for four of the most important reactions

The last column is the **neutronicity** of the reaction, the fraction of the fusion energy released as neutrons. This is an important indicator of the magnitude of the problems associated with neutrons like radiation damage, biological shielding, remote handling, and safety. For the first two reactions it is calculated as $(E_{\text{fus}}-E_{\text{ch}})/E_{\text{fus}}$. For the last two reactions, where this calculation would give zero, the values quoted are rough estimates based on side reactions that produce neutrons in a plasma in thermal equilibrium.

Of course, the reactants should also be mixed in the optimal proportions. This is the case when each reactant ion plus its associated electrons accounts for half the pressure. Assuming that the total pressure is fixed, this means that density of the non-hydrogenic ion is smaller than that of the hydrogenic ion by a factor 2/(Z+1). Therefore, the rate for these reactions is reduced by the same factor, on top of any differences in the values of $\langle \sigma v \rangle / T^2$. On the other hand, because the $^{2}_{1}$D-$^{2}_{1}$D reaction has only one reactant, its rate is twice as high as when the fuel is divided between two different hydrogenic species, thus creating a more efficient reaction.

Thus there is a "penalty" of (2/(Z+1)) for non-hydrogenic fuels arising from the fact that they require more electrons, which take up pressure without participating in the fusion reaction. (It is usually a good assumption that the electron temperature will be nearly equal to the ion temperature. Some authors, however discuss the possibility that the electrons could be maintained substantially colder than the ions. In such a case, known as a "hot ion mode", the "penalty" would not apply.) There is at the same time a "bonus" of a factor 2 for $^{2}_{1}$D-$^{2}_{1}$D because each ion can react with any of the other ions, not just a fraction of them.

We can now compare these reactions in the following table.

The maximum value of $\langle \sigma v \rangle / T^2$ is taken from a previous table. The "penalty/bonus" factor is that related to a non-hydrogenic reactant or a single-species reaction. The values in the column "reactivity" are found by dividing 1.24×10^{-24} by the product of the second and third columns. It indicates the factor by which the other reactions occur more slowly than the $^{2}_{1}$D-$^{3}_{1}$D

1T reaction under comparable conditions. The column "Lawson criterion" weights these results with E_{ch} and gives an indication of how much more difficult it is to achieve ignition with these reactions, relative to the difficulty for the 2
1D-3
1T reaction. The last column is labeled "power density" and weights the practical reactivity with E_{fus}. It indicates how much lower the fusion power density of the other reactions is compared to the 2
1D-3
1T reaction and can be considered a measure of the economic potential.

12.5.4   Bremsstrahlung losses in quasineutral, isotropic plasmas

The ions undergoing fusion in many systems will essentially never occur alone but will be mixed with electrons that in aggregate neutralize the ions' bulk electrical charge and form a plasma. The electrons will generally have a temperature comparable to or greater than that of the ions, so they will collide with the ions and emit x-ray radiation of 10–30 keV energy, a process known as Bremsstrahlung.

The huge size of the Sun and stars means that the x-rays produced in this process will not escape and will deposit their energy back into the plasma. They are said to are opaque to x-rays. But any terrestrial fusion reactor will be optically thin for x-rays of this energy range. X-rays are difficult to reflect but they are effectively absorbed (and converted into heat) in less than mm thickness of stainless steel (which is part of a reactor's shield). This means the bremsstrahlung process is carrying energy out of the plasma, cooling it.

The ratio of fusion power produced to x-ray radiation lost to walls is an important figure of merit. This ratio is generally maximized at a much higher temperature than that which maximizes the power density (see the previous subsection). The following table shows estimates of the optimum temperature and the power ratio at that temperature for several reactions.[26]

The actual ratios of fusion to Bremsstrahlung power will likely be significantly lower for several reasons. For one, the calculation assumes that the energy of the fusion products is transmitted completely to the fuel ions, which then lose energy to the electrons by collisions, which in turn lose energy by Bremsstrahlung. However, because the fusion products move much faster than the fuel ions, they will give up a significant fraction of their energy directly to the electrons. Secondly, the ions in the plasma are assumed to be purely fuel ions. In practice, there will be a significant proportion of impurity ions, which will then lower the ratio. In particular, the fusion products themselves *must* remain in the plasma until they have given up their energy, and *will* remain some time after that in any proposed confinement scheme. Finally, all channels of energy loss other than Bremsstrahlung have been neglected. The last two factors are related. On theoretical and experimental grounds, particle and energy confinement seem to be closely related. In a confinement scheme that does a good job of retaining energy, fusion products will build up. If the fusion products are efficiently ejected, then energy confinement will be poor, too.

The temperatures maximizing the fusion power compared to the Bremsstrahlung are in every case higher than the temperature that maximizes the power density and minimizes the required value of the fusion triple product. This will not change the optimum operating point for 2
1D-3
1T very much because the Bremsstrahlung fraction is low, but it will push the other fuels into regimes where the power density relative to 2
1D-3
1T is even lower and the required confinement even more difficult to achieve. For 2
1D-2
1D and 2
1D-3
2He, Bremsstrahlung losses will be a serious, possibly prohibitive problem. For 3
2He-3
2He, p+-6
3Li and p+-11
5B the Bremsstrahlung losses appear to make a fusion reactor using these fuels with a quasineutral, isotropic plasma impossible. Some ways out of this dilemma are considered—and rejected—in *Fundamental limitations on plasma fusion*

systems not in thermodynamic equilibrium by Todd Rider.[28] This limitation does not apply to non-neutral and anisotropic plasmas; however, these have their own challenges to contend with.

12.6   See also

- Aneutronic fusion
- CNO cycle
- Direct energy conversion
- Inertial electrostatic confinement
- Focus fusion
- Fusenet
- Fusion power
- Fusion rocket
- Helium-3
- Impulse generator
- ITER
- Joint European Torus
- List of fusion experiments
- List of plasma (physics) articles
- National Ignition Facility
- Nuclear fission
- Nuclear physics
- Nuclear reactor
- Nucleosynthesis
- Neutron generator
- Neutron source
- Periodic table
- Polywell
- Proton-proton chain
- Pulsed power
- Teller–Ulam design
- Thermonuclear fusion
- Timeline of nuclear fusion
- Triple-alpha process

12.7   References

[1] Shultis, J.K. and Faw, R.E. (2002). *Fundamentals of nuclear science and engineering*. CRC Press. p. 151. ISBN 0-8247-0834-2.

[2] Physics Flexbook. Ck12.org. Retrieved on 2012-12-19.

[3] Bethe, Hans A. "The Hydrogen Bomb", *Bulletin of the Atomic Scientists*, April 1950, p. 99.

[4] "Progress in Fusion". ITER. Retrieved 2010-02-15.

[5] "ITER – the way to new energy". *ITER*. 2014.

[6] "The National Ignition Facility: Ushering in a new age for high energy density science". National Ignition Facility. Retrieved 2014-03-27.

[7] "DOE looks again at inertial fusion as potential clean-energy source", David Kramer, *Physics Today*, March 2011, p 26

[8] S. Atzeni, J. Meyer-ter-Vehn (2004). Chapter 1: "Nuclear fusion reactions". The Physics of Inertial Fusion. University of Oxford Press. ISBN 978-0-19-856264-1

[9] The Most Tightly Bound Nuclei. Hyperphysics.phy-astr.gsu.edu. Retrieved on 2011-08-17.

[10] What Is The Lawson Criteria, Or How to Make Fusion Power Viable

[11] "Fusor Forums • Index page". Fusor.net. Retrieved 2014-08-24.

[12] "Build a Nuclear Fusion Reactor? No Problem". Clhsonline.net. 2012-03-23. Retrieved 2014-08-24.

[13] "Extreme DIY: Building a homemade nuclear reactor in NYC". *BBC News*. Retrieved 30 October 2014.

[14] Schechner, Sam (2008-08-18). "Nuclear Ambitions: Amateur Scientists Get a Reaction From Fusion – WSJ". Online.wsj.com. Retrieved 2014-08-24.

[15] "Will's Amateur Science and Engineering: Fusion Reactor's First Light!". Tidbit77.blogspot.com. 2010-02-09. Retrieved 2014-08-24.

[16] Park J, Nebel RA, Stange S, Murali SK (2005). "Experimental Observation of a Periodically Oscillating Plasma Sphere in a Gridded Inertial Electrostatic Confinement Device". *Phys Rev Lett* **95** (1): 015003. Bibcode:2005PhRvL..95a5003P. doi:10.1103/PhysRevLett.95.015003. PMID 16090625.

[17] "The Multiple Ambipolar Recirculating Beam Line Experiment" Poster presentation, 2011 US-Japan IEC conference, Dr. Alex Klein

[18] Jones, S.E. (1986). "Muon-Catalysed Fusion Revisited". *Nature* **321** (6066): 127–133. Bibcode:1986Natur.321..127J

[19] Supplementary methods for "Observation of nuclear fusion driven by a pyroelectric crystal". Main article Naranjo, B.; Gimzewski, J.K.; Putterman, S. (2005). "Observation of nuclear fusion driven by a pyroelectric crystal". *Nature* **434** (7037): 1115–1117. Bibcode:2005Natur.434.1115N. doi:10.1038/nature03575. PMID 15858570.

[20] UCLA Crystal Fusion. Rodan.physics.ucla.edu. Retrieved on 2011-08-17. Archived 8 June 2015 at the Wayback Machine

[21] Schewe, Phil and Stein, Ben (2005). "Pyrofusion: A Room-Temperature, Palm-Sized Nuclear Fusion Device". *Physics News Update* **729** (1). Archived from the original on 12 November 2013.

[22] Coming in out of the cold: nuclear fusion, for real. Christiansciencemonitor.com (2005-06-06). Retrieved on 2011-08-17.

[23] fusion on the desktop ... really!. MSNBC (2005-04-27). Retrieved on 2011-08-17.

[24] Gerstner, E. (2009). "Nuclear energy: The hybrid returns". *Nature* **460** (7251): 25–8. doi:10.1038/460025a. PMID 19571861.

[25] FusEdWeb | Fusion Education. Fusedweb.pppl.gov (1998-11-09). Retrieved on 2011-08-17.

[26] Archived January 3, 2006 at the Wayback Machine. Retrieved on 2012-12-19.

[27] Subsection 4.7.4c. Kayelaby.npl.co.uk. Retrieved on 2012-12-19.

[28] Portable Document Format (PDF) Archived 26 March 2006 at the Wayback Machine

12.8   Further reading

- "What is Nuclear Fusion?". NuclearFiles.org.

- S. Atzeni, J. Meyer-ter-Vehn (2004). "Nuclear fusion reactions". *The Physics of Inertial Fusion* (PDF). University of Oxford Press. ISBN 978-0-19-856264-1.

- G. Brumfiel (22 May 2006). "Chaos could keep fusion under control". *Nature*. doi:10.1038/news060522-2.

- R.W. Bussard (9 November 2006). "Should Google Go Nuclear? Clean, Cheap, Nuclear Power". *Google TechTalks*.

- A. Wenisch, R. Kromp, D. Reinberger (November 2007). "Science of Fiction: Is there a Future for Nuclear?" (PDF). Austrian Institute of Ecology.

- W.J. Nuttall (September 2008). "Fusion as an Energy Source: Challenges and Opportunities" (PDF). *Institute of Physics Report*. Institute of Physics.

12.9   External links

- NuclearFiles.org—A repository of documents related to nuclear power.

- Annotated bibliography for nuclear fusion from the Alsos Digital Library for Nuclear Issues

- -NRL Fusion Formulary

Organizations

- Fusion for Energy website

- ITER (International Thermonuclear Experimental Reactor) website

- CCFE (Culham Centre for Fusion Energy) website

- JET (Joint European Torus) website

- Naka Fusion Institute at JAEA (Japan Atomic Energy Agency) website

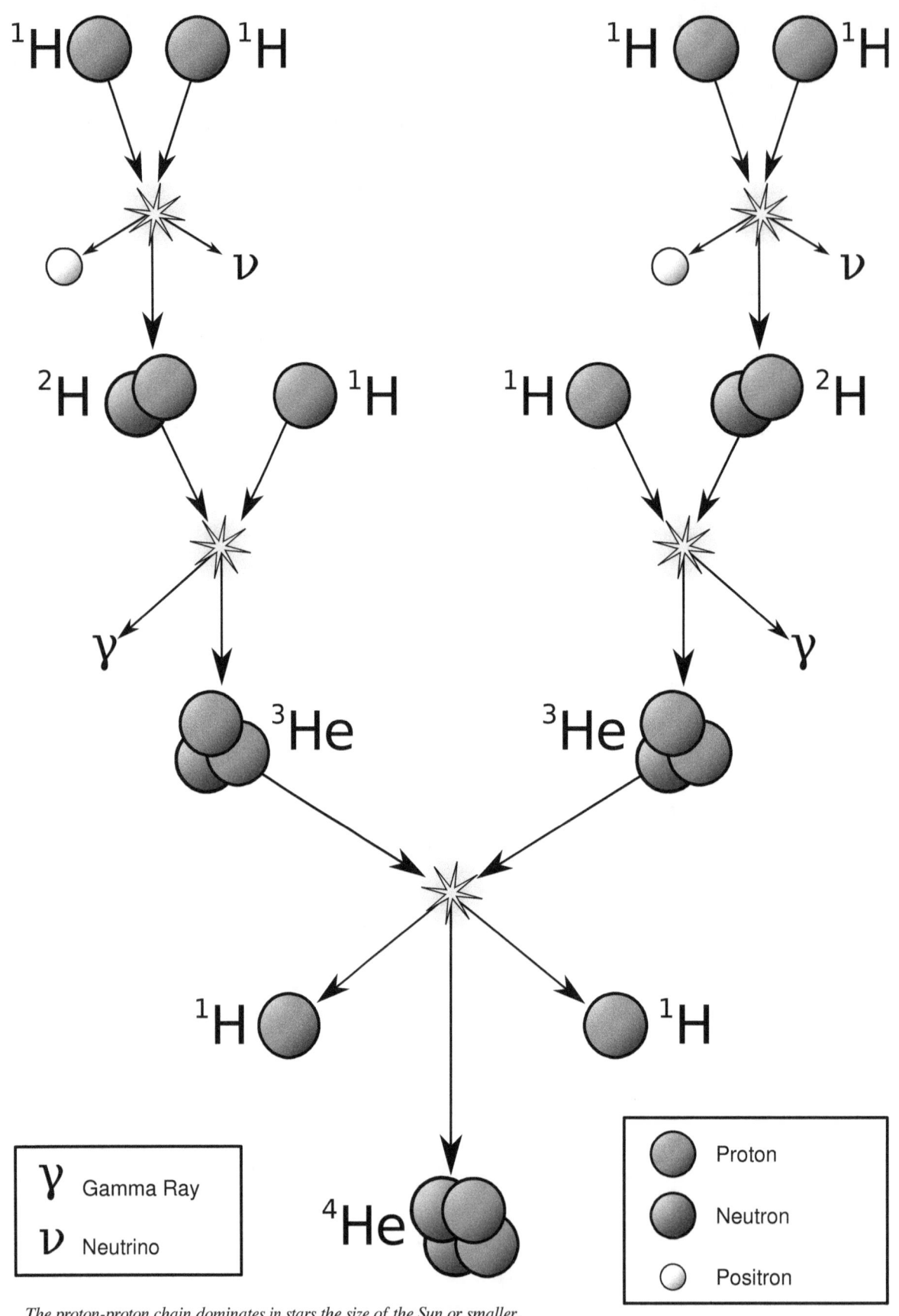

The proton-proton chain dominates in stars the size of the Sun or smaller.

γ Gamma Ray

ν Neutrino

Proton

Neutron

Positron

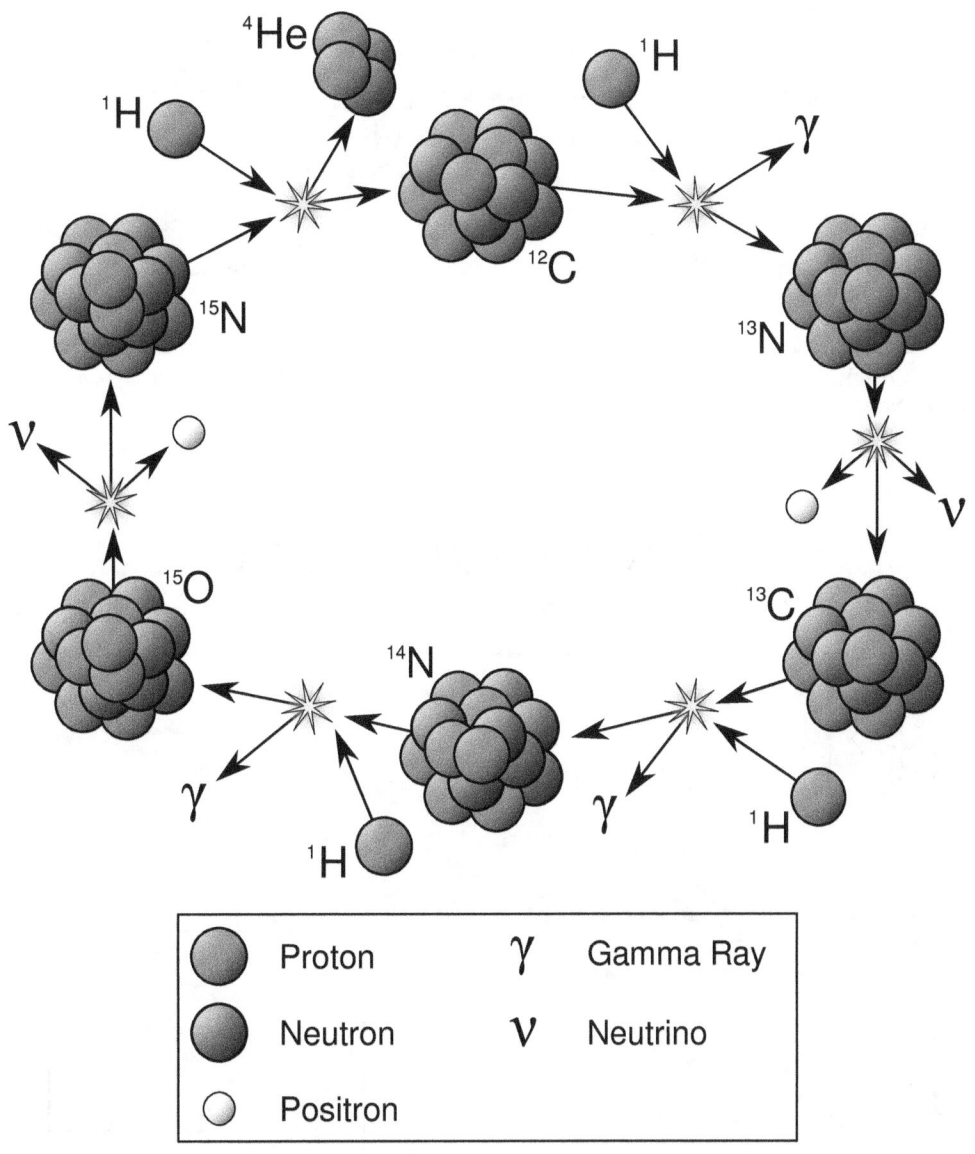

The CNO cycle dominates in stars heavier than the Sun.

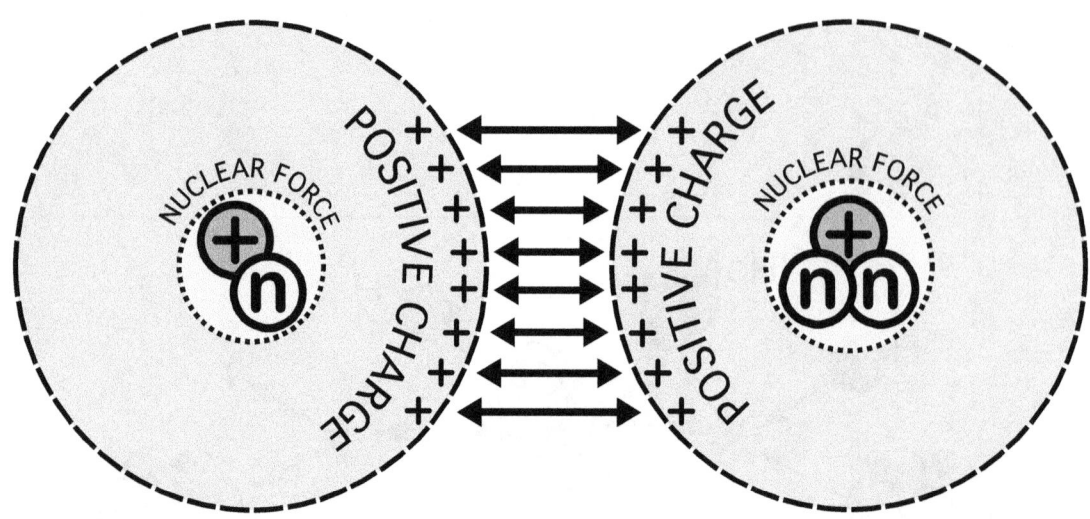

The electrostatic force between the positively charged nuclei is repulsive, but when the separation is small enough, the attractive nuclear force is stronger. Therefore, the prerequisite for fusion is that the nuclei have enough kinetic energy that they can approach each other despite the electrostatic repulsion.

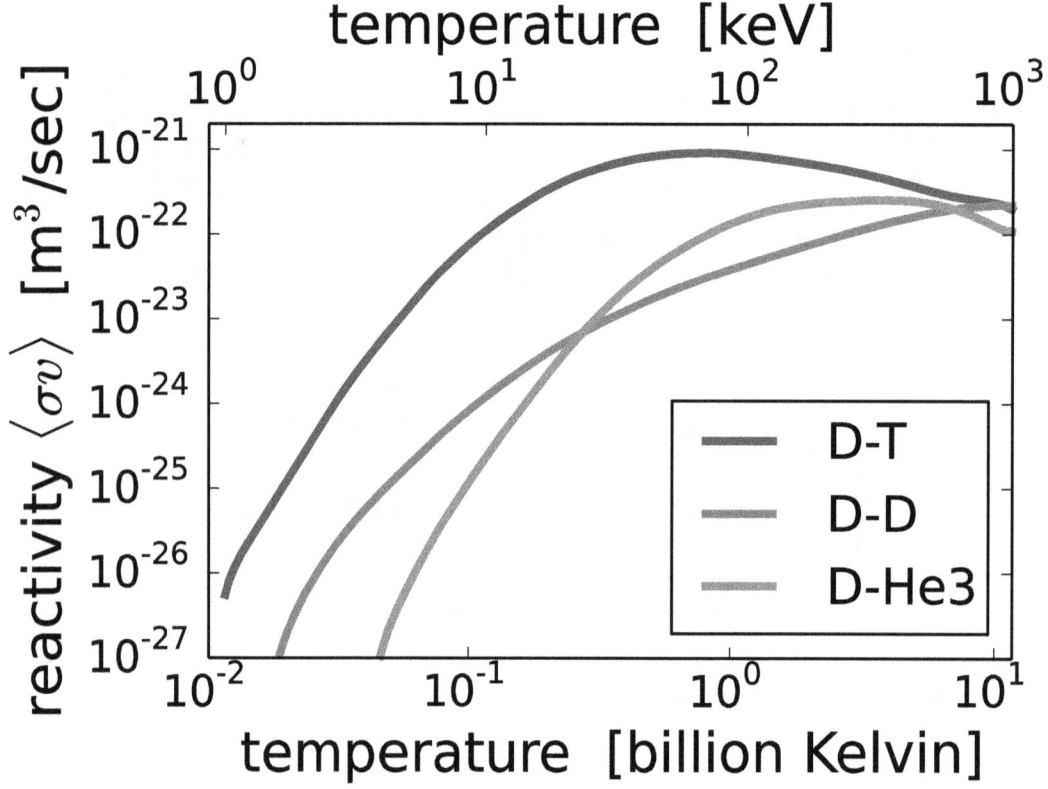

The fusion reaction rate increases rapidly with temperature until it maximizes and then gradually drops off. The DT rate peaks at a lower temperature (about 70 keV, or 800 million kelvin) and at a higher value than other reactions commonly considered for fusion energy.

The Tokamak à configuration variable, *research fusion reactor, at the École Polytechnique Fédérale de Lausanne (Switzerland).*

The only man-made fusion device to achieve ignition to date is the hydrogen bomb. The detonation of the first device, codenamed Ivy Mike, occurred in 1952 and is shown here.

Chapter 13

Nuclear power

"Atomic power" and "Atomic Power" redirect here. For the film, see Atomic Power (film).
This article is about nuclear fission and fusion power sources primarily. For commercial quantities of nuclear energy attained from nuclear decay, see Geothermal energy. For the political term, see List of states with nuclear weapons.

The 1200 MWe, Leibstadt fission-electric power station in Switzerland. The boiling water reactor(BWR), located inside the dome capped cylindrical structure, is dwarfed in size by its cooling tower. The station produces a yearly average of 25 million kilowatt-hours per day, sufficient to power a city the size of Boston.[1]

The Palo Verde Nuclear Generating Station, the largest in the US with 3 pressurized water reactors(PWRs), is situated in the Arizona desert. It uses sewage from cities as its cooling water in 9 squat mechanical draft cooling towers.[2][3] Its total spent fuel/"waste" inventory produced since 1986, is contained in dry cask storage cylinders located between the artificial body of water and the electrical switchyard.

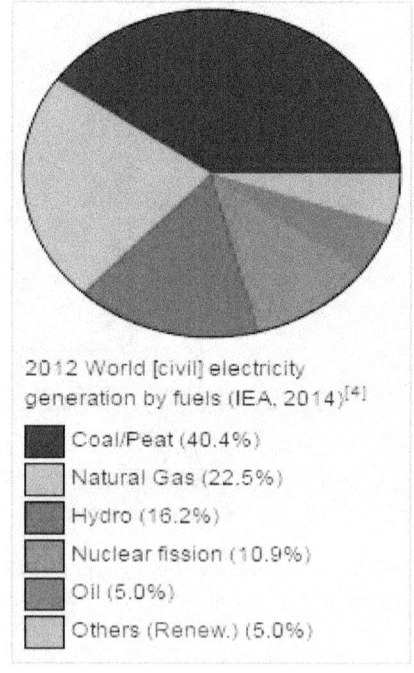

2012 World [civil] electricity
generation by fuels (IEA, 2014)[4]

- Coal/Peat (40.4%)
- Natural Gas (22.5%)
- Hydro (16.2%)
- Nuclear fission (10.9%)
- Oil (5.0%)
- Others (Renew.) (5.0%)

U.S. nuclear powered ships,(top to bottom) cruisers USS Bainbridge, *the USS* Long Beach *and the* USS Enterprise, *the longest ever naval vessel, and the first nuclear-powered aircraft carrier. Picture taken in 1964 during a record setting voyage of 26,540 nmi (49,190 km) around the world in 65 days without refueling. Crew members are spelling out Einstein's mass-energy equivalence formula* $E = mc^2$ *on the flight deck.*

Others (Renew.) (5.0%)

Nuclear power is the use of nuclear reactions that release nuclear energy[5] to generate heat, which most frequently is then used in steam turbines to produce electricity in a nuclear power station. The term includes nuclear fission, nuclear decay and nuclear fusion. Presently, the nuclear fission of elements in the actinide series of the periodic table produce the vast majority of nuclear energy in the direct service of humankind, with nuclear decay processes, primarily in the form of geothermal energy, and radioisotope thermoelectric generators, in niche uses making up the rest.

Nuclear (fission) power stations, excluding the contribution from naval nuclear fission reactors, provided 11% of the world's electricity in 2012,[6] somewhat less than that generated by hydro-electric stations at 16%. Since electricity accounts for about 25% of humanity's energy usage with the majority of the rest coming from fossil fuel reliant sectors such as transport, manufacture and home heating, nuclear fission's contribution to the global final energy consumption is about 2.5%,[7] a little more than the combined global electricity production from "new renewables"; wind, solar, biofuel and geothermal power, which together provided 2% of global final energy consumption in 2014.[8]

Regional differences in the use of fission energy are large. Fission energy generation, with a 20% share of the U.S. electricity production, is the single largest deployed technology among current low-carbon power sources in the country.[9] In addition, two-thirds of the European Union's twenty-seven nations's low-carbon energy is produced by fission.[10] Some of these nations have banned its generation, such as Italy, which ended the use of fission-electric generation, which started in 1963, in 1990. France is the largest user of nuclear energy, deriving 75% of its electricity from fission.

The Russian nuclear-powered icebreaker NS Yamal on a joint scientific expedition with the NSF in 1994.

Along with other sustainable energy sources, nuclear fission power is a low carbon power generation method of producing electricity, meaning that it is in the renewable energy family of low associated greenhouse gas emissions per unit of energy generated.[11] As all electricity supplying technologies use cement etc., during construction, emissions are yet to be brought to zero. A 2014 analysis of the carbon footprint literature by the Intergovernmental Panel on Climate Change (IPCC) reported that fission electricities embodied total life-cycle emission intensity value of 12 g CO_2 eq/kWh is the lowest out of all commercial Baseload energy sources,[12][13] and second lowest out of all commercial electricity technologies known, after wind power which is an Intermittent energy source with embodied greenhouse gas emissions, per unit of energy generated of 11 g CO_2eq/kWh. Each result is contrasted with coal & fossil gas at 820 and 490 g CO_2 eq/kWh.[12][13] With this translating into, from the beginning of Fission-electric power station commercialization in the 1970s, having prevented the emission of about 64 billion tonnes of carbon dioxide equivalent, greenhouse gases that would have otherwise resulted from the burning of fossil fuels in thermal power stations.[14]

In 2013, the IAEA reported that there are 437 operational civil fission-electric reactors[15] in 31 countries,[16] although not every reactor is producing electricity.[17] In addition, there are approximately 140 naval vessels using nuclear propulsion in operation, powered by some 180 reactors.[18][19][20] As of 2013, attaining a net energy gain from sustained nuclear fusion reactions, excluding natural fusion power sources such as the Sun, remains an ongoing area of international physics and engineering research. With commercial fusion power production remaining unlikely before 2050.[21]

In 2015, the IAEA report that worldwide there were 67 civil fission-electric power reactors under construction in 15 countries including Gulf states such as the United Arab Emirates (UAE).[15] Over half of the 67 total being built are in Asia, with 28 in the Peoples Republic of China (PRC), with the most recently completed fission-electric reactor to be connected to the electrical grid, as of August 2015, occurring in Wolseong Nuclear Power Plant in the Republic of Korea.[22] Five other new grid connections were completed by the PRC so far this year.[23] In the USA, four new Generation III reactors are under construction at Vogtle and Summer station, while a fifth is nearing completion at Watts Bar station, all five are expected to enter service before 2020.[24] In 2013, four aging uncompetitive U.S reactors were closed.[25][26]

There is a social debate about nuclear power.

[Proponents, such as the World Nuclear Association and Environmentalists for Nuclear Energy, contend that nuclear power is a safe, sustainable energy source that reduces carbon emissions.[30] Opponents, such as Greenpeace International and NIRS, contend that nuclear power poses many threats to people and the environment.[31][32][33]

Far reaching fission power reactor accidents, or accidents that resulted in medium to long-lived fission product contamination of inhabited areas, have occurred in Generation I & II reactor designs, blueprinted between 1950 and 1980. These include the Chernobyl disaster which occurred in 1986, the Fukushima Daiichi nuclear disaster (2011), and the more contained Three Mile Island accident (1979).[34] There have also been some nuclear submarine accidents.[34][35][36] In terms of lives lost per unit of energy generated, analysis has determined that fission-electric reactors have caused less fatalities per unit of energy generated than the other major sources of energy generation. Energy production from coal, petroleum, natural gas and hydroelectricity has caused a greater number of fatalities per unit of energy generated due to air pollution and energy accident effects.[37][38][39][40][41] However, the economic costs of nuclear power accidents is high, and meltdowns can render areas uninhabitable for very long periods. The human costs of evacuations of affected populations and lost livelihoods is also significant.[42][43]

Japan's 2011 Fukushima Daiichi nuclear disaster, which occurred in a reactor design from the 1960s, prompted a re-examination of nuclear safety and nuclear energy policy in many countries.[44] Germany plans to close all its reactors by 2022, and Italy has re-affirmed its ban on electric utilities generating, but not importing, fission derived electricity.[44] In 2011 the International Energy Agency halved its prior estimate of new generating capacity to be built by 2035.[45][46] In 2013 Japan signed a deal worth $22 billion, in which Mitsubishi Heavy Industries would build four modern *Atmea* reactors for Turkey.[47] In August 2015, following 4 years of near zero fission-electricity generation, Japan began restarting its fission fleet, after safety upgrades were completed, beginning with Sendai fission-electric station.[48]

13.1 Use

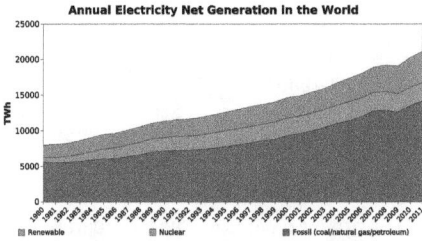

Net electrical generation by source and growth from 1980 to 2010. (Brown) - fossil fuels.(Red) - Fission.(Green)- "all renewables". In terms of energy generated between 1980 and 2010, the contribution from fission grew the fastest.

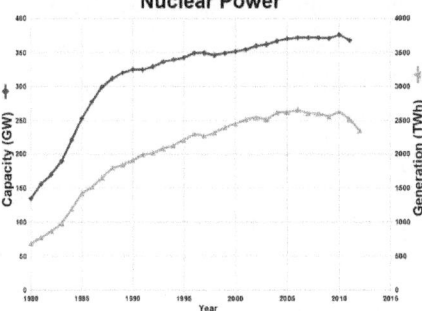

Worldwide civilian fission-electric power, installed nameplate capacity (in blue) in units of GW and actual electrical generation (in red) in units of TWh. 1980 to 2010 (EIA)

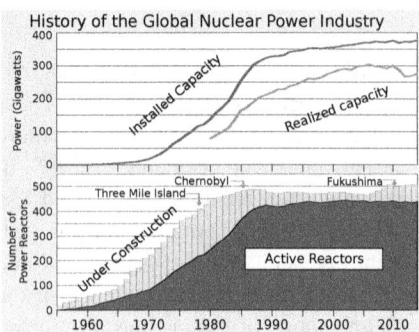

The rate of new construction builds for civilian fission-electric reactors essentially halted in the late 1980s, with the effects of accidents having a chilling effect. Increased capacity factor realizations in existing reactors was primarily responsible for the continuing increase in electrical energy produced during this period. The halting of new builds c. 1985, resulted in greater fossil fuel generation, see above graph.

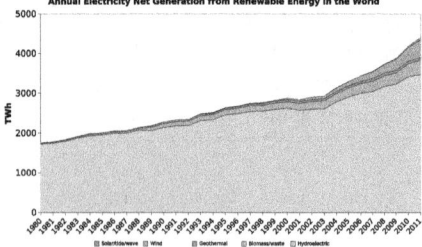

Hydropower(blue) is the dominant electricity generating "renewable" source and more new hydro-electricity was generated from 1980 to 2010, than all other renewable sources combined.

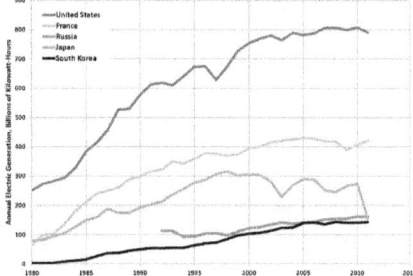

Electricitiy generation trends in the top five fission-energy producing countries (US EIA data)
  See also: Nuclear power by country and List of nuclear reactors

In 2011 nuclear power provided 10% of the world's electricity[6] In 2007, the IAEA reported there were 439 nuclear power reactors in operation in the world,[49] operating in 31 countries.[16] However, many have now ceased operation in the wake of the Fukushima nuclear disaster while they are assessed for safety. In 2011 worldwide nuclear output fell by 4.3%, the largest decline on record, on the back of sharp declines in Japan (−44.3%) and Germany (−23.2%).[50]

Since commercial nuclear energy began in the mid-1950s, 2008 was the first year that no new nuclear power plant was connected to the grid, although two were connected in 2009.[51][52]

Annual generation of nuclear power has been on a slight downward trend since 2007, decreasing 1.8% in 2009 to 2558 TWh with nuclear power meeting 13–14% of the world's electricity demand.[53] One factor in the nuclear power percentage decrease since 2007 has been the prolonged shutdown of large reactors at the Kashiwazaki-Kariwa Nuclear Power Plant in Japan following the Niigata-Chuetsu-Oki earthquake.[53]

The United States produces the most nuclear energy, with nuclear power providing 19%[54] of the electricity it consumes, while France produces the highest percentage of its electrical energy from nuclear reactors—80% as of 2006.[55] In the European Union as a whole, nuclear energy provides 30% of the electricity.[56] Nuclear energy policy differs among European Union countries, and some, such as Austria, Estonia, Ireland and Italy, have no active nuclear power stations. In comparison, France has a large number of these plants, with 16 multi-unit stations in current use.

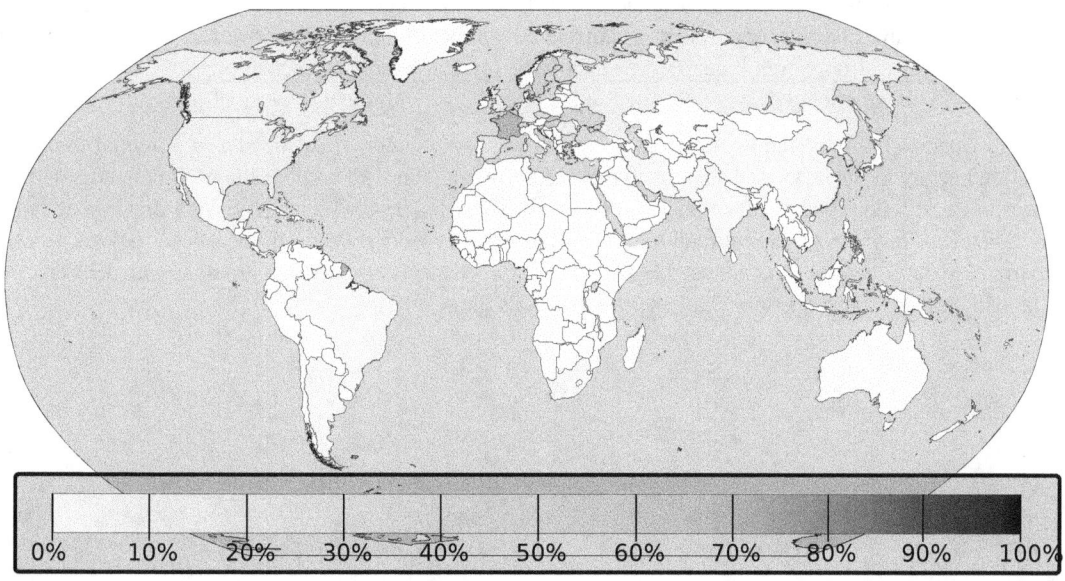

Percentage of a nations electricity, produced by fission-electric power stations.

In the US, while the coal and gas electricity industry is projected to be worth \$85 billion by 2013, nuclear power generators are forecast to be worth \$18 billion.[57]

Many military and some civilian (such as some icebreaker) ships use nuclear marine propulsion, a form of nuclear propulsion.[58] A few space vehicles have been launched using full-fledged nuclear reactors: 33 reactors belong to the Soviet RORSAT series and one was the American SNAP-10A.

International research is continuing into safety improvements such as passively safe plants,[59] the use of nuclear fusion, and additional uses of process heat such as hydrogen production (in support of a hydrogen economy), for desalinating sea water, and for use in district heating systems.

13.1.1 Use in space

Main article: Nuclear power in space

Both fission and fusion appear promising for space propulsion applications, generating higher mission velocities with less reaction mass. This is due to the much higher energy density of nuclear reactions: some 7 orders of magnitude (10,000,000 times) more energetic than the chemical reactions which power the current generation of rockets.

Radioactive decay has been used on a relatively small scale (few kW), mostly to power space missions and experiments by using radioisotope thermoelectric generators such as those developed at Idaho National Laboratory.

13.2 History

13.2.1 Origins

See also: Nuclear fission § History and Atomic Age

The pursuit of nuclear energy for electricity generation began soon after the discovery in the early 20th century that radioactive elements, such as radium, released immense amounts of energy, according to the principle of mass–energy equivalence. However, means of harnessing such energy was impractical, because intensely radioactive elements were, by their very nature, short-lived (high energy release is correlated with short half-lives). However, the dream of harnessing "atomic energy" was quite strong, even though it was dismissed by such fathers of nuclear physics like Ernest Rutherford as "moonshine."[60] This situation, however, changed in the late 1930s, with the discovery of nuclear fission.

In 1932, James Chadwick discovered the neutron,[61] which was immediately recognized as a potential tool for nuclear experimentation because of its lack of an electric charge. Experimentation with bombardment of materials with neutrons led Frédéric and Irène Joliot-Curie to discover induced radioactivity in 1934, which allowed the creation of radium-like elements at much less the price of natural radium.[62] Further work by Enrico Fermi in the 1930s focused on using slow neutrons to increase the effectiveness of induced radioactivity. Experiments bombarding uranium with neutrons led Fermi to believe he had created a new, transuranic element, which was dubbed hesperium.[63]

December 2, 1942. A depiction of the scene when scientists observed the world's first man made nuclear reactor, the Chicago Pile-1, as it became self-sustaining/critical at the University of Chicago.

But in 1938, German chemists Otto Hahn[64] and Fritz Strassmann, along with Austrian physicist Lise Meitner[65] and Meitner's nephew, Otto Robert Frisch,[66] conducted experiments with the products of neutron-bombarded uranium, as a means of further investigating Fermi's claims. They determined that the relatively tiny neutron split the nucleus of the massive uranium atoms into two roughly equal pieces, contradicting Fermi.[63] This was an extremely surprising result: all other forms of nuclear decay involved only small changes to the mass of the nucleus, whereas this process—dubbed "fission" as a reference to biology—involved a complete rupture of the nucleus. Numerous scientists, including Leó Szilárd, who was one of the first, recognized that if fission reactions released additional neutrons, a self-sustaining nuclear chain reaction could result. Once this was experimentally confirmed and announced by Frédéric Joliot-Curie in 1939, scientists in many countries (including the United States, the United Kingdom, France, Germany, and the Soviet Union) petitioned their governments for support of nuclear fission research, just on the cusp of World War II, for the development of a nuclear weapon.[67]

In the United States, where Fermi and Szilárd had both emigrated, this led to the creation of the first man-made reactor, known as Chicago Pile-1, which achieved criticality on December 2, 1942. This work became part of the Manhattan Project, which made enriched uranium and built large reactors to breed plutonium for use in the first nuclear weapons, which were used on the cities of Hiroshima and Nagasaki.

Unexpectedly high costs in the U.S. nuclear weapons program, along with competition with the Soviet Union and a desire to spread democracy through the world, created "...pressure on federal officials to develop a civilian nuclear power industry that could help justify the government's considerable expenditures."[68] In 1945, the pocketbook *The Atomic Age*

The first light bulbs ever lit by electricity generated by nuclear power at EBR-1 at Argonne National Laboratory-West, December 20, 1951.

heralded the untapped atomic power in everyday objects and depicted a future where fossil fuels would go unused. One science writer, David Dietz, wrote that instead of filling the gas tank of your car two or three times a week, you will travel for a year on a pellet of atomic energy the size of a vitamin pill. Glenn Seaborg, who chaired the Atomic Energy Commission, wrote "there will be nuclear powered earth-to-moon shuttles, nuclear powered artificial hearts, plutonium heated swimming pools for SCUBA divers, and much more". These overly optimistic predications remain unfulfilled.[69]

United Kingdom, Canada,[70] and USSR proceeded over the course of the late 1940s and early 1950s. Electricity was generated for the first time by a nuclear reactor on December 20, 1951, at the EBR-I experimental station near Arco, Idaho, which initially produced about 100 kW.[71][72] Work was also strongly researched in the US on nuclear marine propulsion, with a test reactor being developed by 1953 (eventually, the USS Nautilus, the first nuclear-powered submarine, would launch in 1955).[73] In 1953, US President Dwight Eisenhower gave his "Atoms for Peace" speech at the United Nations, emphasizing the need to develop "peaceful" uses of nuclear power quickly. This was followed by the 1954 Amendments to the Atomic Energy Act which allowed rapid declassification of U.S. reactor technology and encouraged development by the private sector. This involved a significant learning phase, with many early partial core meltdowns and accidents at experimental reactors and research facilities.[74]

13.2.2   Early years

On June 27, 1954, the USSR's Obninsk Nuclear Power Plant became the world's first nuclear power plant to generate electricity for a power grid, and produced around 5 megawatts of electric power.[75][76]

Later in 1954, Lewis Strauss, then chairman of the United States Atomic Energy Commission (U.S. AEC, forerunner of the U.S. Nuclear Regulatory Commission and the United States Department of Energy) spoke of electricity in the future being "too cheap to meter".[77] Strauss was very likely referring to hydrogen fusion[78] —which was secretly being developed as part of Project Sherwood at the time—but Strauss's statement was interpreted as a promise of very cheap energy from nuclear fission. The U.S. AEC itself had issued far more realistic testimony regarding nuclear fission to the U.S. Congress only months before, projecting that "costs can be brought down... [to]... about the same as the cost of electricity from conventional sources..."[79] Significant disappointment would develop later on, when the new nuclear plants did not provide energy "too cheap to meter."

In 1955 the United Nations' "First Geneva Conference", then the world's largest gathering of scientists and engineers, met to explore the technology. In 1957 EURATOM was launched alongside the European Economic Community (the latter is now the European Union). The same year also saw the launch of the International Atomic Energy Agency (IAEA).

The Shippingport Atomic Power Station in Shippingport, Pennsylvania was the first commercial reactor in the USA and was opened in 1957.

The world's first commercial nuclear power station, Calder Hall at Windscale, England, was opened in 1956 with an initial capacity of 50 MW (later 200 MW).[80][81] The first commercial nuclear generator to become operational in the United States was the Shippingport Reactor (Pennsylvania, December 1957).

One of the first organizations to develop nuclear power was the U.S. Navy, for the purpose of propelling submarines and aircraft carriers. The first nuclear-powered submarine, USS *Nautilus* (SSN-571), was put to sea in December 1954.[82] Two U.S. nuclear submarines, USS *Scorpion* and USS *Thresher*, have been lost at sea. Eight Soviet and Russian nuclear submarines have been lost at sea. This includes the Soviet submarine K-19 reactor accident in 1961 which resulted in 8 deaths and more than 30 other people were over-exposed to radiation.[35] The Soviet submarine K-27 reactor accident in 1968 resulted in 9 fatalities and 83 other injuries.[36] Moreover, Soviet submarine K-429 sank twice, but was raised after each incident. Several serious nuclear and radiation accidents have involved nuclear submarine mishaps.[34][36]

The U.S. Army also had a nuclear power program, beginning in 1954. The SM-1 Nuclear Power Plant, at Fort Belvoir, Virginia, was the first power reactor in the U.S. to supply electrical energy to a commercial grid (VEPCO), in April 1957, before Shippingport. The SL-1 was a U.S. Army experimental nuclear power reactor at the National Reactor Testing Station in eastern Idaho. It underwent a steam explosion and meltdown in January 1961, which killed its three operators.[83] In Soviet Union in The Mayak Production Association there were a number of accidents including an explosion that released 50-100 tonnes of high-level radioactive waste, contaminating a huge territory in the eastern Urals and causing numerous deaths and injuries. The Soviet regime kept this accident secret for about 30 years. The event was eventually rated at 6 on the seven-level INES scale (third in severity only to the disasters at Chernobyl and Fukushima).

13.2.3 Development

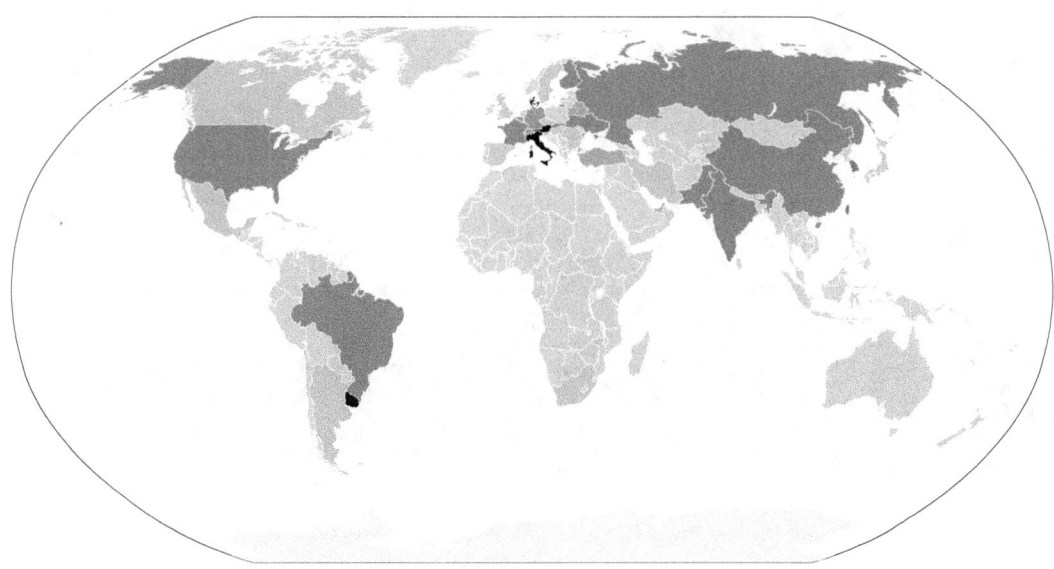

The status of nuclear power globally
(click image for legend)

Installed nuclear capacity initially rose relatively quickly, rising from less than 1 gigawatt (GW) in 1960 to 100 GW in the late 1970s, and 300 GW in the late 1980s. Since the late 1980s worldwide capacity has risen much more slowly, reaching 366 GW in 2005. Between around 1970 and 1990, more than 50 GW of capacity was under construction (peaking at over 150 GW in the late 1970s and early 1980s) — in 2005, around 25 GW of new capacity was planned. More than two-thirds of all nuclear plants ordered after January 1970 were eventually cancelled.[82] A total of 63 nuclear units were canceled in the USA between 1975 and 1980.[84]

During the 1970s and 1980s rising economic costs (related to extended construction times largely due to regulatory changes and pressure-group litigation)[85] and falling fossil fuel prices made nuclear power plants then under construction less attractive. In the 1980s (U.S.) and 1990s (Europe), flat load growth and electricity liberalization also made the addition of large new baseload capacity unattractive.

The 1973 oil crisis had a significant effect on countries, such as France and Japan, which had relied more heavily on oil for electric generation (39%[86] and 73% respectively) to invest in nuclear power.[87]

Some local opposition to nuclear power emerged in the early 1960s,[88] and in the late 1960s some members of the scientific community began to express their concerns.[89] These concerns related to nuclear accidents, nuclear proliferation, high cost of nuclear power plants, nuclear terrorism and radioactive waste disposal.[90] In the early 1970s, there were large protests about a proposed nuclear power plant in Wyhl, Germany. The project was cancelled in 1975 and anti-nuclear success at Wyhl inspired opposition to nuclear power in other parts of Europe and North America.[91][92] By the mid-1970s anti-nuclear activism had moved beyond local protests and politics to gain a wider appeal and influence, and nuclear

Washington Public Power Supply System Nuclear Power Plants 3 and 5 were never completed.

power became an issue of major public protest.[93] Although it lacked a single co-ordinating organization, and did not have uniform goals, the movement's efforts gained a great deal of attention.[94] In some countries, the nuclear power conflict "reached an intensity unprecedented in the history of technology controversies".[95]

In France, between 1975 and 1977, some 175,000 people protested against nuclear power in ten demonstrations.[96] In West Germany, between February 1975 and April 1979, some 280,000 people were involved in seven demonstrations at nuclear sites. Several site occupations were also attempted. In the aftermath of the Three Mile Island accident in 1979, some 120,000 people attended a demonstration against nuclear power in Bonn.[96] In May 1979, an estimated 70,000 people, including then governor of California Jerry Brown, attended a march and rally against nuclear power in Washington, D.C.[97] Anti-nuclear power groups emerged in every country that has had a nuclear power programme. Some of these anti-nuclear power organisations are reported to have developed considerable expertise on nuclear power and energy issues.[98]

Health and safety concerns, the 1979 accident at Three Mile Island, and the 1986 Chernobyl disaster played a part in stopping new plant construction in many countries,[89] although the public policy organization, the Brookings Institution states that new nuclear units, at the time of publishing in 2006, had not been built in the U.S. because of soft demand for electricity, and cost overruns on nuclear plants due to regulatory issues and construction delays.[99] By the end of the 1970s it became clear that nuclear power would not grow nearly as dramatically as once believed. Eventually, more than 120 reactor orders in the U.S. were ultimately cancelled[100] and the construction of new reactors ground to a halt. A cover story in the February 11, 1985, issue of *Forbes* magazine commented on the overall failure of the U.S. nuclear power program, saying it "ranks as the largest managerial disaster in business history".[101]

Unlike the Three Mile Island accident, the much more serious Chernobyl accident did not increase regulations affecting Western reactors since the Chernobyl reactors were of the problematic RBMK design only used in the Soviet Union,

120,000 people attended an anti-nuclear protest in Bonn, Germany, on October 14, 1979, following the Three Mile Island accident.[96]

for example lacking "robust" containment buildings.[102] Many of these RBMK reactors are still in use today. However, changes were made in both the reactors themselves (use of a safer enrichment of uranium) and in the control system (prevention of disabling safety systems), amongst other things, to reduce the possibility of a duplicate accident.[103]

An international organization to promote safety awareness and professional development on operators in nuclear facilities was created: WANO; World Association of Nuclear Operators.

Opposition in Ireland and Poland prevented nuclear programs there, while Austria (1978), Sweden (1980) and Italy (1987) (influenced by Chernobyl) voted in referendums to oppose or phase out nuclear power. In July 2009, the Italian Parliament passed a law that cancelled the results of an earlier referendum and allowed the immediate start of the Italian nuclear program.[104] After the Fukushima Daiichi nuclear disaster a one-year moratorium was placed on nuclear power development,[105] followed by a referendum in which over 94% of voters (turnout 57%) rejected plans for new nuclear power.[106]

13.3 Nuclear power plant

Main article: Nuclear power plant

Just as many conventional thermal power stations generate electricity by harnessing the thermal energy released from burning fossil fuels, nuclear power plants convert the energy released from the nucleus of an atom via nuclear fission that takes place in a nuclear reactor. The heat is removed from the reactor core by a cooling system that uses the heat to generate steam, which drives a steam turbine connected to a generator producing electricity.

The abandoned city of Pripyat with Chernobyl plant in the distance.

13.4   Life cycle

Main article: Nuclear fuel cycle

A nuclear reactor is only part of the life-cycle for nuclear power. The process starts with mining (see *Uranium mining*). Uranium mines are underground, open-pit, or in-situ leach mines. In any case, the uranium ore is extracted, usually converted into a stable and compact form such as yellowcake, and then transported to a processing facility. Here, the yellowcake is converted to uranium hexafluoride, which is then enriched using various techniques. At this point, the enriched uranium, containing more than the natural 0.7% U-235, is used to make rods of the proper composition and geometry for the particular reactor that the fuel is destined for. The fuel rods will spend about 3 operational cycles (typically 6 years total now) inside the reactor, generally until about 3% of their uranium has been fissioned, then they will be moved to a spent fuel pool where the short lived isotopes generated by fission can decay away. After about 5 years in a spent fuel pool the spent fuel is radioactively and thermally cool enough to handle, and it can be moved to dry storage casks or reprocessed.

13.4.1   Conventional fuel resources

Main articles: Uranium market and Energy development - Nuclear energy

 Uranium is a fairly common element in the Earth's crust. Uranium is approximately as common as tin or germanium in the Earth's crust, and is about 40 times more common than silver.[107] Uranium is a constituent of most rocks, dirt, and of the oceans. The fact that uranium is so spread out is a problem because mining uranium is only economically feasible

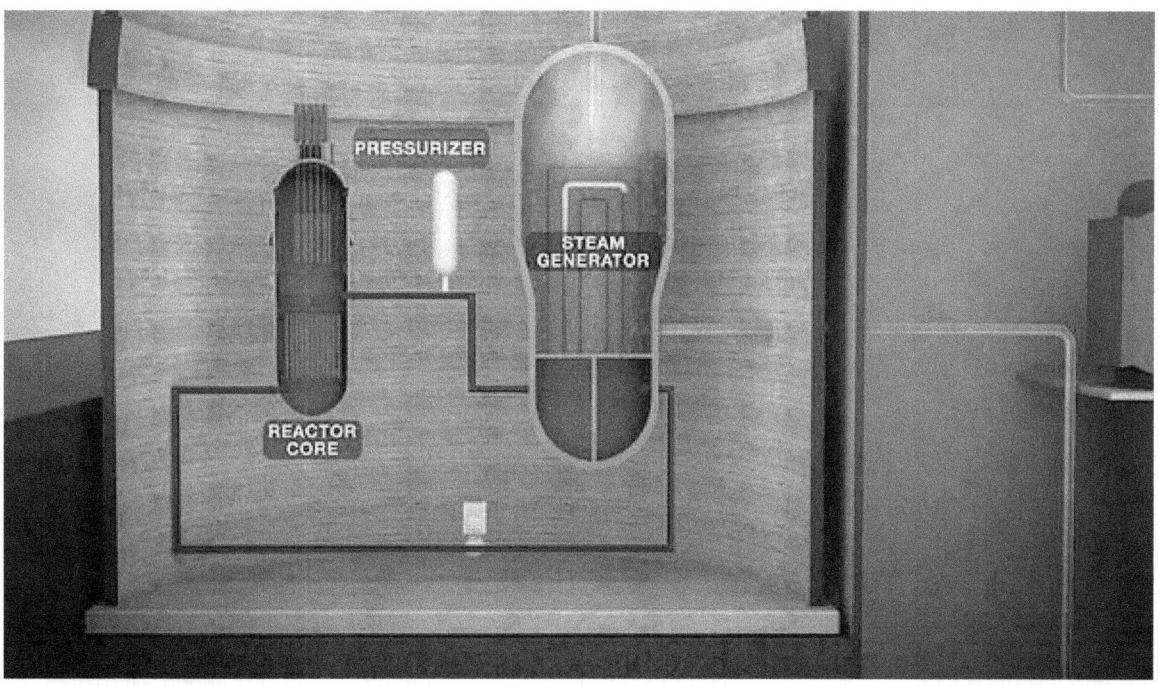

An animation of a Pressurized water reactor in operation.

Unlike fossil fuel power plants, the only substance leaving the cooling towers of nuclear power plants is water vapour and thus does not pollute the air or cause global warming.

where there is a large concentration. Still, the world's present measured resources of uranium, economically recoverable at a price of 130 USD/kg, are enough to last for between 70 and 100 years.[108][109][110]

According to the OECD in 2006, there is an expected 85 years worth of uranium in identified resources, when that uranium is used in present reactor technology, with 670 years of economically recoverable uranium in total conventional resources and phosphate ores, while also using present reactor technology, a resource that is recoverable from between 60-100 US$/kg of Uranium.[111] The OECD have noted that:

> Even if the nuclear industry expands significantly, sufficient fuel is available for centuries. If advanced breeder reactors could be designed in the future to efficiently utilize recycled or depleted uranium and all

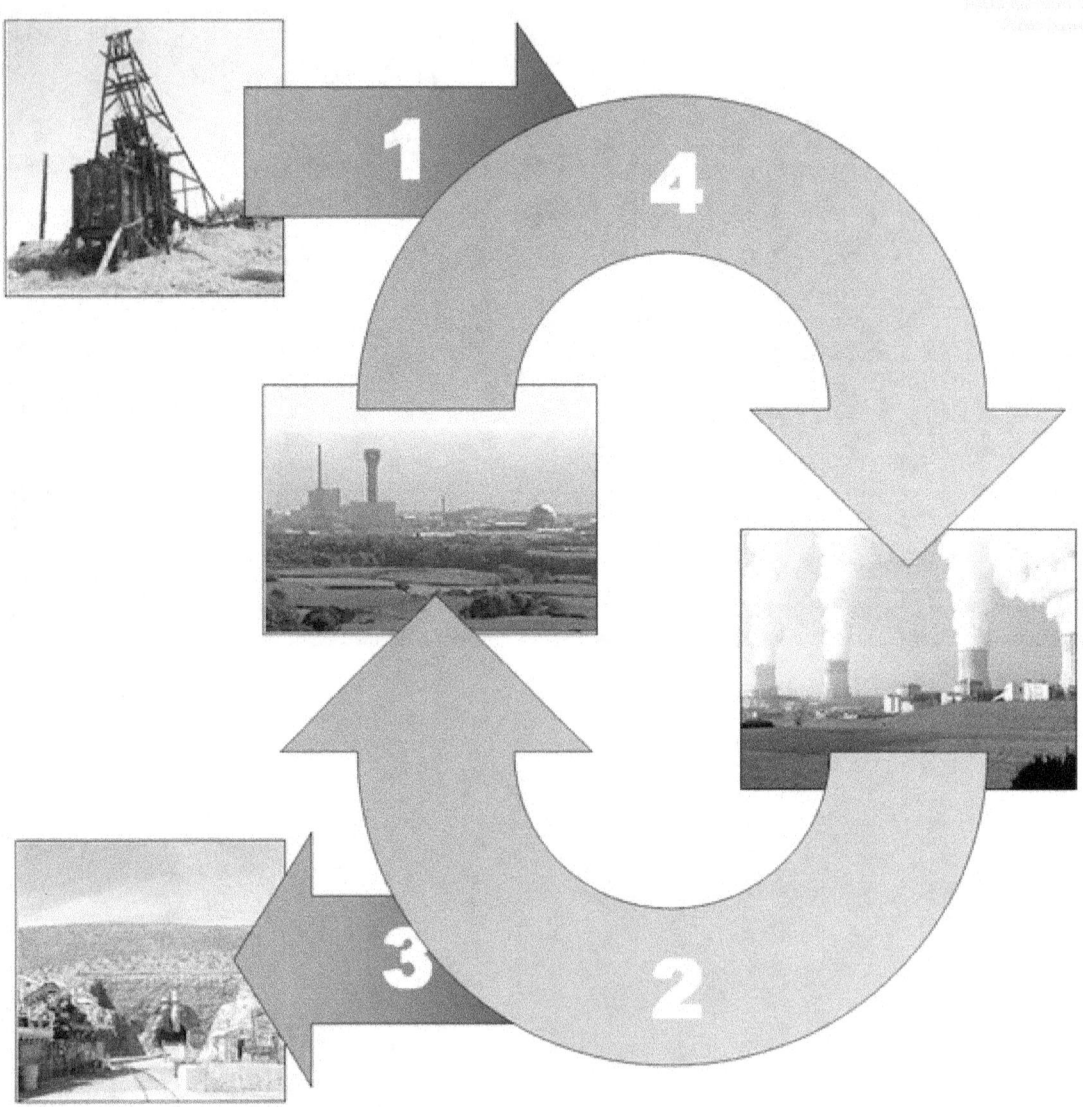

The nuclear fuel cycle begins when uranium is mined, enriched, and manufactured into nuclear fuel, (1) which is delivered to a nuclear power plant. After usage in the power plant, the spent fuel is delivered to a reprocessing plant (2) or to a final repository (3) for geological disposition. In reprocessing 95% of spent fuel can potentially be recycled to be returned to usage in a power plant (4).

actinides, then the resource utilization efficiency would be further improved by an additional factor of eight.

For example, the OECD have determined that with a pure fast reactor fuel cycle with a burn up of, and recycling of, all the Uranium and actinides, actinides which presently make up the most hazardous substances in nuclear waste, there is 160,000 years worth of Uranium in total conventional resources and phosphate ore.[112] According to the OECD's red book in 2011, due to increased exploration, known uranium resources have grown by 12.5% since 2008, with this increase translating into greater than a century of uranium available if the metals usage rate were to continue at the 2011 level.[113][114]

Current light water reactors make relatively inefficient use of nuclear fuel, fissioning only the very rare uranium-235 isotope. Nuclear reprocessing can make this waste reusable, and more efficient reactor designs, such as the currently under construction Generation III reactors achieve a higher efficiency burn up of the available resources, than the current vintage generation II reactors, which make up the vast majority of reactors worldwide.[115]

Breeding

Main articles: Breeder reactor and Nuclear power proposed as renewable energy

As opposed to current light water reactors which use uranium-235 (0.7% of all natural uranium), fast breeder reactors use uranium-238 (99.3% of all natural uranium). It has been estimated that there is up to five billion years' worth of uranium-238 for use in these power plants.[116]

Breeder technology has been used in several reactors, but the high cost of reprocessing fuel safely, at 2006 technological levels, requires uranium prices of more than 200 USD/kg before becoming justified economically.[117] Breeder reactors are still however being pursued as they have the potential to burn up all of the actinides in the present inventory of nuclear waste while also producing power and creating additional quantities of fuel for more reactors via the breeding process.[118][119] In 2005, there were two breeder reactors producing power: the Phénix in France, which has since powered down in 2009 after 36 years of operation, and the BN-600 reactor, a reactor constructed in 1980 Beloyarsk, Russia which is still operational as of 2013. The electricity output of BN-600 is 600 MW — Russia plans to expand the nation's use of breeder reactors with the BN-800 reactor, was scheduled to become operational in 2014,[120] but due to delays is not scheduled to produce power until 2017.[121] The technical design of a yet larger breeder, the BN-1200 reactor was originally scheduled to be finalized in 2013, with construction slated for 2015 but has also been delayed.[122] Japan's Monju breeder reactor restarted (having been shut down in 1995) in 2010 for 3 months, but shut down again after equipment fell into the reactor during reactor checkups, it is planned to become re-operational in late 2013.[123] Both China and India are building breeder reactors. With the Indian 500 MWe Prototype Fast Breeder Reactor scheduled to become operational in 2014, with plans to build five more by 2020.[124] The China Experimental Fast Reactor began producing power in 2011.[125]

Another alternative to fast breeders is thermal breeder reactors that use uranium-233 bred from thorium as fission fuel in the thorium fuel cycle. Thorium is about 3.5 times more common than uranium in the Earth's crust, and has different geographic characteristics. This would extend the total practical fissionable resource base by 450%.[126] India's three-stage nuclear power programme features the use of a thorium fuel cycle in the third stage, as it has abundant thorium reserves but little uranium.

13.4.2   Solid waste

For more details on this topic, see Radioactive waste.
See also: List of nuclear waste treatment technologies

The most important waste stream from nuclear power plants is spent nuclear fuel. It is primarily composed of unconverted uranium as well as significant quantities of transuranic actinides (plutonium and curium, mostly). In addition, about 3% of it is fission products from nuclear reactions. The actinides (uranium, plutonium, and curium) are responsible for the bulk of the long-term radioactivity, whereas the fission products are responsible for the bulk of the short-term radioactivity.[127]

High-level radioactive waste

Main article: High-level radioactive waste management

 High-level radioactive waste management concerns management and disposal of highly radioactive materials created during production of nuclear power. The technical issues in accomplishing this are daunting, due to the extremely long periods radioactive wastes remain deadly to living organisms. Of particular concern are two long-lived fission products, Technetium-99 (half-life 220,000 years) and Iodine-129 (half-life 15.7 million years),[130] which dominate spent nuclear fuel radioactivity after a few thousand years. The most troublesome transuranic elements in spent fuel are Neptunium-237 (half-life two million years) and Plutonium-239 (half-life 24,000 years).[131] Consequently, high-level radioactive waste requires sophisticated treatment and management to successfully isolate it from the biosphere. This usually necessitates treatment, followed by a long-term management strategy involving permanent storage, disposal or transformation of the waste into a non-toxic form.[132]

Governments around the world are considering a range of waste management and disposal options, usually involving deep-geologic placement, although there has been limited progress toward implementing long-term waste management solutions.[133] This is partly because the timeframes in question when dealing with radioactive waste range from 10,000 to millions of years,[134][135] according to studies based on the effect of estimated radiation doses.[136]

Some proposed nuclear reactor designs however such as the American Integral Fast Reactor and the Molten salt reactor can use the nuclear waste from light water reactors as a fuel, transmutating it to isotopes that would be safe after hundreds, instead of tens of thousands of years. This offers a potentially more attractive alternative to deep geological disposal.[137][138][139]

Another possibility is the use of thorium in a reactor especially designed for thorium (rather than mixing in thorium with uranium and plutonium (i.e. in existing reactors). Used thorium fuel remains only a few hundreds of years radioactive, instead of tens of thousands of years.[140]

Since the fraction of a radioisotope's atoms decaying per unit of time is inversely proportional to its half-life, the relative radioactivity of a quantity of buried human radioactive waste would diminish over time compared to natural radioisotopes (such as the decay chains of 120 trillion tons of thorium and 40 trillion tons of uranium which are at relatively trace concentrations of parts per million each over the crust's $3 * 10^{19}$ ton mass).[141][142][143] For instance, over a timeframe of thousands of years, after the most active short half-life radioisotopes decayed, burying U.S. nuclear waste would increase the radioactivity in the top 2000 feet of rock and soil in the United States (10 million km^2) by $\approx$ 1 part in 10 million over the cumulative amount of natural radioisotopes in such a volume, although the vicinity of the site would have a far higher concentration of artificial radioisotopes underground than such an average.[144]

Low-level radioactive waste

See also: Low-level waste

The nuclear industry also produces a large volume of low-level radioactive waste in the form of contaminated items like clothing, hand tools, water purifier resins, and (upon decommissioning) the materials of which the reactor itself is built. In the US, the Nuclear Regulatory Commission has repeatedly attempted to allow low-level materials to be handled as normal waste: landfilled, recycled into consumer items, etcetera.

Comparing radioactive waste to industrial toxic waste

In countries with nuclear power, radioactive wastes comprise less than 1% of total industrial toxic wastes, much of which remains hazardous for long periods.[115] Overall, nuclear power produces far less waste material by volume than fossil-fuel based power plants.[145] Coal-burning plants are particularly noted for producing large amounts of toxic and mildly radioactive ash due to concentrating naturally occurring metals and mildly radioactive material from the coal.[146] A 2008 report from Oak Ridge National Laboratory concluded that coal power actually results in more radioactivity being released into the environment than nuclear power operation, and that the population effective dose equivalent, or dose to the public from radiation from coal plants is 100 times as much as from the ideal operation of nuclear plants.[147] Indeed, coal ash is much less radioactive than spent nuclear fuel on a weight per weight basis, but coal ash is produced in much higher quantities per unit of energy generated, and this is released directly into the environment as fly ash, whereas nuclear plants use shielding to protect the environment from radioactive materials, for example, in dry cask storage vessels.[148]

Waste disposal

Disposal of nuclear waste is often said to be the Achilles' heel of the industry.[149] Presently, waste is mainly stored at individual reactor sites and there are over 430 locations around the world where radioactive material continues to accumulate. Some experts suggest that centralized underground repositories which are well-managed, guarded, and monitored, would be a vast improvement.[149] There is an "international consensus on the advisability of storing nuclear waste in deep geological repositories",[150] with the lack of movement of nuclear waste in the 2 billion year old natural nuclear fission reactors in Oklo, Gabon being cited as "a source of essential information today."[151][152]

As of 2009 there were no commercial scale purpose built underground repositories in operation.[150][153][154][155] The Waste Isolation Pilot Plant in New Mexico has been taking nuclear waste since 1999 from production reactors, but as the name suggests is a research and development facility.

13.4.3 Reprocessing

For more details on this topic, see Nuclear reprocessing.

Reprocessing can potentially recover up to 95% of the remaining uranium and plutonium in spent nuclear fuel, putting it into new mixed oxide fuel. This produces a reduction in long term radioactivity within the remaining waste, since this is largely short-lived fission products, and reduces its volume by over 90%. Reprocessing of civilian fuel from power reactors is currently done in Britain, France and (formerly) Russia, soon will be done in China and perhaps India, and is being done on an expanding scale in Japan. The full potential of reprocessing has not been achieved because it requires breeder reactors, which are not commercially available. France is generally cited as the most successful reprocessor, but it presently only recycles 28% (by mass) of the yearly fuel use, 7% within France and another 21% in Russia.[156]

Reprocessing is not allowed in the U.S.[157] The Obama administration has disallowed reprocessing of nuclear waste, citing nuclear proliferation concerns.[158] In the U.S., spent nuclear fuel is currently all treated as waste.[159]

Depleted uranium

Main article: Depleted uranium

Uranium enrichment produces many tons of depleted uranium (DU) which consists of U-238 with most of the easily fissile U-235 isotope removed. U-238 is a tough metal with several commercial uses—for example, aircraft production, radiation shielding, and armor—as it has a higher density than lead. Depleted uranium is also controversially used in munitions; DU penetrators (bullets or APFSDS tips) "self sharpen", due to uranium's tendency to fracture along shear bands.[160][161]

13.5 Economics

Main article: Economics of new nuclear power plants
Internationally the price of nuclear plants rose 15% annually in 1970-1990. Total costs rose tenfold. The nuclear plant construction time doubled. According to Al Gore if intended plan does not hold, the delay cost a billion dollars a year.[164] Yet, nuclear power has total costs in 2012 of about $96 per megawatt hour (MWh), most of which involves capital construction costs, compared with solar power at $130 per MWh, and natural gas at the low end at $64 per MWh.[165]

The economics of new nuclear power plants is a controversial subject, since there are diverging views on this topic, and multibillion-dollar investments ride on the choice of an energy source. Nuclear power plants typically have high capital costs for building the plant, but low fuel costs. Therefore, comparison with other power generation methods is strongly dependent on assumptions about construction timescales and capital financing for nuclear plants as well as the future costs of fossil fuels and renewables as well as for energy storage solutions for intermittent power sources. Cost estimates also need to take into account plant decommissioning and nuclear waste storage costs. On the other hand, measures to mitigate global warming, such as a carbon tax or carbon emissions trading, may favor the economics of nuclear power.

In 2015, the *Bulletin of the Atomic Scientists* unveiled the Nuclear Fuel Cycle Cost Calculator, an online tool that estimates the full cost of electricity produced by three configurations of the nuclear fuel cycle. Two years in the making, this interactive calculator is the first generally accessible model to provide a nuanced look at the economic costs of nuclear power; it lets users test how sensitive the price of electricity is to a full range of components—more than 60 parameters that can be adjusted for the three configurations of the nuclear fuel cycle considered by this tool (once-through, limited-recycle, full-recycle). Users can select the fuel cycle they would like to examine, change cost estimates for each component of that cycle, and even choose uncertainty ranges for the cost of particular components. This approach allows users around the world to compare the cost of different nuclear power approaches in a sophisticated way, while taking account of prices relevant to their own countries or regions.

In recent years there has been a slowdown of electricity demand growth and financing has become more difficult, which has an impact on large projects such as nuclear reactors, with very large upfront costs and long project cycles which carry a large variety of risks.[166] In Eastern Europe, a number of long-established projects are struggling to find finance, notably Belene in Bulgaria and the additional reactors at Cernavoda in Romania, and some potential backers have pulled out.[166] Where the electricity market is competitive, cheap natural gas is available, and its future supply relatively secure, this also poses a major problem for nuclear projects[166] and existing plants.[167]

Analysis of the economics of nuclear power must take into account who bears the risks of future uncertainties. To date all operating nuclear power plants were developed by state-owned or regulated utility monopolies[168] where many of the risks associated with construction costs, operating performance, fuel price, accident liability and other factors were borne by consumers rather than suppliers. In addition, because the potential liability from a nuclear accident is so great, the full cost of liability insurance is generally limited/capped by the government, which the U.S. Nuclear Regulatory Commission concluded constituted a significant subsidy.[169] Many countries have now liberalized the electricity market where these risks, and the risk of cheaper competitors emerging before capital costs are recovered, are borne by plant suppliers and operators rather than consumers, which leads to a significantly different evaluation of the economics of new nuclear power plants.[170]

Following the 2011 Fukushima Daiichi nuclear disaster, costs are expected to increase for currently operating and new nuclear power plants, due to increased requirements for on-site spent fuel management and elevated design basis threats.[171]

13.6   Accidents and safety, the human and financial costs

See also: Energy accidents, Nuclear safety, Nuclear and radiation accidents and Lists of nuclear disasters and radioactive incidents

Some serious nuclear and radiation accidents have occurred. Benjamin K. Sovacool has reported that worldwide there have been 99 accidents at nuclear power plants.[174] Fifty-seven accidents have occurred since the Chernobyl disaster, and 57% (56 out of 99) of all nuclear-related accidents have occurred in the USA.[174][175]

Nuclear power plant accidents include the Chernobyl accident (1986) with approximately 60 deaths so far attributed to the accident and a predicted, eventual total death toll, of from 4000 to 25,000 latent cancers deaths. The Fukushima Daiichi nuclear disaster (2011), has not caused any radiation related deaths, with a predicted, eventual total death toll, of from 0 to 1000, and the Three Mile Island accident (1979), no causal deaths, cancer or otherwise, have been found in follow up studies of this accident.[34] Nuclear-powered submarine mishaps include the K-19 reactor accident (1961),[35] the K-27 reactor accident (1968),[36] and the K-431 reactor accident (1985).[34] International research is continuing into safety improvements such as passively safe plants,[59] and the possible future use of nuclear fusion.

In terms of lives lost per unit of energy generated, nuclear power has caused fewer accidental deaths per unit of energy generated than all other major sources of energy generation. Energy produced by coal, petroleum, natural gas and hydropower has caused more deaths per unit of energy generated, from air pollution and energy accidents. This is found in the following comparisons, when the immediate nuclear related deaths from accidents are compared to the immediate deaths from these other energy sources,[38] when the latent, or predicted, indirect cancer deaths from nuclear energy accidents are compared to the immediate deaths from the above energy sources,[40][41][176] and when the combined immediate and indirect fatalities from nuclear power and all fossil fuels are compared, fatalities resulting from the mining of the necessary natural resources to power generation and to air pollution.[177] With these data, the use of nuclear power has been calculated to have prevented in the region of 1.8 million deaths between 1971 and 2009, by reducing the proportion of energy that would otherwise have been generated by fossil fuels, and is projected to continue to do so.[178][179]

Although according to Benjamin K. Sovacool, fission energy accidents ranked first in terms of their total economic cost, accounting for 41 percent of all property damage attributed to energy accidents.[180] Analysis presented in the international Journal, *Human and Ecological Risk Assessment* found that coal, oil, Liquid petroleum gas and hydroelectric accidents(primarily due to the Banqiao dam burst) have resulted in greater economic impacts than nuclear power accidents.[181]

Following the 2011 Japanese Fukushima nuclear disaster, authorities shut down the nation's 54 nuclear power plants, but it has been estimated that if Japan had never adopted nuclear power, accidents and pollution from coal or gas plants would have caused more lost years of life.[182] As of 2013, the Fukushima site remains highly radioactive, with some 160,000 evacuees still living in temporary housing, and some land will be unfarmable for centuries. The difficult Fukushima disaster cleanup will take 40 or more years, and cost tens of billions of dollars.[42][43]

Forced evacuation from a nuclear accident may lead to social isolation, anxiety, depression, psychosomatic medical problems, reckless behavior, even suicide. Such was the outcome of the 1986 Chernobyl nuclear disaster in Ukraine. A comprehensive 2005 study concluded that "the mental health impact of Chernobyl is the largest public health problem unleashed by the accident to date".[183] Frank N. von Hippel, a U.S. scientist, commented on the 2011 Fukushima nuclear disaster, saying that "fear of ionizing radiation could have long-term psychological effects on a large portion of the population in the contaminated areas".[184]

13.7 Nuclear proliferation

Many technologies and materials associated with the creation of a nuclear power program have a dual-use capability, in that they can be used to make nuclear weapons if a country chooses to do so. When this happens a nuclear power program can become a route leading to a nuclear weapon or a public annex to a "secret" weapons program. The concern over Iran's nuclear activities is a case in point.[185]

A fundamental goal for American and global security is to minimize the nuclear proliferation risks associated with the expansion of nuclear power. If this development is "poorly managed or efforts to contain risks are unsuccessful, the nuclear future will be dangerous".[185] The Global Nuclear Energy Partnership is one such international effort to create a distribution network in which developing countries in need of energy, would receive nuclear fuel at a discounted rate, in exchange for that nation agreeing to forgo their own indigenous develop of a uranium enrichment program. The France-based Eurodif/*European Gaseous Diffusion Uranium Enrichment Consortium* was/is one such program that successfully implemented this concept, with Spain and other countries without enrichment facilities buying a share of the fuel produced at the French controlled enrichment facility, but without a transfer of technology.[188] Iran was an early participant from 1974, and remains a shareholder of Eurodif via Sofidif.

According to Benjamin K. Sovacool, a "number of high-ranking officials, even within the United Nations, have argued that they can do little to stop states using nuclear reactors to produce nuclear weapons".[189] A 2009 United Nations report said that:

> the revival of interest in nuclear power could result in the worldwide dissemination of uranium enrichment
> and spent fuel reprocessing technologies, which present obvious risks of proliferation as these technologies
> can produce fissile materials that are directly usable in nuclear weapons.[189]

On the other hand, one factor influencing the support of power reactors is due to the appeal that these reactors have at reducing nuclear weapons arsenals through the Megatons to Megawatts Program, a program which eliminated 425 metric tons of highly enriched uranium(HEU), the equivalent of 17,000 nuclear warheads, by diluting it with natural uranium making it equivalent to low enriched uranium(LEU), and thus suitable as nuclear fuel for commercial fission reactors. This is the single most successful non-proliferation program to date.[186]

The Megatons to Megawatts Program, the brainchild of Thomas Neff of MIT,[190][191] was hailed as a major success by anti-nuclear weapon advocates as it has largely been the driving force behind the sharp reduction in the quantity of nuclear weapons worldwide since the cold war ended.[186] However without an increase in nuclear reactors and greater demand for fissile fuel, the cost of dismantling and down blending has dissuaded Russia from continuing their disarmament.

Currently, according to Harvard professor Matthew Bunn: "The Russians are not remotely interested in extending the program beyond 2013. We've managed to set it up in a way that costs them more and profits them less than them just making new low-enriched uranium for reactors from scratch. But there are other ways to set it up that would be very profitable for them and would also serve some of their strategic interests in boosting their nuclear exports."[192]

Up to 2005, the Megatons to Megawatts Program had processed $8 billion of HEU/weapons grade uranium into LEU/reactor grade uranium, with that corresponding to the elimination of 10,000 nuclear weapons.[193]

For approximately two decades, this material generated nearly 10 percent of all the electricity consumed in the United States (about half of all US nuclear electricity generated) with a total of around 7 trillion kilowatt-hours of electricity produced.[194] Enough energy to energize the entire United States electric grid for about two years.[190] In total it is estimated to have cost $17 billion, a "bargain for US ratepayers", with Russia profiting $12 billion from the deal.[194] Much needed profit for the Russian nuclear oversight industry, which after the collapse of the Soviet economy, had difficulties paying for the maintenance and security of the Russian Federations Highly enriched uranium and warheads.[190]

In April 2012 there were thirty one countries that have civil nuclear power plants,[195] of which nine have nuclear weapons, with the vast majority of these nuclear weapons states having first produced weapons, before commercial fission electricity stations. Moreover, the re-purposing of civilian nuclear industries for military purposes would be a breach of the Non-proliferation treaty, of which 190 countries adhere to.

13.8   Environmental issues

Main articles: Environmental effects of nuclear power and Comparisons of life-cycle greenhouse gas emissions

Life cycle analysis (LCA) of carbon dioxide emissions show nuclear power as comparable to renewable energy sources. Emissions from burning fossil fuels are many times higher.[196][198][199]

According to the United Nations (UNSCEAR), regular nuclear power plant operation including the nuclear fuel cycle causes radioisotope releases into the environment amounting to 0.0002 mSv (milli-Sievert) per year of public exposure as a global average.[200] (Such is small compared to variation in natural background radiation, which averages 2.4 mSv/a globally but frequently varies between 1 mSv/a and 13 mSv/a depending on a person's location as determined by UNSCEAR).[200] As of a 2008 report, the remaining legacy of the worst nuclear power plant accident (Chernobyl) is 0.002 mSv/a in global average exposure (a figure which was 0.04 mSv per person averaged over the entire populace of the Northern Hemisphere in the year of the accident in 1986, although far higher among the most affected local populations and recovery workers).[200]

13.8.1   Climate change

Climate change causing weather extremes such as heat waves, reduced precipitation levels and droughts can have a significant impact on nuclear energy infrastructure.[201] Seawater is corrosive and so nuclear energy supply is likely to be negatively affected by the fresh water shortage.[201] This generic problem may become increasingly significant over time.[201] This can force nuclear reactors to be shut down, as happened in France during the 2003 and 2006 heat waves. Nuclear power supply was severely diminished by low river flow rates and droughts, which meant rivers had reached the maximum temperatures for cooling reactors.[201] During the heat waves, 17 reactors had to limit output or shut down. 77% of French electricity is produced by nuclear power and in 2009 a similar situation created a 8GW shortage and forced the French government to import electricity.[201] Other cases have been reported from Germany, where extreme temperatures have reduced nuclear power production 9 times due to high temperatures between 1979 and 2007.[201] In particular:

- the Unterweser nuclear power plant reduced output by 90% between June and September 2003[201]

- the Isar nuclear power plant cut production by 60% for 14 days due to excess river temperatures and low stream flow in the river Isar in 2006[201]

Similar events have happened elsewhere in Europe during those same hot summers.[201] If global warming continues, this disruption is likely to increase.

13.9 Nuclear decommissioning

The price of energy inputs and the environmental costs of every nuclear power plant continue long after the facility has finished generating its last useful electricity. Once no longer economically viable, nuclear reactors and uranium enrichment facilities are generally decommissioned, returning the facility and its parts to a safe enough level to be entrusted for other uses, such as greenfield status. After a cooling-off period that may last decades, reactor core materials are dismantled and cut into small pieces to be packed in containers for interim storage or transmutation experiments. The process is expensive, time-consuming, dangerous for workers and potentially hazardous to the natural environment as it presents opportunities for human error, accidents or sabotage.[202]

The total energy required for decommissioning can be as much as 50% more than the energy needed for the original construction. In most cases, the decommissioning process costs between US $300 million to US$5.6 billion. Decommissioning at nuclear sites which have experienced a serious accident are the most expensive and time-consuming. In the U.S. in 2011, there are 13 reactors that had permanently shut down and are in some phase of decommissioning.[202] With Yankee Rowe Nuclear Power Station having completed the process in 2007, after ceasing commercial electricity production in 1992. The majority of the 15 years, was used to allow the station to naturally cool-down on its own, which makes the manual disassembly process both safer and cheaper.

13.10 Debate on nuclear power

Main article: Nuclear power debate
See also: Nuclear energy policy and Anti-nuclear movement

The nuclear power debate concerns the controversy[28][29][94] which has surrounded the deployment and use of nuclear fission reactors to generate electricity from nuclear fuel for civilian purposes. The debate about nuclear power peaked during the 1970s and 1980s, when it "reached an intensity unprecedented in the history of technology controversies", in some countries.[95][203]

Proponents of nuclear energy contend that nuclear power is a sustainable energy source that reduces carbon emissions and increases energy security by decreasing dependence on imported energy sources.[30] Proponents claim that nuclear power produces virtually no conventional air pollution, such as greenhouse gases and smog, in contrast to the chief viable alternative of fossil fuel.[204] Nuclear power can produce base-load power unlike many renewables which are intermittent energy sources lacking large-scale and cheap ways of storing energy.[205] M. King Hubbert saw oil as a resource that would run out, and proposed nuclear energy as a replacement energy source.[206] Proponents claim that the risks of storing waste are small and can be further reduced by using the latest technology in newer reactors, and the operational safety record in the Western world is excellent when compared to the other major kinds of power plants.[207]

Opponents believe that nuclear power poses many threats to people and the environment.[31][32][33] These threats include the problems of processing, transport and storage of radioactive nuclear waste, the risk of nuclear weapons proliferation and terrorism, as well as health risks and environmental damage from uranium mining.[208][209] They also contend that reactors themselves are enormously complex machines where many things can and do go wrong; and there have been serious nuclear accidents.[210][211] Critics do not believe that the risks of using nuclear fission as a power source can be fully offset through the development of new technology. They also argue that when all the energy-intensive stages of the nuclear fuel chain are considered, from uranium mining to nuclear decommissioning, nuclear power is neither a low-carbon nor an economical electricity source.[212][213][214]

Arguments of economics and safety are used by both sides of the debate.

13.11 Comparison with renewable energy

See also: Renewable energy debate, Nuclear power proposed as renewable energy and 100% renewable energy

As of 2013, the World Nuclear Association has said "There is unprecedented interest in renewable energy, particularly solar and wind energy, which provide electricity without giving rise to any carbon dioxide emission. Harnessing these for electricity depends on the cost and efficiency of the technology, which is constantly improving, thus reducing costs per peak kilowatt".[215]

Renewable electricity production, from sources such as wind power and solar power, is sometimes criticized for being intermittent or variable.[216][217] However, the International Energy Agency concluded that deployment of renewable technologies (RETs), when it increases the diversity of electricity sources, contributes to the flexibility of the system. However, the report also concluded (p. 29): "At high levels of grid penetration by RETs the consequences of unmatched demand and supply can pose challenges for grid management. This characteristic may affect how, and the degree to which, RETs can displace fossil fuels and nuclear capacities in power generation."[218]

Renewable electricity supply in the 20-50+% range has already been implemented in several European systems, albeit in the context of an integrated European grid system.[219] In 2012, the share of electricity generated by renewable sources in Germany was 21.9%, compared to 16.0% for nuclear power after Germany shut down 7-8 of its 18 nuclear reactors in 2011.[220] In the United Kingdom, the amount of energy produced from renewable energy is expected to exceed that from nuclear power by 2018,[221] and Scotland plans to obtain all electricity from renewable energy by 2020.[222] The majority of installed renewable energy across the world is in the form of hydro power.

The IPCC has said that if governments were supportive, and the full complement of renewable energy technologies were deployed, renewable energy supply could account for almost 80% of the world's energy use within forty years.[223] Rajendra Pachauri, chairman of the IPCC, said the necessary investment in renewables would cost only about 1% of global GDP annually. This approach could contain greenhouse gas levels to less than 450 parts per million, the safe level beyond which climate change becomes catastrophic and irreversible.[223]

The cost of nuclear power has followed an increasing trend whereas the cost of electricity is declining for wind power.[224] In about 2011, wind power became as inexpensive as natural gas, and anti-nuclear groups have suggested that in 2010 solar power became cheaper than nuclear power.[225][226] Data from the EIA in 2011 estimated that in 2016, solar will have a levelized cost of electricity almost twice that of nuclear (21¢/kWh for solar, 11.39¢/kWh for nuclear), and wind somewhat less (9.7¢/kWh).[227] However, the US EIA has also cautioned that levelized costs of intermittent sources such as wind and solar are not directly comparable to costs of "dispatchable" sources (those that can be adjusted to meet demand).[228]

From a safety stand point, nuclear power, in terms of lives lost per unit of electricity delivered, is comparable to and in some cases, lower than many renewable energy sources.[37][38][229] There is however no radioactive spent fuel that needs to be stored or reprocessed with conventional renewable energy sources.[230] A nuclear plant needs to be disassembled and removed. Much of the disassembled nuclear plant needs to be stored as low level nuclear waste.[231]

13.12   Nuclear renaissance

Main article: Nuclear renaissance

Since about 2001 the term *nuclear renaissance* has been used to refer to a possible nuclear power industry revival, driven by rising fossil fuel prices and new concerns about meeting greenhouse gas emission limits.[237] However, the World Nuclear Association has reported that nuclear electricity generation in 2012 was at its lowest level since 1999.[238]

In March 2011 the nuclear emergencies at Japan's Fukushima I Nuclear Power Plant and shutdowns at other nuclear facilities raised questions among some commentators over the future of the renaissance.[239][240][241][242][243] Platts has reported that "the crisis at Japan's Fukushima nuclear plants has prompted leading energy-consuming countries to review the safety of their existing reactors and cast doubt on the speed and scale of planned expansions around the world".[244] In 2011 Siemens exited the nuclear power sector following the Fukushima disaster and subsequent changes to German energy policy, and supported the German government's planned energy transition to renewable energy technologies.[245] China, Germany, Switzerland, Israel, Malaysia, Thailand, United Kingdom, Italy[246] and the Philippines have reviewed their nuclear power programs. Indonesia and Vietnam still plan to build nuclear power plants.[247][248][249][250] Countries such as Australia, Austria, Denmark, Greece, Ireland, Latvia, Liechtenstein, Luxembourg, Portugal, Israel, Malaysia, New Zealand, and Norway remain opposed to nuclear power. Following the Fukushima I nuclear accidents, the International

Energy Agency halved its estimate of additional nuclear generating capacity built by 2035.[45]

The World Nuclear Association has said that "nuclear power generation suffered its biggest ever one-year fall through 2012 as the bulk of the Japanese fleet remained offline for a full calendar year". Data from the International Atomic Energy Agency showed that nuclear power plants globally produced 2346 TWh of electricity in 2012 – seven per cent less than in 2011. The figures illustrate the effects of a full year of 48 Japanese power reactors producing no power during the year. The permanent closure of eight reactor units in Germany was also a factor. Problems at Crystal River, Fort Calhoun and the two San Onofre units in the USA meant they produced no power for the full year, while in Belgium Doel 3 and Tihange 2 were out of action for six months. Compared to 2010, the nuclear industry produced 11% less electricity in 2012.[238]

13.13   Future of the industry

See also: List of prospective nuclear units in the United States, Nuclear power in the United States, Nuclear energy policy and Mitigation of global warming
 As already noted, the nuclear power industry in western nations has a history of construction delays, cost overruns, plant cancellations, and nuclear safety issues despite significant government subsidies and support.[101][252][253][254] In December 2013, *Forbes* magazine reported that, in developed countries, "reactors are not a viable source of new power".[255] Even in developed nations where they make economic sense, they are not feasible because nuclear's "enormous costs, political and popular opposition, and regulatory uncertainty".[255] This view echoes the statement of former Exelon CEO John Rowe, who said in 2012 that new nuclear plants "don't make any sense right now" and won't be economically viable in the foreseeable future.[255] John Quiggin, economics professor, also says the main problem with the nuclear option is that it is not economically-viable. Quiggin says that we need more efficient energy use and more renewable energy commercialization.[162] Former NRC member Peter Bradford and Professor Ian Lowe have recently made similar statements.[256][257] However, some "nuclear cheerleaders" and lobbyists in the West continue to champion reactors, often with proposed new but largely untested designs, as a source of new power.[255][256][258][259][260][261][262]

Much more new build activity is occurring in developing countries like South Korea, India and China. China has 25 reactors under construction, with plans to build more,[263][264] However, according to a government research unit, China must not build "too many nuclear power reactors too quickly", in order to avoid a shortfall of fuel, equipment and qualified plant workers.[265]

In the USA, licenses of almost half its reactors have been extended to 60 years,[266][267] Two new Generation III reactors are under construction at Vogtle, a dual construction project which marks the end of a 34-year period of stagnation in the US construction of civil nuclear power reactors. The station operator licenses of almost half the present 104 power reactors in the US, as of 2008, have been given extensions to 60 years.[266] As of 2012, U.S. nuclear industry officials expect five new reactors to enter service by 2020, all at existing plants.[24] In 2013, four aging, uncompetitive, reactors were permanently closed.[25][26] Relevant state legislatures are trying to close Vermont Yankee and Indian Point Nuclear Power Plant.[26]

The U.S. NRC and the U.S. Department of Energy have initiated research into Light water reactor sustainability which is hoped will lead to allowing extensions of reactor licenses beyond 60 years, provided that safety can be maintained, as the loss in non-CO_2-emitting generation capacity by retiring reactors "may serve to challenge U.S. energy security, potentially resulting in increased greenhouse gas emissions, and contributing to an imbalance between electric supply and demand."[268]

There is a possible impediment to production of nuclear power plants as only a few companies worldwide have the capacity to forge single-piece reactor pressure vessels,[269] which are necessary in the most common reactor designs. Utilities across the world are submitting orders years in advance of any actual need for these vessels. Other manufacturers are examining various options, including making the component themselves, or finding ways to make a similar item using alternate methods.[270]

According to the World Nuclear Association, globally during the 1980s one new nuclear reactor started up every 17 days on average, and by the year 2015 this rate could increase to one every 5 days.[271] As of 2007, Watts Bar 1 in Tennessee, which came on-line on February 7, 1996, was the last U.S. commercial nuclear reactor to go on-line. This is often quoted as evidence of a successful worldwide campaign for nuclear power phase-out.[272] Electricity shortages, fossil fuel price increases, global warming, and heavy metal emissions from fossil fuel use, new technology such as passively safe plants, and national energy security may renew the demand for nuclear power plants.

13.13.1 Nuclear phase out

Main article: Nuclear power phase-out

Following the Fukushima Daiichi nuclear disaster, the International Energy Agency halved its estimate of additional nuclear generating capacity to be built by 2035.[45][46] Platts has reported that "the crisis at Japan's Fukushima nuclear plants has prompted leading energy-consuming countries to review the safety of their existing reactors and cast doubt on the speed and scale of planned expansions around the world".[244] In 2011, *The Economist* reported that nuclear power "looks dangerous, unpopular, expensive and risky", and that "it is replaceable with relative ease and could be forgone with no huge structural shifts in the way the world works".[273]

In early April 2011, analysts at Swiss-based investment bank UBS said: "At Fukushima, four reactors have been out of control for weeks, casting doubt on whether even an advanced economy can master nuclear safety . . .. We believe the Fukushima accident was the most serious ever for the credibility of nuclear power".[274]

In 2011, Deutsche Bank analysts concluded that "the global impact of the Fukushima accident is a fundamental shift in public perception with regard to how a nation prioritizes and values its populations health, safety, security, and natural environment when determining its current and future energy pathways". As a consequence, "renewable energy will be a clear long-term winner in most energy systems, a conclusion supported by many voter surveys conducted over the past few weeks. At the same time, we consider natural gas to be, at the very least, an important transition fuel, especially in those regions where it is considered secure".[275]

In September 2011, German engineering giant Siemens announced it will withdraw entirely from the nuclear industry, as a response to the Fukushima nuclear disaster in Japan, and said that it would no longer build nuclear power plants anywhere in the world. The company's chairman, Peter Löscher, said that "Siemens was ending plans to cooperate with Rosatom, the Russian state-controlled nuclear power company, in the construction of dozens of nuclear plants throughout Russia over the coming two decades".[276][277] Also in September 2011, IAEA Director General Yukiya Amano said the Japanese nuclear disaster "caused deep public anxiety throughout the world and damaged confidence in nuclear power".[278]

In February 2012, the United States Nuclear Regulatory Commission approved the construction of two additional reactors at the Vogtle Electric Generating Plant, the first reactors to be approved in over 30 years since the Three Mile Island accident,[279] but NRC Chairman Gregory Jaczko cast a dissenting vote citing safety concerns stemming from Japan's 2011 Fukushima nuclear disaster, and saying "I cannot support issuing this license as if Fukushima never happened".[24] One week after Southern received the license to begin major construction on the two new reactors, a dozen environmental and anti-nuclear groups are suing to stop the Plant Vogtle expansion project, saying "public safety and environmental problems since Japan's Fukushima Daiichi nuclear reactor accident have not been taken into account".[280]

Countries such as Australia, Austria, Denmark, Greece, Ireland, Italy, Latvia, Liechtenstein, Luxembourg, Malta, Portugal, Israel, Malaysia, New Zealand, and Norway have no nuclear power reactors and remain opposed to nuclear power.[273][281] However, by contrast, some countries remain in favor, and financially support nuclear fusion research, including EU wide funding of the ITER project.[282][283]

Worldwide wind power has been increasing at 26%/year, and solar power at 58%/year, from 2006 to 2011, as a replacement for thermal generation of electricity.[284]

13.13.2 Advanced concepts

Main article: Generation IV reactor

Current fission reactors in operation around the world are second or third generation systems, with most of the first-generation systems having been retired some time ago. Research into advanced generation IV reactor types was officially started by the Generation IV International Forum (GIF) based on eight technology goals, including to improve nuclear safety, improve proliferation resistance, minimize waste, improve natural resource utilization, the ability to consume existing nuclear waste in the production of electricity, and decrease the cost to build and run such plants. Most of these reactors differ significantly from current operating light water reactors, and are generally not expected to be available for commercial construction before 2030.[285]

The nuclear reactors to be built at Vogtle are new AP1000 third generation reactors, which are said to have safety improvements over older power reactors.[279] However, John Ma, a senior structural engineer at the NRC, is concerned that some parts of the AP1000 steel skin are so brittle that the "impact energy" from a plane strike or storm driven projectile could shatter the wall.[286] Edwin Lyman, a senior staff scientist at the Union of Concerned Scientists, is concerned about the strength of the steel containment vessel and the concrete shield building around the AP1000.[286][287]

The Union of Concerned Scientists has referred to the EPR (nuclear reactor), currently under construction in China, Finland and France, as the only new reactor design under consideration in the United States that "...appears to have the potential to be significantly safer and more secure against attack than today's reactors."[288]

One disadvantage of any new reactor technology is that safety risks may be greater initially as reactor operators have little experience with the new design. Nuclear engineer David Lochbaum has explained that almost all serious nuclear accidents have occurred with what was at the time the most recent technology. He argues that "the problem with new reactors and accidents is twofold: scenarios arise that are impossible to plan for in simulations; and humans make mistakes".[289] As one director of a U.S. research laboratory put it, "fabrication, construction, operation, and maintenance of new reactors will face a steep learning curve: advanced technologies will have a heightened risk of accidents and mistakes. The technology may be proven, but people are not".[289]

13.13.3 Hybrid nuclear fusion-fission

Hybrid nuclear power is a proposed means of generating power by use of a combination of nuclear fusion and fission processes. The concept dates to the 1950s, and was briefly advocated by Hans Bethe during the 1970s, but largely remained unexplored until a revival of interest in 2009, due to delays in the realization of pure fusion. When a sustained nuclear fusion power plant is built, it has the potential to be capable of extracting all the fission energy that remains in spent fission fuel, reducing the volume of nuclear waste by orders of magnitude, and more importantly, eliminating all actinides present in the spent fuel, substances which cause security concerns.[290]

13.13.4 Nuclear fusion

Main articles: Nuclear fusion and Fusion power

Nuclear fusion reactions have the potential to be safer and generate less radioactive waste than fission.[291][292] These reactions appear potentially viable, though technically quite difficult and have yet to be created on a scale that could be used in a functional power plant. Fusion power has been under theoretical and experimental investigation since the 1950s.

Construction of the ITER facility began in 2007, but the project has run into many delays and budget overruns. The facility is now not expected to begin operations until the year 2027 – 11 years after initially anticipated.[293] A follow on commercial nuclear fusion power station, DEMO, has been proposed.[21][294] There is also suggestions for a power plant based upon a different fusion approach, that of a Inertial fusion power plant.

Fusion powered electricity generation was initially believed to be readily achievable, as fission-electric power had been. However, the extreme requirements for continuous reactions and plasma containment led to projections being extended by several decades. In 2010, more than 60 years after the first attempts, commercial power production was still believed to be unlikely before 2050.[21]

13.14 Nuclear power organizations

There are multiple organizations which have taken a position on nuclear power – some are proponents, and some are opponents.

13.14.1 Proponents

Main article: List of nuclear power groups

- Environmentalists for Nuclear Energy (International)
- Nuclear Industry Association (United Kingdom)
- World Nuclear Association, a confederation of companies connected with nuclear power production. (International)
- International Atomic Energy Agency (IAEA)
- Nuclear Energy Institute (United States)
- American Nuclear Society (United States)
- United Kingdom Atomic Energy Authority (United Kingdom)
- EURATOM (Europe)
- European Nuclear Education Network (Europe)
- Atomic Energy of Canada Limited (Canada)

13.14.2 Opponents

Main article: List of anti-nuclear power groups

- Friends of the Earth International, a network of environmental organizations.[295]
- Greenpeace International, a non-governmental organization[296]
- Nuclear Information and Resource Service (International)
- World Information Service on Energy (International)
- Sortir du nucléaire (France)
- Pembina Institute (Canada)
- Institute for Energy and Environmental Research (United States)
- Sayonara Nuclear Power Plants (Japan)

13.15   See also

- Alsos Digital Library for Nuclear Issues
- German nuclear energy project
- Linear no-threshold model
- Nuclear power in France
- Nuclear weapons debate
- Harry Shearer's Le Show— a weekly radio show and podcast featuring "Clean, Safe, Too Cheap to Meter", a series of regular updates on nuclear power
- Uranium mining debate
- World energy consumption

13.16 References

[1] Nuclear Energy: Statistics, Dr. Elizabeth Ervin

[2] An oasis filled with grey water 25 June 2013

[3] Topical issues of infrastructure development IAEA 2012

[4] "2014 Key World Energy Statistics" (PDF). *http://www.iea.org/publications/freepublications/*. IEA. 2014. p. 24. Archived from the original on 5 May 2014.

[5] "Nuclear Energy". *Energy Education is an interactive curriculum supplement for secondary-school science students, funded by the U. S. Department of Energy and the Texas State Energy Conservation Office (SECO)*. U. S. Department of Energy and the Texas State Energy Conservation Office (SECO). July 2010. Retrieved 2010-07-10.

[6] "Key World Energy Statistics 2012" (PDF). International Energy Agency. 2012. Retrieved 2012-12-16.

[7] Nicola Armaroli, Vincenzo Balzani, *Towards an electricity-powered world*. In: *Energy and Environmental Science* 4, (2011), 3193-3222, p. 3200, doi:10.1039/c1ee01249e.

[8] REN 21. RENEWABLES 2014 GLOBAL STATUS REPORT

[9] Issues in Science & Technology Online; "Promoting Low-Carbon Electricity Production"

[10] The European Strategic Energy Technology Plan SET-Plan Towards a low-carbon future 2010. Nuclear power provides "2/3 of the EU's low carbon energy" pg 6.

[11] "Collectively, life cycle assessment literature shows that nuclear power is similar to other renewable and much lower than fossil fuel in total life cycle GHG emissions."". Nrel.gov. 2013-01-24. Retrieved 2013-06-22.

[12] "IPCC Working Group III – Mitigation of Climate Change, Annex II I: Technology - specific cost and performance parameters" (PDF). IPCC. 2014. p. 10. Retrieved 1 August 2014.

[13] "IPCC Working Group III – Mitigation of Climate Change, Annex II Metrics and Methodology. pg 37 to 40,41" (PDF).

[14] "Prevented Mortality and Greenhouse Gas Emissions from Historical and Projected Nuclear Power - global nuclear power has prevented an average of 1.84 million air pollution-related deaths and 64 gigatonnes of CO2-equivalent (GtCO2-eq) greenhouse gas (GHG) emissions that would have resulted from fossil fuel burning". Pubs.acs.org. Bibcode:2013EnST...47.4889K. doi:10.1021/es3051197.

[15] "PRIS - Home". Iaea.org. Retrieved 2013-06-14.

[16] "World Nuclear Power Reactors 2007-08 and Uranium Requirements". World Nuclear Association. 2008-06-09. Archived from the original on March 3, 2008. Retrieved 2008-06-21.

[17] "Japan approves two reactor restarts". Taipei Times. 2013-06-07. Retrieved 2013-06-14.

[18] "What is Nuclear Power Plant - How Nuclear Power Plants work | What is Nuclear Power Reactor - Types of Nuclear Power Reactors". EngineersGarage. Retrieved 2013-06-14.

[19] "Nuclear-Powered Ships | Nuclear Submarines". World-nuclear.org. Retrieved 2013-06-14.

[20] http://www.ewp.rpi.edu/hartford/~{}ernesto/F2010/EP2/Materials4Students/Misiaszek/NuclearMarinePropulsion.pdf Naval Nuclear Propulsion, Magdi Ragheb. *As of 2001, about 235 naval reactors had been built*

[21] "Beyond ITER". *The ITER Project*. Information Services, Princeton Plasma Physics Laboratory. Archived from the original on 7 November 2006. Retrieved 5 February 2011. - Projected fusion power timeline

[22] South Korea's Shin-Wolsong-2 Enters Commercial Operation

[23] "Grid Connection for Fuqing-2 in China 7 August 2015". Worldnuclearreport.org. Retrieved 2015-08-12.

[24] Ayesha Rascoe (Feb 9, 2012). "U.S. approves first new nuclear plant in a generation". *Reuters*.

[25] Mark Cooper (18 June 2013). "Nuclear aging: Not so graceful". *Bulletin of the Atomic Scientists*.

[26] Matthew Wald (June 14, 2013). "Nuclear Plants, Old and Uncompetitive, Are Closing Earlier Than Expected". *New York Times.*

[27] Union-Tribune Editorial Board (March 27, 2011). "The nuclear controversy". *Union-Tribune.*

[28] James J. MacKenzie. Review of The Nuclear Power Controversy by Arthur W. Murphy *The Quarterly Review of Biology*, Vol. 52, No. 4 (Dec., 1977), pp. 467-468.

[29] In February 2010 the nuclear power debate played out on the pages of the *New York Times*, see A Reasonable Bet on Nuclear Power and Revisiting Nuclear Power: A Debate and A Comeback for Nuclear Power?

[30] U.S. Energy Legislation May Be 'Renaissance' for Nuclear Power.

[31] Share. "Nuclear Waste Pools in North Carolina". Projectcensored.org. Retrieved 2010-08-24.

[32] "Nuclear Power". Nc Warn. Retrieved 2013-06-22.

[33] Sturgis, Sue. "Investigation: Revelations about Three Mile Island disaster raise doubts over nuclear plant safety". Southern-studies.org. Retrieved 2010-08-24.

[34] iPad iPhone Android TIME TV Populist The Page (2009-03-25). "The Worst Nuclear Disasters". Time.com. Retrieved 2013-06-22.

[35] Strengthening the Safety of Radiation Sources p. 14.

[36] Johnston, Robert (September 23, 2007). "Deadliest radiation accidents and other events causing radiation casualties". Database of Radiological Incidents and Related Events.

[37] Markandya, A.; Wilkinson, P. (2007). "Electricity generation and health". *Lancet* **370** (9591): 979–990. doi:10.1016/S0140-6736(07)61253-7. PMID 17876910.

[38] "Dr. MacKay *Sustainable Energy without the hot air". Data from studies by the Paul Scherrer Institute including non EU data.* p. 168. Retrieved 15 September 2012.

[39] http://www.forbes.com/sites/jamesconca/2012/06/10/energys-deathprint-a-price-always-paid/ with Chernobyl's total predicted linear no-threshold cancer deaths included, nuclear power is safer when compared to many alternative energy sources' immediate, death rate.

[40] Brendan Nicholson (2006-06-05). "Nuclear power 'cheaper, safer' than coal and gas". Melbourne: The Age. Retrieved 2008-01-18.

[41] Burgherr, P.; Hirschberg, S. (2008). "A Comparative Analysis of Accident Risks in Fossil, Hydro, and Nuclear Energy Chains". *Human and Ecological Risk Assessment: an International Journal* **14** (5): 947. doi:10.1080/10807030802387556. If you cannot access the paper via the above link, the following link is open to the public, credit to the authors. http://gabe.web.psi.ch/pdfs/_2012_LEA_Audit/TA01.pdf Page 962 to 965. Comparing Nuclear's *latent* cancer deaths, such as cancer with other energy sources *immediate* deaths per unit of energy generated(GWeyr). This study does not include Fossil fuel related cancer and other indirect deaths created by the use of fossil fuel consumption in its "severe accident", an accident with more than 5 fatalities, classification.

[42] Richard Schiffman (12 March 2013). "Two years on, America hasn't learned lessons of Fukushima nuclear disaster". *The Guardian* (London).

[43] Martin Fackler (June 1, 2011). "Report Finds Japan Underestimated Tsunami Danger". *New York Times.*

[44] Sylvia Westall and Fredrik Dahl (June 24, 2011). "IAEA Head Sees Wide Support for Stricter Nuclear Plant Safety". *Scientific American.* Archived from the original on April 14, 2014.

[45] "Gauging the pressure". The Economist. 28 April 2011.

[46] European Environment Agency (Jan 23, 2013). "Late lessons from early warnings: science, precaution, innovation: Full Report". p. 476.

[47] "Turkey Prepares to Host First ATMEA 1 Nuclear Reactors". *PowerMag* (Electric Power). Retrieved 24 May 2015.

[48] August 11, 2015, Kyushu Electric Power Company Inc. Startup of Sendai Nuclear Power Unit No.1

[49] "Nuclear Power Plants Information. Number of Reactors Operation Worldwide". International Atomic Energy Agency. Retrieved 2008-06-21.

[50] "BP Statistical Review of World Energy June 2012" (PDF). BP. Retrieved 2012-12-16.

[51] Trevor Findlay (2010). The Future of Nuclear Energy to 2030 and its Implications for Safety, Security and Nonproliferation: Overview, The Centre for International Governance Innovation (CIGI), Waterloo, Ontario, Canada, pp. 10-11.

[52] Mycle Schneider, Steve Thomas, Antony Froggatt, and Doug Koplow (August 2009). The World Nuclear Industry Status Report 2009 Commissioned by German Federal Ministry of Environment, Nature Conservation and Reactor Safety, p. 5. Archived July 7, 2014 at the Wayback Machine

[53] World Nuclear Association. Another drop in nuclear generation *World Nuclear News*, 05 May 2010.

[54] "Summary status for the US". Energy Information Administration. 2010-01-21. Retrieved 2010-02-18.

[55] Eleanor Beardsley (2006). "France Presses Ahead with Nuclear Power". NPR. Retrieved 2006-11-08.

[56] "Gross electricity generation, by fuel used in power-stations". Eurostat. 2006. Retrieved 2007-02-03.

[57] *Nuclear Power Generation, US Industry Report"* IBISWorld, August 2008

[58] "Nuclear Icebreaker Lenin". Bellona. 2003-06-20. Retrieved 2007-11-01.

[59] David Baurac (2002). "Passively safe reactors rely on nature to keep them cool". *Logos* (Argonne National Laboratory) **20** (1). Retrieved 2012-07-25.

[60] "Moonshine". Atomicarchive.com. Retrieved 2013-06-22.

[61] "The Atomic Solar System". Atomicarchive.com. Retrieved 2013-06-22.

[62] taneya says:. "What do you mean by Induced Radioactivity?". Thebigger.com. Retrieved 2013-06-22.

[63] "Neptunium". Vanderkrogt.net. Retrieved 2013-06-22.

[64] "Otto Hahn, The Nobel Prize in Chemistry, 1944". http://www.nobelprize.org. Retrieved 2007-11-01.

[65] "Otto Hahn, Fritz Strassmann, and Lise Meitner". http://www.chemheritage.org. Retrieved 2007-11-01.

[66] "Otto Robert Frisch". http://www.nuclearfiles.org. Retrieved 2007-11-01.

[67] "The Einstein Letter". Atomicarchive.com. Retrieved 2013-06-22.

[68] John Byrne and Steven M. Hoffman (1996). *Governing the Atom: The Politics of Risk*, Transaction Publishers, p. 136.

[69] Benjamin K. Sovacool, *The National Politics of Nuclear Power*, Routledge, p. 68.

[70] Bain, Alastair S.; et al. (1997). *Canada enters the nuclear age: a technical history of Atomic Energy of Canada*. Magill-Queen's University Press. p. ix. ISBN 0-7735-1601-8.

[71] "Reactors Designed by Argonne National Laboratory: Fast Reactor Technology". U.S. Department of Energy, Argonne National Laboratory. 2012. Retrieved 2012-07-25.

[72] "Reactor Makes Electricity." *Popular Mechanics*, March 1952, p. 105.

[73] "STR (Submarine Thermal Reactor) in "Reactors Designed by Argonne National Laboratory: Light Water Reactor Technology Development"". U.S. Department of Energy, Argonne National Laboratory. 2012. Retrieved 2012-07-25.

[74] Benjamin K. Sovacool. The costs of failure: A preliminary assessment of major energy accidents, 1907–2007, *Energy Policy* 36 (2008), p. 1808.

[75] "From Obninsk Beyond: Nuclear Power Conference Looks to Future". *International Atomic Energy Agency*. Retrieved 2006-06-27.

[76] "Nuclear Power in Russia". *World Nuclear Association*. Retrieved 2006-06-27.

[77] "This Day in Quotes: SEPTEMBER 16 - Too cheap to meter: the great nuclear quote debate". This day in quotes. 2009. Retrieved 2009-09-16.

[78] Pfau, Richard (1984) *No Sacrifice Too Great: The Life of Lewis L. Strauss* University Press of Virginia, Charlottesville, Virginia, p. 187, ISBN 978-0-8139-1038-3

[79] David Bodansky (2004). Nuclear Energy: Principles, Practices, and Prospects. Springer. p. 32. ISBN 978-0-387-20778-0. Retrieved 2008-01-31.

[80] Kragh, Helge (1999). *Quantum Generations: A History of Physics in the Twentieth Century.* Princeton NJ: Princeton University Press. p. 286. ISBN 0-691-09552-3.

[81] "On This Day: October 17". BBC News. 1956-10-17. Retrieved 2006-11-09.

[82] "50 Years of Nuclear Energy" (PDF). International Atomic Energy Agency. Retrieved 2006-11-09.

[83] McKeown, William (2003). *Idaho Falls: The Untold Story of America's First Nuclear Accident.* Toronto: ECW Press. ISBN 978-1-55022-562-4.

[84] The Changing Structure of the Electric Power Industry p. 110.

[85] Bernard L. Cohen. "THE NUCLEAR ENERGY OPTION". Plenum Press. Retrieved December 2007.

[86] Evolution of Electricity Generation by Fuel PDF (39.4 KB)

[87] Sharon Beder, 'The Japanese Situation', English version of conclusion of Sharon Beder, "Power Play: The Fight to Control the World's Electricity", Soshisha, Japan, 2006.

[88] Paula Garb. Review of Critical Masses, *Journal of Political Ecology*, Vol 6, 1999.

[89] Rüdig, Wolfgang, ed. (1990). *Anti-nuclear Movements: A World Survey of Opposition to Nuclear Energy.* Detroit, MI: Longman Current Affairs. p. 1. ISBN 0-8103-9000-0.

[90] Brian Martin. Opposing nuclear power: past and present, *Social Alternatives*, Vol. 26, No. 2, Second Quarter 2007, pp. 43-47.

[91] Stephen Mills and Roger Williams (1986). Public Acceptance of New Technologies Routledge, pp. 375-376.

[92] Robert Gottlieb (2005). Forcing the Spring: The Transformation of the American Environmental Movement, Revised Edition, Island Press, USA, p. 237.

[93] Jim Falk (1982). *Global Fission: The Battle Over Nuclear Power*, Oxford University Press, pp. 95-96.

[94] Walker, J. Samuel (2004). *Three Mile Island: A Nuclear Crisis in Historical Perspective* (Berkeley: University of California Press), pp. 10-11.

[95] Herbert P. Kitschelt. Political Opportunity and Political Protest: Anti-Nuclear Movements in Four Democracies *British Journal of Political Science*, Vol. 16, No. 1, 1986, p. 57.

[96] Herbert P. Kitschelt. Political Opportunity and Political Protest: Anti-Nuclear Movements in Four Democracies *British Journal of Political Science*, Vol. 16, No. 1, 1986, p. 71.

[97] Social Protest and Policy Change p. 45.

[98] Lutz Mez, Mycle Schneider and Steve Thomas (Eds.) (2009). *International Perspectives of Energy Policy and the Role of Nuclear Power*, Multi-Science Publishing Co. Ltd, p. 279.

[99] "The Political Economy of Nuclear Energy in the United States" (PDF). *Social Policy*. The Brookings Institution. 2004. Retrieved 2006-11-09.

[100] Nuclear Power: Outlook for New U.S. Reactors p. 3.

[101] "Nuclear Follies", a February 11, 1985 cover story in *Forbes* magazine.

[102] "Backgrounder on Chernobyl Nuclear Power Plant Accident". *Nuclear Regulatory Commission*. Retrieved 2006-06-28.

[103] "RBMK Reactors | reactor bolshoy moshchnosty kanalny | Positive void coefficient". World-nuclear.org. 2009-09-07. Retrieved 2013-06-14.

[104] "Italy rejoins the nuclear family". World Nuclear News. 2009-07-10. Retrieved 2009-07-17.

[105] "Italy puts one year moratorium on nuclear". 2011-03-13.

[106] "Italy nuclear: Berlusconi accepts referendum blow". *BBC News.* 2011-06-14.

[107] "uranium Facts, information, pictures | Encyclopedia.com articles about uranium". Encyclopedia.com. 2001-09-11. Retrieved 2013-06-14.

[108] "Second Thoughts About Nuclear Power" (PDF). *A Policy Brief - Challenges Facing Asia.* January 2011. Archived from the original (PDF) on April 16, 2015.

[109] "Uranium resources sufficient to meet projected nuclear energy requirements long into the future". Nuclear Energy Agency (NEA). June 3, 2008. Retrieved 2008-06-16.

[110] NEA, IAEA: Uranium 2007 – Resources, Production and Demand. OECD Publishing, June 10, 2008, ISBN 978-92-64-04766-2.

[111] https://www.ipcc.ch/pdf/assessment-report/ar4/wg3/ar4-wg3-chapter4.pdf table 4.10 and page 271

[112] https://www.ipcc.ch/pdf/assessment-report/ar4/wg3/ar4-wg3-chapter4.pdf figure 4.10 and page 271

[113] "Uranium 2011 - OECD Online Bookshop". Oecdbookshop.org. Retrieved 2013-06-14.

[114] "Global Uranium Supply Ensured For Long Term, New Report Shows". Oecd-nea.org. 2012-07-26. Retrieved 2013-06-14.

[115] "Waste Management in the Nuclear Fuel Cycle". *Information and Issue Briefs.* World Nuclear Association. 2006. Retrieved 2006-11-09.

[116] John McCarthy (2006). "Facts From Cohen and Others". *Progress and its Sustainability.* Stanford. Retrieved 2006-11-09. Citing Breeder reactors: A renewable energy source, *American Journal of Physics*, vol. 51, (1), Jan. 1983.

[117] "Advanced Nuclear Power Reactors". *Information and Issue Briefs.* World Nuclear Association. 2006. Retrieved 2006-11-09.

[118] http://www.worldenergy.org/documents/p001515.pdf

[119] rebecca kessler. "Are Fast-Breeder Reactors A Nuclear Power Panacea? by Fred Pearce: Yale Environment 360". E360.yale.edu. Retrieved 2013-06-14.

[120] "Sodium coolant arrives at Beloyarsk". World-nuclear-news.org. 2013-01-24. Retrieved 2013-06-14.

[121] (Russian) http://www.atominfo.ru/newsl/s0420.htm

[122] "Large fast reactor approved for Beloyarsk". World-nuclear-news.org. 2012-06-27. Retrieved 2013-06-14.

[123] "Atomic agency plans to restart Monju prototype fast breeder reactor - AJW by The Asahi Shimbun". Ajw.asahi.com. Retrieved 2013-06-14.

[124] "India's breeder reactor to be commissioned in 2013". Hindustan Times. Retrieved 2013-06-14.

[125] "China makes nuclear power development - Xinhua | English.news.cn". News.xinhuanet.com. Retrieved 2013-06-14.

[126] "Thorium". *Information and Issue Briefs.* World Nuclear Association. 2006. Retrieved 2006-11-09.

[127] M. I. Ojovan, W.E. Lee. *An Introduction to Nuclear Waste Immobilisation*, Elsevier Science Publishers B.V., Amsterdam, 315pp. (2005).

[128] "NRC: Dry Cask Storage". Nrc.gov. 2013-03-26. Retrieved 2013-06-22.

[129] "Yankee Nuclear Power Plant". Yankeerowe.com. Retrieved 2013-06-22.

[130] "Environmental Surveillance, Education and Research Program". Idaho National Laboratory. Archived from the original on 2008-11-21. Retrieved 2009-01-05.

[131] Vandenbosch 2007, p. 21.

[132] Ojovan, M. I.; Lee, W.E. (2005). *An Introduction to Nuclear Waste Immobilisation*. Amsterdam: Elsevier Science Publishers. p. 315. ISBN 0-08-044462-8.

[133] Brown, Paul (2004-04-14). "Shoot it at the sun. Send it to Earth's core. What to do with nuclear waste?". *The Guardian* (London).

[134] National Research Council (1995). *Technical Bases for Yucca Mountain Standards*. Washington, D.C.: National Academy Press. p. 91. ISBN 0-309-05289-0.

[135] "The Status of Nuclear Waste Disposal". The American Physical Society. January 2006. Retrieved 2008-06-06.

[136] "Public Health and Environmental Radiation Protection Standards for Yucca Mountain, Nevada; Proposed Rule" (PDF). United States Environmental Protection Agency. 2005-08-22. Retrieved 2008-06-06.

[137] Duncan Clark (2012-07-09). "Nuclear waste-burning reactor moves a step closer to reality | Environment | guardian.co.uk". London: Guardian. Retrieved 2013-06-14.

[138] "George Monbiot – A Waste of Waste". Monbiot.com. Retrieved 2013-06-14.

[139] "Energy From Thorium: A Nuclear Waste Burning Liquid Salt Thorium Reactor". YouTube. 2009-07-23. Retrieved 2013-06-14.

[140] NWT magazine, oktober 2012

[141] Sevior M. (2006). "Considerations for nuclear power in Australia" (PDF). *International Journal of Environmental Studies* **63** (6): 859–872. doi:10.1080/00207230601047255.

[142] Thorium Resources In Rare Earth Elements

[143] American Geophysical Union, Fall Meeting 2007, abstract #V33A-1161. Mass and Composition of the Continental Crust

[144] Interdisciplinary Science Reviews 23:193-203;1998. Dr. Bernard L. Cohen, University of Pittsburgh. Perspectives on the High Level Waste Disposal Problem

[145] "The Challenges of Nuclear Power".

[146] "Coal Ash Is More Radioactive than Nuclear Waste". December 13, 2007.

[147] Alex Gabbard (February 5, 2008). "Coal Combustion: Nuclear Resource or Danger". Oak Ridge National Laboratory. Retrieved 2008-01-31.

[148] "Coal ash is *not* more radioactive than nuclear waste". CE Journal. 2008-12-31.

[149] Montgomery, Scott L. (2010). *The Powers That Be*, University of Chicago Press, p. 137.

[150] Al Gore (2009). *Our Choice*, Bloomsbury, pp. 165-166.

[151] "international Journal of Environmental Studies, The Solutions for Nuclear waste, December 2005" (PDF). Retrieved 2013-06-22.

[152] "Oklo: Natural Nuclear Reactors". U.S. Department of Energy Office of Civilian Radioactive Waste Management, Yucca Mountain Project, DOE/YMP-0010. November 2004. Archived from the original on August 25, 2009. Retrieved September 15, 2009.

[153] "A Nuclear Power Renaissance?". *Scientific American*. April 28, 2008. Retrieved 2008-05-15.

[154] von Hippel, Frank N. (April 2008). "Nuclear Fuel Recycling: More Trouble Than It's Worth". *Scientific American*. Retrieved 2008-05-15.

[155] Is the Nuclear Renaissance Fizzling?

[156] IEEE Spectrum: Nuclear Wasteland. Retrieved on 2007-04-22

[157] "Nuclear Fuel Reprocessing: U.S. Policy Development" (PDF). Retrieved 2009-07-25.

[158] "Adieu to nuclear recycling". *Nature* **460** (7252): 152. 2009. Bibcode:2009Natur.460R.152.. doi:10.1038/460152b.

[159] Processing of Used Nuclear Fuel for Recycle. WNA

[160] Hambling, David (July 30, 2003). "'Safe' alternative to depleted uranium revealed". *New Scientist.* Retrieved 2008-07-16.

[161] Stevens, J. B.; R. C. Batra. "Adiabatic Shear Banding in Axisymmetric Impact and Penetration Problems". Virginia Polytechnic Institute and State University. Retrieved 2008-07-16.

[162] John Quiggin (8 November 2013). "Reviving nuclear power debates is a distraction. We need to use less energy". *The Guardian.*

[163] Loan Program for Reactors Is Fizzling

[164] Al Gore: Our Choice,A plan to solve the climate crises, Bloomsbury 2009

[165] http://northdenvernews.com/what-does-nuclear-power-actually-cost-peakoil/

[166] Kidd, Steve (January 21, 2011). "New reactors—more or less?". *Nuclear Engineering International.*

[167] Henry Fountain (December 22, 2014). "Nuclear: Carbon Free, but Not Free of Unease". *The New York Times* (The Times Company). Retrieved December 23, 2014. the plant had become unprofitable in recent years, a victim largely of lower energy prices resulting from a glut of natural gas used to fire electricity plants

[168] Ed Crooks (12 September 2010). "Nuclear: New dawn now seems limited to the east". Financial Times. Retrieved 12 September 2010.

[169] United States Nuclear Regulatory Commission, 1983. The Price-Anderson Act: the Third Decade, NUREG-0957

[170] *The Future of Nuclear Power.* Massachusetts Institute of Technology. 2003. ISBN 0-615-12420-8. Retrieved 2006-11-10.

[171] Massachusetts Institute of Technology (2011). "The Future of the Nuclear Fuel Cycle" (PDF). p. xv.

[172] Tomoko Yamazaki and Shunichi Ozasa (June 27, 2011). "Fukushima Retiree Leads Anti-Nuclear Shareholders at Tepco Annual Meeting". *Bloomberg.*

[173] Mari Saito (May 7, 2011). "Japan anti-nuclear protesters rally after PM call to close plant". *Reuters.*

[174] Benjamin K. Sovacool. A Critical Evaluation of Nuclear Power and Renewable Electricity in Asia *Journal of Contemporary Asia*, Vol. 40, No. 3, August 2010, pp. 393–400.

[175] Benjamin K. Sovacool (2009). The Accidental Century - Prominent Energy Accidents in the Last 100 Years Archived August 21, 2012 at the Wayback Machine

[176] http://www.forbes.com/sites/jamesconca/2012/06/10/energys-deathprint-a-price-always-paid/ with and without Chernobyl's total predicted, by the Linear no-threshold, cancer deaths included.

[177] Markandya, A.; Wilkinson, P. (2007). "Electricity generation and health". *Lancet* **370** (9591): 979–990. doi:10.1016/S0140-6736(07)61253-7. PMID 17876910. - *Nuclear power has lower electricity related health risks than Coal, Oil, & gas. ...the health burdens are appreciably smaller for generation from natural gas, and lower still for nuclear power.* This study includes the latent or indirect fatalities, for example those caused by the inhalation of fossil fuel created particulate matter, smog induced Cardiopulmonary events, black lung etc. in its comparison.)

[178] "Nuclear Power Prevents More Deaths Than It Causes | Chemical & Engineering News". Cen.acs.org. Retrieved 2014-01-24.

[179] Kharecha, P. A.; Hansen, J. E. (2013). "Prevented Mortality and Greenhouse Gas Emissions from Historical and Projected Nuclear Power". *Environmental Science & Technology* **47** (9): 4889. doi:10.1021/es3051197.

[180] Sovacool, B. K. (2008). "The costs of failure: A preliminary assessment of major energy accidents, 1907–2007". *Energy Policy* **36** (5): 1802–1820. doi:10.1016/j.enpol.2008.01.040.

[181] Burgherr, P.; Hirschberg, S. (2008). "A Comparative Analysis of Accident Risks in Fossil, Hydro, and Nuclear Energy Chains". *Human and Ecological Risk Assessment: an International Journal* **14** (5): 947. doi:10.1080/10807030802387556.

[182] Dennis Normile (27 July 2012). "Is Nuclear Power Good for You?". *Science* **337** (6093): 395. doi:10.1126/science.337.6093.395-b.

[183] Andrew C. Revkin (March 10, 2012). "Nuclear Risk and Fear, from Hiroshima to Fukushima". *New York Times.*

[184] Frank N. von Hippel (September–October 2011). "The radiological and psychological consequences of the Fukushima Daiichi accident". *Bulletin of the Atomic Scientists* **67** (5): 27–36. doi:10.1177/0096340211421588.

[185] Steven E. Miller & Scott D. Sagan (Fall 2009). "Nuclear power without nuclear proliferation?". *Dædalus* **138**

[186] "The Bulletin of atomic scientists support the megatons to megawatts program". Archived from the original on July 8, 2011. Retrieved 15 September 2012.

[187] "home". usec.com. 2013-05-24. Retrieved 2013-06-14.

[188] http://www.world-nuclear.org/info/Nuclear-Fuel-Cycle/Conversion-Enrichment-and-Fabrication/Uranium-Enrichment/

[189] Benjamin K. Sovacool (2011). *Contesting the Future of Nuclear Power: A Critical Global Assessment of Atomic Energy*, World Scientific, p. 190.

[190] A Farewell to Arms, 2014.

[191] From Warheads to Cheap Energy, Thomas L. Neff's Idea Turned Russian Warheads Into American Electricity, Jan 2014

[192] All Things Considered (2009-12-05). "Future Unclear For 'Megatons To Megawatts' Program". Npr.org. Retrieved 2013-06-22.

[193] "Megatons to Megawatts Eliminates Equivalent of 10,000 Nuclear Warheads". Usec.com. 2005-09-21. Retrieved 2013-06-22.

[194] 02/21/2014 - 09:16 *More megatons to megawatts* Dawn Stover

[195] "Nuclear Power in the World Today". World-nuclear.org. Retrieved 2013-06-22.

[196] Benjamin K. Sovacool. Valuing the greenhouse gas emissions from nuclear power: A critical survey. *Energy Policy*, Vol. 36, 2008, p. 2950.

[197] Warner, E. S.; Heath, G. A. (2012). "Life Cycle Greenhouse Gas Emissions of Nuclear Electricity Generation". *Journal of Industrial Ecology* **16**: S73. doi:10.1111/j.1530-9290.2012.00472.x.

[198] "Energy Balances and CO2 Implications". World Nuclear Association. November 2005. Retrieved 2014-01-24.

[199] "Life-cycle emissions analyses". Nei.org. Retrieved 2010-08-24.

[200] "UNSCEAR 2008 Report to the General Assembly" (PDF). United Nations Scientific Committee on the Effects of Atomic Radiation. 2008.

[201] Dr. Frauke Urban and Dr. Tom Mitchell 2011. Climate change, disasters and electricity generation. London: Overseas Development Institute and Institute of Development Studies

[202] Benjamin K. Sovacool (2011). *Contesting the Future of Nuclear Power: A Critical Global Assessment of Atomic Energy*, World Scientific, p. 118-119.

[203] Jim Falk (1982). *Global Fission: The Battle Over Nuclear Power*, Oxford University Press.

[204] Patterson, Thom (November 3, 2013). "Climate change warriors: It's time to go nuclear". *CNN*.

[205] "Renewable Energy and Electricity". World Nuclear Association. June 2010. Retrieved 2010-07-04.

[206] M. King Hubbert (June 1956). "Nuclear Energy and the Fossil Fuels 'Drilling and Production Practice'" (PDF). API. p. 36. Retrieved 2008-04-18.

[207] Bernard Cohen. "The Nuclear Energy Option". Retrieved 2009-12-09.

[208] Greenpeace International and European Renewable Energy Council (January 2007). *Energy Revolution: A Sustainable World Energy Outlook*, p. 7.

[209] Giugni, Marco (2004). *Social Protest and Policy Change: Ecology, Antinuclear, and Peace Movements*.

[210] Benjamin K. Sovacool. The costs of failure: A preliminary assessment of major energy accidents, 1907–2007, *Energy Policy* 36 (2008), pp. 1802-1820.

[211] Stephanie Cooke (2009). *In Mortal Hands: A Cautionary History of the Nuclear Age*, Black Inc., p. 280.

[212] Kurt Kleiner. Nuclear energy: assessing the emissions *Nature Reports*, Vol. 2, October 2008, pp. 130-131.

[213] Mark Diesendorf (2007). *Greenhouse Solutions with Sustainable Energy*, University of New South Wales Press, p. 252.

[214] Mark Diesendorf. Is nuclear energy a possible solution to global warming? Archived July 22, 2012 at the Wayback Machine

[215] World Nuclear Association (September 2013). "Renewable Energy and Electricity".

[216] Kloor, Keith (11 January 2013). "The Pro-Nukes Environmental Movement". *Slate.com "The Big Questions" Blog* (The Slate Group). Retrieved 11 March 2013.

[217] Smil, Vaclav (2012-06-28). "A Skeptic Looks at Alternative Energy - IEEE Spectrum". Spectrum.ieee.org. Retrieved 2014-01-24.

[218] International Energy Agency (2007). Contribution of Renewables to Energy Security IEA Information Paper, p. 5.

[219] Amory Lovins (2011). *Reinventing Fire*, Chelsea Green Publishing, p. 199.

[220] Entwicklungen in der deutschen Strom- und Gaswirtschaft 2012 BDEW (german) Archived January 17, 2015 at the Wayback Machine

[221] Harvey, Fiona (2012-10-30). "Renewable energy will overtake nuclear power by 2018, research says". *The Guardian* (London).

[222] Steve Colquhoun (2012-10-31). "Scotland aims for 100% renewable energy by 2020". The Sydney Morning Herald. Retrieved 2014-01-24.

[223] Fiona Harvey (9 May 2011). "Renewable energy can power the world, says landmark IPCC study". *The Guardian* (London).

[224] Archived October 21, 2012 at the Wayback Machine

[225] "Is solar power cheaper than nuclear power?". August 9, 2010. Retrieved 2013-01-04.

[226] "Solar and Nuclear Costs — The Historic Crossover" (PDF). July 2010. Retrieved 2013-01-16.

[227] "Solar and Nuclear Costs — The Historic Crossover". July 2010. Retrieved 2013-01-16.

[228] Chris Namovicz, Assessing the Economic Value of New Utility-Scale Renewable Generation Projects US Energy Information Administration Energy Conference, 17 June 2013.

[229] Nils Starfelt; Carl-Erik Wikdahl. "Economic Analysis of Various Options of Electricity Generation - Taking into Account Health and Environmental Effects" (PDF). Retrieved 2012-09-08.

[230] David Biello (2009-01-28). "Spent Nuclear Fuel: A Trash Heap Deadly for 250,000 Years or a Renewable Energy Source?". Scientificamerican.com. Retrieved 2014-01-24.

[231] "Closing and Decommissioning Nuclear Power Plants" (PDF). March 7, 2012.

[232] "Olkiluoto pipe welding 'deficient', says regulator". World Nuclear News. 16 October 2009. Retrieved 8 June 2010.

[233] Kinnunen, Terhi (2010-07-01). "Finnish parliament agrees plans for two reactors". Reuters. Retrieved 2010-07-02.

[234] "Olkiluoto 3 delayed beyond 2014". World Nuclear News. 17 July 2012. Retrieved 24 July 2012.

[235] "Finland's Olkiluoto 3 nuclear plant delayed again". BBC. 16 July 2012. Retrieved 10 August 2012.

[236] http://www.iaea.org/PRIS/WorldStatistics/WorldTrendinElectricalProduction.aspx International Atomic Energy Agency, March 2014

[237] "The Nuclear Renaissance". World Nuclear Association. Retrieved 2014-01-24.

[238] WNA (20 June 2013). "Nuclear power down in 2012". *World Nuclear News*.

[239] Nuclear Renaissance Threatened as Japan's Reactor Struggles Bloomberg, published March 2011, accessed 2011-03-14

[240] Analysis: Nuclear renaissance could fizzle after Japan quake Reuters, published 2011-03-14, accessed 2011-03-14

[241] Japan nuclear woes cast shadow over U.S. energy policy Reuters, published 2011-03-13, accessed 2011-03-14

[242] Nuclear winter? Quake casts new shadow on reactors MarketWatch, published 2011-03-14, accessed 2011-03-14

[243] Will China's nuclear nerves fuel a boom in green energy? Channel 4, published 2011-03-17, accessed 2011-03-17

[244] "NEWS ANALYSIS: Japan crisis puts global nuclear expansion in doubt". Platts. 21 March 2011.

[245] "Siemens to quit nuclear industry". *BBC News.* September 18, 2011.

[246] "Italy announces nuclear moratorium". World Nuclear News. 24 March 2011. Retrieved 23 May 2011.

[247] Jo Chandler (March 19, 2011). "Is this the end of the nuclear revival?". *The Sydney Morning Herald.*

[248] Aubrey Belford (March 17, 2011). "Indonesia to Continue Plans for Nuclear Power". *New York Times.*

[249] Israel Prime Minister Netanyahu: Japan situation has "caused me to reconsider" nuclear power Piers Morgan on CNN, published 2011-03-17, accessed 2011-03-17

[250] Israeli PM cancels plan to build nuclear plant xinhuanet.com, published 2011-03-18, accessed 2011-03-17

[251] "Bruce Power's Unit 2 sends electricity to Ontario grid for first time in 17 years". Bruce Power. 2012-10-16. Retrieved 2014-01-24.

[252] James Kanter. In Finland, Nuclear Renaissance Runs Into Trouble *New York Times*, May 28, 2009.

[253] James Kanter. Is the Nuclear Renaissance Fizzling? *Green*, 29 May 2009.

[254] Rob Broomby. Nuclear dawn delayed in Finland *BBC News*, 8 July 2009.

[255] Jeff McMahon (10 November 2013). "New-Build Nuclear Is Dead: Morningstar". *Forbes.*

[256] Hannah Northey (18 March 2011). "Former NRC Member Says Renaissance is Dead, for Now". *New York Times.*

[257] Ian Lowe (March 20, 2011). "No nukes now, or ever". *The Age* (Melbourne).

[258] Leo Hickman (28 November 2012). "Nuclear lobbyists wined and dined senior civil servants, documents show". *The Guardian* (London).

[259] Diane Farseta (September 1, 2008). "The Campaign to Sell Nuclear". *Bulletin of the Atomic Scientists* **64** (4). pp. 38–56.

[260] Jonathan Leake. " The Nuclear Charm Offensive" *New Statesman*, 23 May 2005.

[261] Union of Concerned Scientists. Nuclear Industry Spent Hundreds of Millions of Dollars Over the Last Decade to Sell Public, Congress on New Reactors, New Investigation Finds News Center, February 1, 2010. Archived November 27, 2013 at the Wayback Machine

[262] Nuclear group spent $460,000 lobbying in 4Q *Business Week*, March 19, 2010.

[263] World Nuclear Association (December 10, 2010). Nuclear Power in China

[264] China is Building the World's Largest Nuclear Capacity 21cbh.com, 21. Sep. 2010

[265] "China Should Control Pace of Reactor Construction, Outlook Says". *Bloomberg News.* January 11, 2011.

[266] "Nuclear Power in the USA". World Nuclear Association. June 2008. Retrieved 2008-07-25.

[267] Matthew L. Wald (December 7, 2010). Nuclear 'Renaissance' Is Short on Largess *The New York Times.*

[268] "NRC/DOE Life After 60 Workshop Report" (PDF). 2008. Retrieved 2009-04-01.

[269] New nuclear build – sufficient supply capability? Steve Kid, Nuclear Engineering International, 3/3/2009

[270] Bloomberg exclusive: Samurai-Sword Maker's Reactor Monopoly May Cool Nuclear Revival By Yoshifumi Takemoto and Alan Katz, bloomberg.com, 3/13/08.

[271] Plans For New Reactors Worldwide, World Nuclear Association

[272] "Nuclear Energy's Role in Responding to the Energy Challenges of the 21st Century" (PDF). *Idaho National Engineering and Environmental Laboratory.* Retrieved 2008-06-21.

[273] "Nuclear power: When the steam clears". *The Economist.* March 24, 2011.

[274] Paton J (April 4, 2011). "Fukushima crisis worse for atomic power than Chernobyl, USB says". *Bloomberg.com.* Retrieved 17 August 2014.

[275] Deutsche Bank Group (2011). The 2011 inflection point for energymarkets: Health, safety, security and the environment. *DB Climate Change Advisors,* May 2.

[276] John Broder (October 10, 2011). "The Year of Peril and Promise in Energy Production". *New York Times.*

[277] "Siemens to quit nuclear industry". *BBC News.* 18 September 2011.

[278] "IAEA sees slow nuclear growth post Japan". *UPI.* September 23, 2011.

[279] Hsu, Jeremy (February 9, 2012). "First Next-Gen US Reactor Designed to Avoid Fukushima Repeat". Live Science (hosted on Yahoo!). Retrieved February 9, 2012.

[280] Kristi E. Swartz (February 16, 2012). "Groups sue to stop Vogtle expansion project". *The Atlanta Journal-Constitution.*

[281] Duroyan Fertl (June 5, 2011). "Germany: Nuclear power to be phased out by 2022". *Green Left.*

[282] "Science/Nature | France gets nuclear fusion plant". BBC News. 2005-06-28. Retrieved 2014-01-24.

[283] "NCPST Homepage | DCU". Ncpst.ie. Retrieved 2014-01-24.

[284] Renewables 2012 Global Status Report p. 21

[285] "4th Generation Nuclear Power — OSS Foundation". Ossfoundation.us. Retrieved 2014-01-24.

[286] Adam Piore (June 2011). "Nuclear energy: Planning for the Black Swan". *Scientific American.*

[287] Matthew L. Wald. Critics Challenge Safety of New Reactor Design *New York Times,* April 22, 2010.

[288] "Nuclear Power in a Warming World" (PDF). *Union of Concerned Scientists.* Retrieved 1 October 2008.

[289] Benjamin K. Sovacool. *A Critical Evaluation of Nuclear Power and Renewable Electricity in Asia,* Journal of Contemporary Asia, *Vol. 40, No. 3, August 2010, p. 381.*

[290] Gerstner, E. (2009). "Nuclear energy: The hybrid returns" (PDF). *Nature* **460** (7251): 25–8. doi:10.1038/460025a. PMID 19571861.

[291] *Introduction to Fusion Energy,* J. Reece Roth, 1986.

[292] T. Hamacher and A.M. Bradshaw (October 2001). "Fusion as a Future Power Source: Recent Achievements and Prospects" (PDF). World Energy Council. Archived from the original (PDF) on 2004-05-06.

[293] W Wayt Gibbs (30 December 2013). "Triple-threat method sparks hope for fusion". *Nature.*

[294] "Overview of EFDA Activities". *EFDA.* European Fusion Development Agreement. Archived from the original on 2006-10-01. Retrieved 2006-11-11.

[295] "About Friends of the Earth International". Friends of the Earth International. Retrieved 2009-06-25.

[296] "United Nations, Department of Public Information, Non-Governmental Organizations". Un.org. 2006-02-23. Retrieved 2010-08-24.

13.17   Further reading

See also: List of books about nuclear issues and List of films about nuclear issues

- Clarfield, Gerald H. and William M. Wiecek (1984). *Nuclear America: Military and Civilian Nuclear Power in the United States 1940-1980*, Harper & Row.

- Cooke, Stephanie (2009). *In Mortal Hands: A Cautionary History of the Nuclear Age*, Black Inc.

- Cravens, Gwyneth (2007). *Power to Save the World: the Truth about Nuclear Energy*. New York: Knopf. ISBN 0-307-26656-7.

- Elliott, David (2007). *Nuclear or Not? Does Nuclear Power Have a Place in a Sustainable Energy Future?*, Palgrave.

- Falk, Jim (1982). *Global Fission: The Battle Over Nuclear Power*, Oxford University Press.

- Ferguson, Charles D., (2007). *Nuclear Energy: Balancing Benefits and Risks* Council on Foreign Relations.

- Herbst, Alan M. and George W. Hopley (2007). *Nuclear Energy Now: Why the Time has come for the World's Most Misunderstood Energy Source*, Wiley.

- Schneider, Mycle, Steve Thomas, Antony Froggatt, Doug Koplow (2012). *The World Nuclear Industry Status Report*, German Federal Ministry of Environment, Nature Conservation and Reactor Safety.

- Walker, J. Samuel (1992). *Containing the Atom: Nuclear Regulation in a Changing Environment, 1993-1971*, Berkeley: University of California Press.

- Weart, Spencer R. *The Rise of Nuclear Fear*. Cambridge, MA: Harvard University Press, 2012. ISBN 0-674-05233-1

13.18   External links

- Alsos Digital Library for Nuclear Issues — Annotated Bibliography on Nuclear Power

- An entry to nuclear power through an educational discussion of reactors

- Argonne National Laboratory

- A cost comparison of nuclear energy to other commercial energy sources

- Briefing Papers from the Australian EnergyScience Coalition

- British Energy — Understanding Nuclear Energy / Nuclear Power

- Coal Combustion: Nuclear Resource or Danger?

- Congressional Research Service report on Nuclear Energy Policy PDF (94.0 KB)

- Energy Information Administration provides lots of statistics and information

- How Nuclear Power Works

- IAEA Website The International Atomic Energy Agency

    - IAEA's Power Reactor Information System (PRIS)

- Nuclear Power: Climate Fix or Folly? (2009)

- Nuclear Power Education

- Nuclear Tourist.com, nuclear power information

- The World Nuclear Industry Status Reports website

- TED Talk - Bill Gates on energy: Innovating to zero!

- LFTR in 5 Minutes - Creative Commons Film Compares PWR to Th-MSR/LFTR Nuclear Power. on YouTube

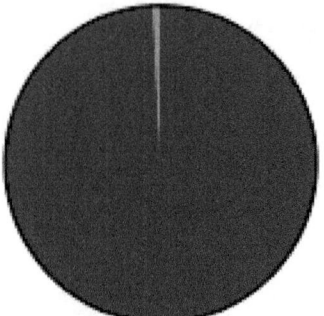

Natural uranium
> 99.2% U-238
0.72% U-235

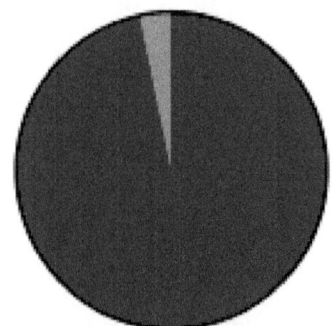

Low-enriched uranium
(reactor grade)
3-4% U-235

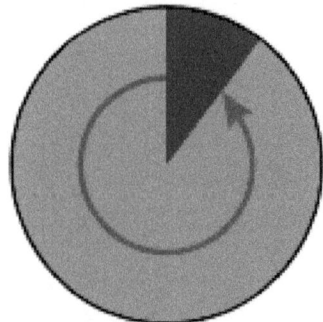

Highly enriched uranium
(weapons grade)
90% U-235

Proportions of the isotopes, uranium-238 (blue) and uranium-235 (red) found naturally, versus grades that are enriched. light water reactors require fuel enriched to (3-4%), while others such as the CANDU reactor uses natural uranium.

A nuclear fuel rod assembly bundle being inspected before entering a reactor.

Following interim storage in a spent fuel pool, the bundles of used fuel assemblies of a typical nuclear power station are often stored on site in the likes of the eight dry cask storage vessels pictured above.[128] At Yankee Rowe Nuclear Power Station, which generated 44 billion kilowatt hours of electricity over its lifetime, its complete spent fuel inventory is contained within sixteen casks.[129]

George W. Bush signing the Energy Policy Act of 2005, which was designed to promote the US nuclear power industry, through incentives and subsidies, including cost-overrun support up to a total of $2 billion for six new nuclear plants.[162] However, as of 2014 some electric utilities have rebuffed the loan package, including South Carolina Electric and Gas which operates Summer Station(the location of 2 new builds), noting instead that "it was easier to raise [loan] money commercially."[163]

The Ikata Nuclear Power Plant, a pressurized water reactor that cools by utilizing a secondary coolant heat exchanger with a large body of water, an alternative cooling approach to large cooling towers.

The 2011 Fukushima Daiichi nuclear disaster, the world's worst nuclear accident since 1986, displaced 50,000 households after radiation leaked into the air, soil and sea.[172] *Radiation checks led to bans of some shipments of vegetables and fish.*[173]

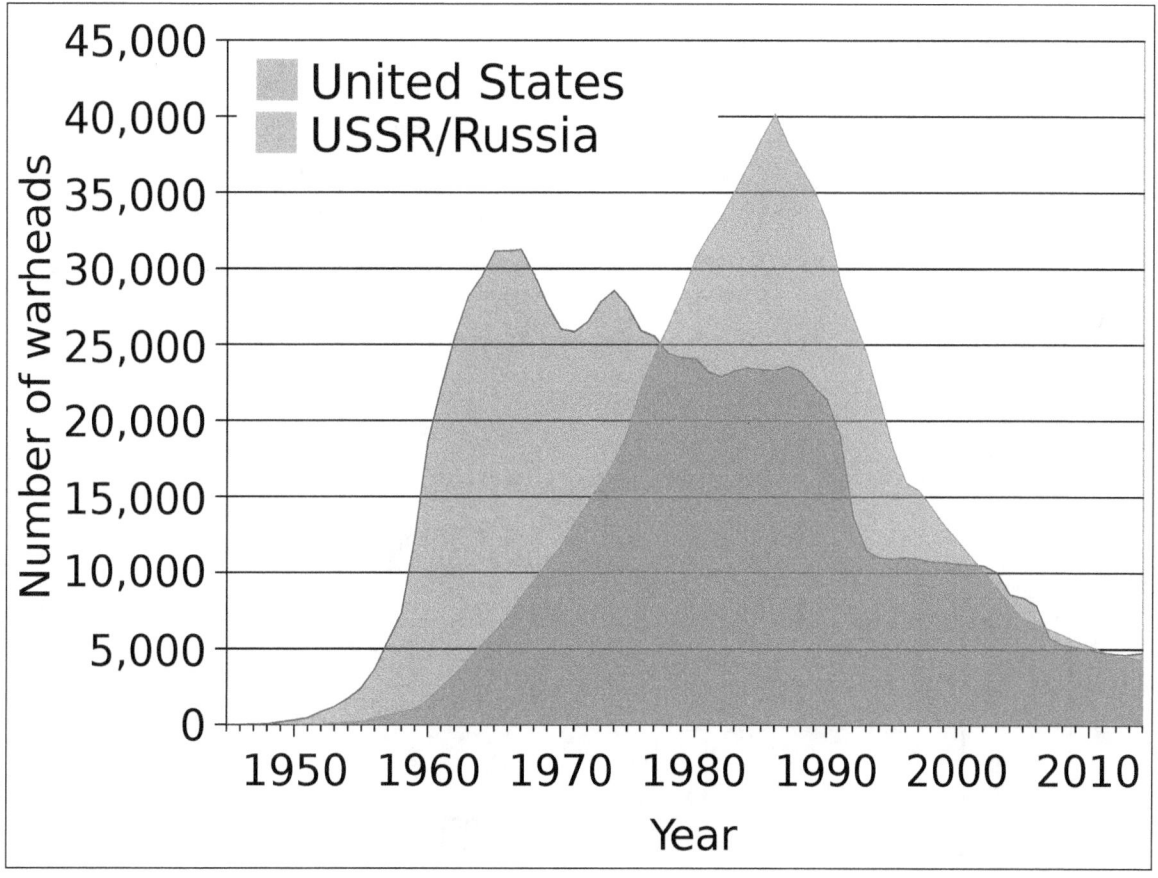

United States and USSR/Russian nuclear weapons stockpiles, 1945-2006. The Megatons to Megawatts Program was the main driving force behind the sharp reduction in the quantity of nuclear weapons worldwide since the cold war ended.[186][187] However without an increase in nuclear reactors and greater demand for fissile fuel, the cost of dismantling has dissuaded Russia from continuing their disarmament.

How much usable nuclear energy is placed in 1 (Minuteman) ICBM ?

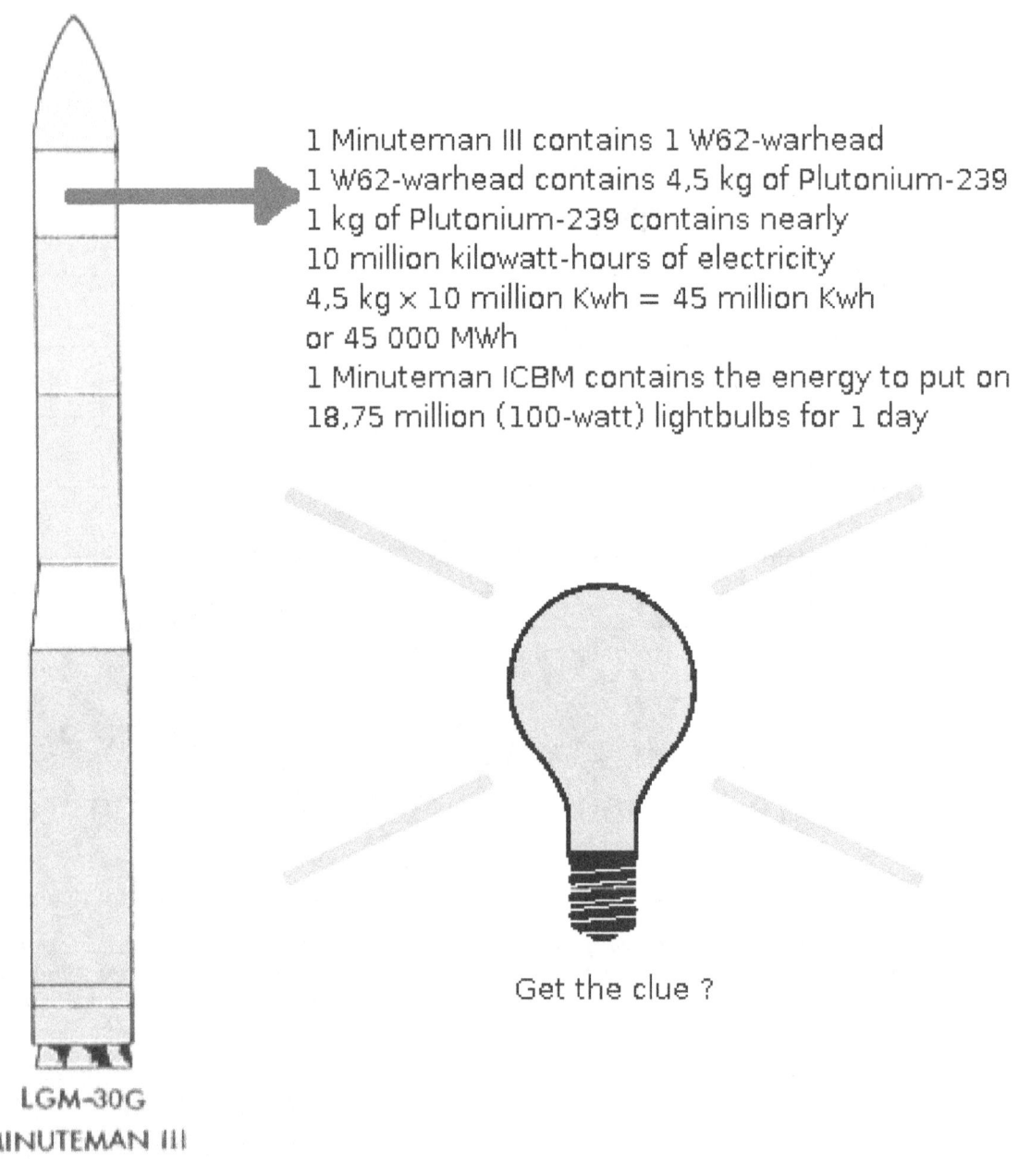

1 Minuteman III contains 1 W62-warhead
1 W62-warhead contains 4,5 kg of Plutonium-239
1 kg of Plutonium-239 contains nearly
10 million kilowatt-hours of electricity
4,5 kg × 10 million Kwh = 45 million Kwh
or 45 000 MWh
1 Minuteman ICBM contains the energy to put on
18,75 million (100-watt) lightbulbs for 1 day

Get the clue ?

LGM–30G
MINUTEMAN III

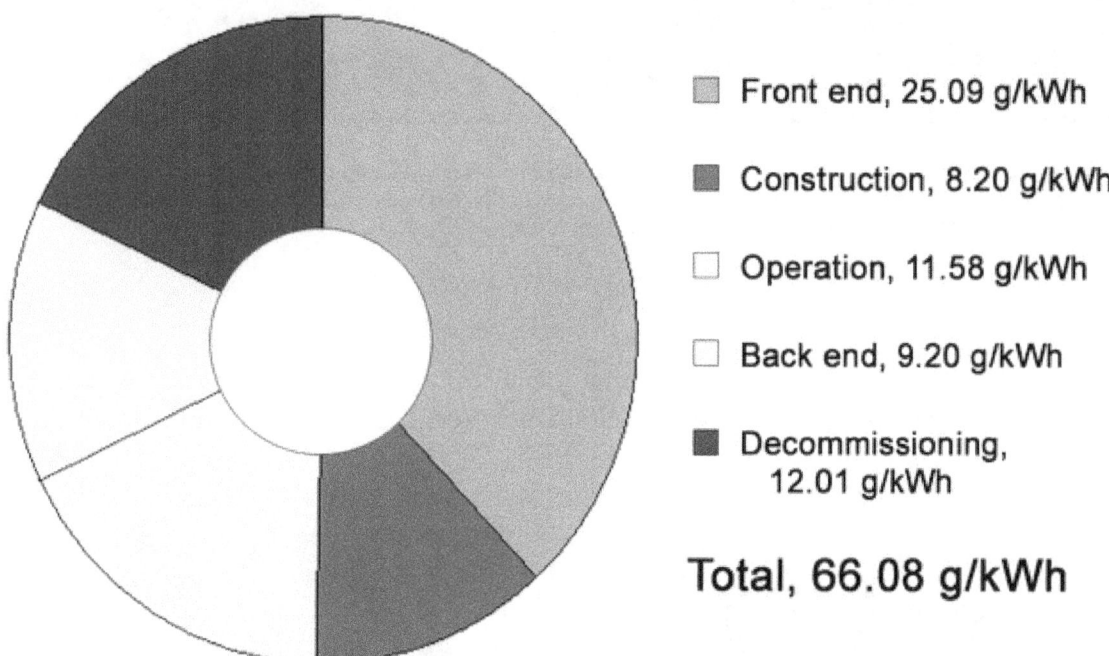

Carbon emissions from nuclear power
Sovacool life cycle study survey, 2008

- Front end, 25.09 g/kWh
- Construction, 8.20 g/kWh
- Operation, 11.58 g/kWh
- Back end, 9.20 g/kWh
- Decommissioning, 12.01 g/kWh

Total, 66.08 g/kWh

Mean value of carbon dioxide emissions from qualified life cycle studies among 103 surveyed. Includes results of 1997 Vattenfall study.

A 2008 synthesis of 103 studies, published by Benjamin K. Sovacool, estimated that the value of CO_2 emissions for nuclear power over the lifecycle of a plant was 66.08 g/kW·h. Comparative results for various renewable power sources were 9–32 g/kW·h.[196] A 2012 study by Yale University arrived at a different value, with the mean value, depending on which Reactor design was analyzed, ranging from 11 to 25 g/kW·h of total life cycle nuclear power CO_2 emissions.[197]

Olkiluoto 3 under construction in 2009. It is the first EPR design, but problems with workmanship and supervision have created costly delays which led to an inquiry by the Finnish nuclear regulator STUK.[232] In December 2012, Areva estimated that the full cost of building the reactor will be about €8.5 billion, or almost three times the original delivery price of €3 billion.[233][234][235]

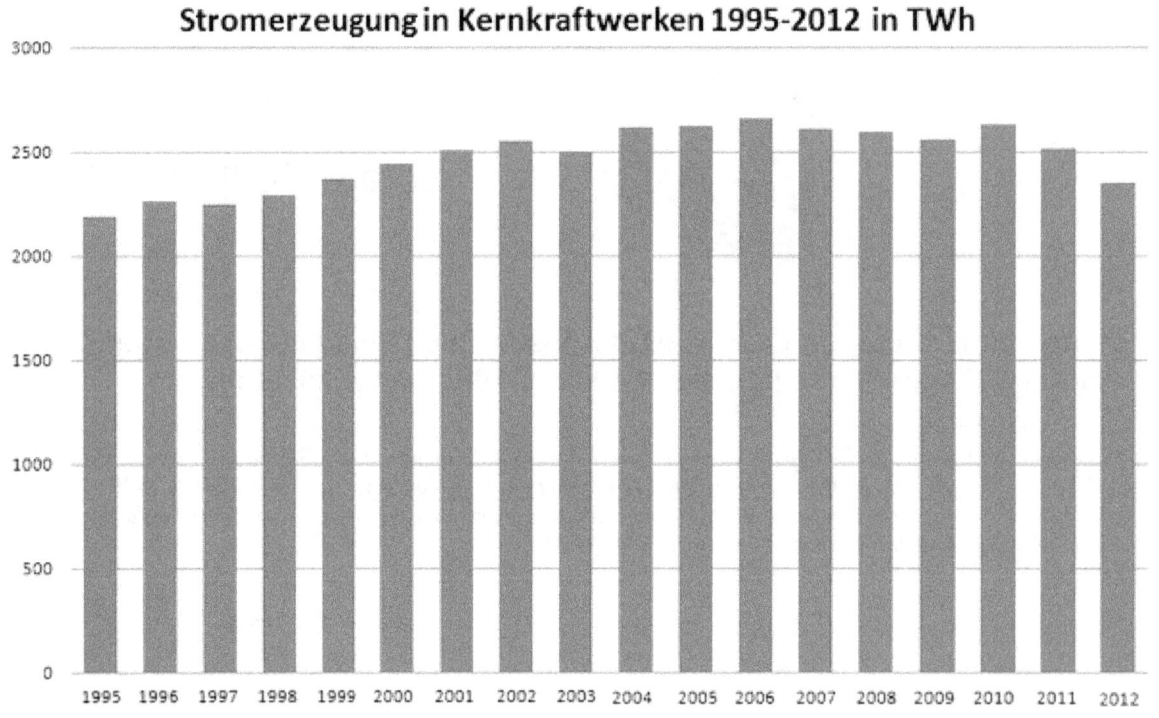

Nuclear power production 1995-2012 in TWh[236]

Brunswick Nuclear Plant discharge canal

The Bruce Nuclear Generating Station, the largest nuclear power facility in the world[251]

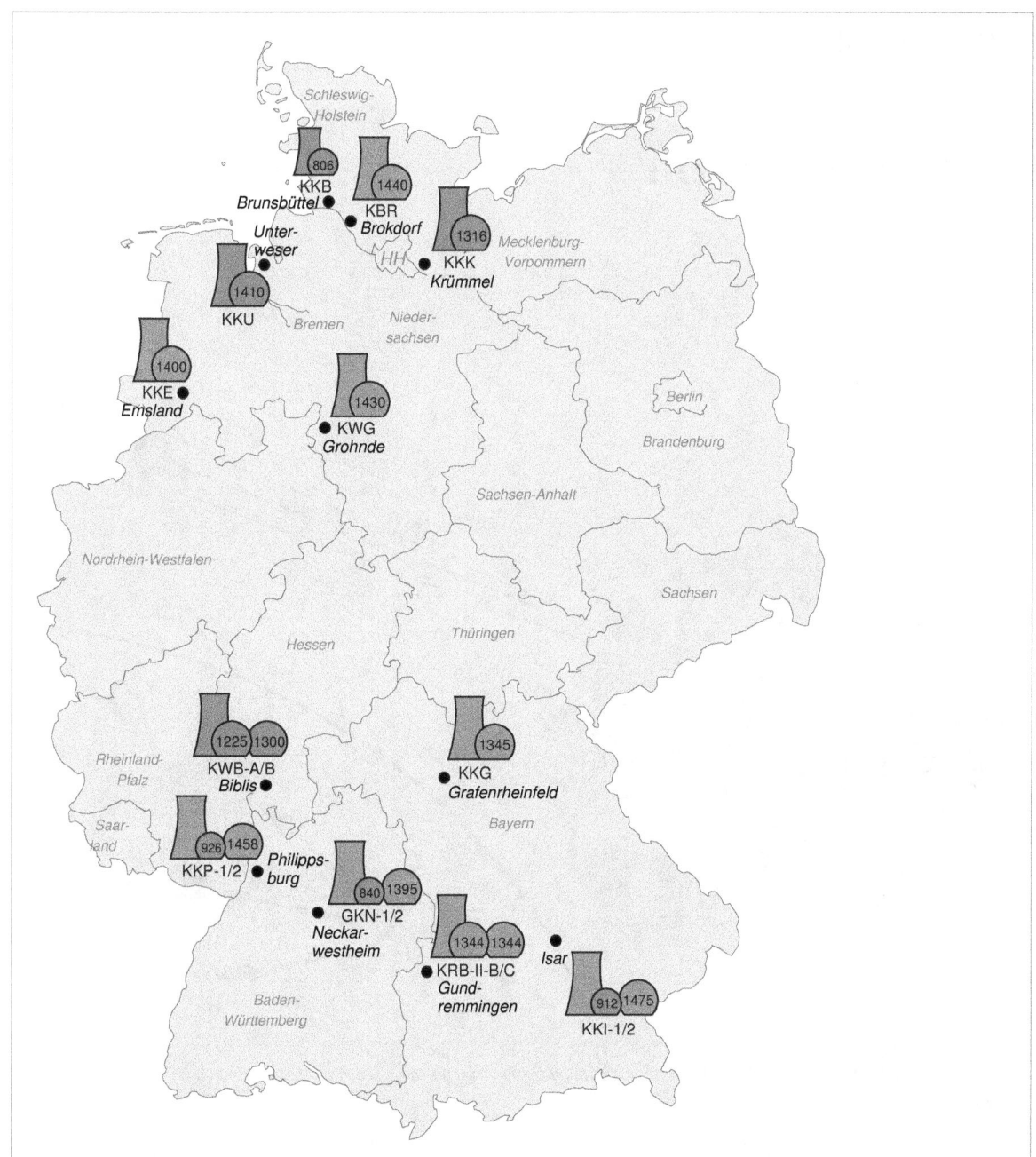

Eight of the seventeen operating reactors in Germany were permanently shut down following the March 2011 Fukushima nuclear disaster.

Chapter 14

Weak interaction

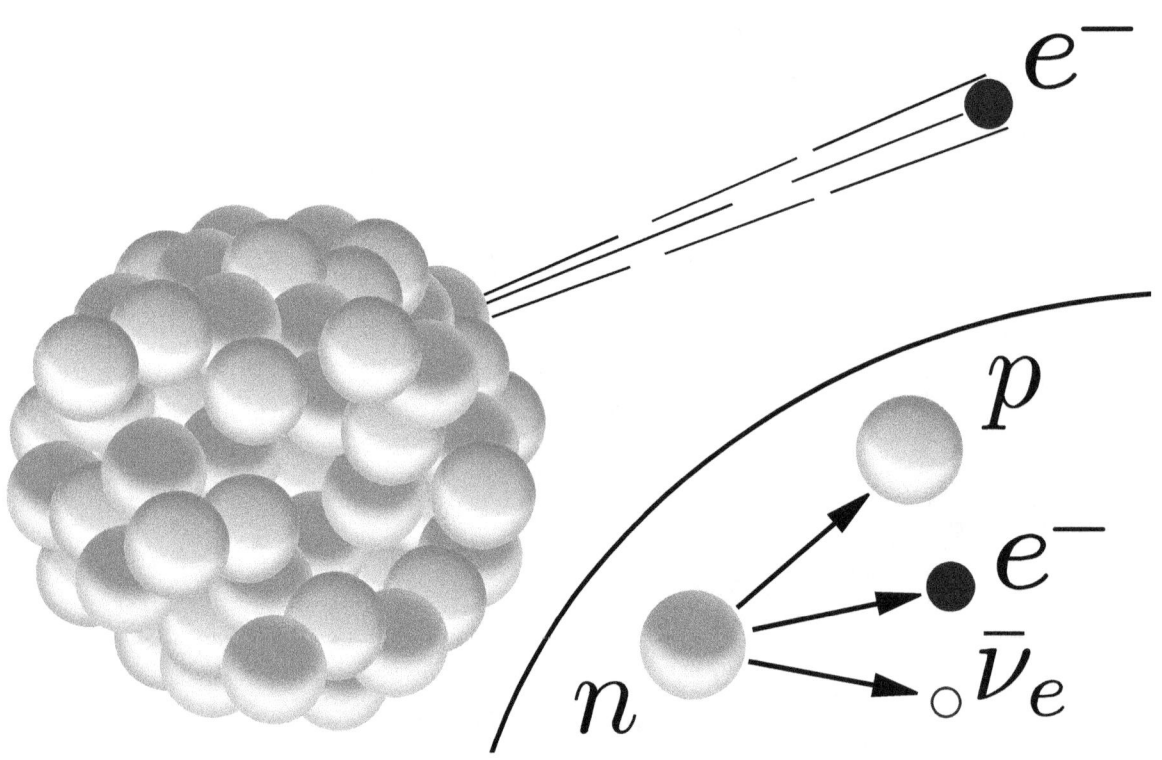

The radioactive beta decay is possible due to the weak interaction, which transforms a neutron into: a proton, an electron, and an electron antineutrino.

In particle physics, the **weak interaction** is the mechanism responsible for the **weak force** or **weak nuclear force**, one of the four known fundamental interactions of nature, alongside the strong interaction, electromagnetism, and gravitation. The weak interaction is responsible for the radioactive decay of subatomic particles, and it plays an essential role in nuclear fission. The theory of the weak interaction is sometimes called **quantum flavordynamics** (**QFD**), in analogy with the terms QCD and QED, but the term is rarely used because the weak force is best understood in terms of electro-weak theory (EWT).[1]

In the Standard Model of particle physics, the weak interaction is caused by the emission or absorption of W and Z bosons. All known fermions interact through the weak interaction. Fermions are particles that have half-integer spin (one of the fundamental properties of particles). A fermion can be an elementary particle, such as the electron, or it can be a composite particle, such as the proton. The masses of W^+, W^-, and Z bosons are each far greater than that of protons or neutrons, consistent with the short range of the weak force. The force is termed *weak* because its field strength over a given distance is typically several orders of magnitude less than that of the strong nuclear force and electromagnetic force.

During the quark epoch, the electroweak force split into the electromagnetic and weak forces. Important examples of weak interaction include beta decay, and the production, from hydrogen, of deuterium needed to power the sun's thermonuclear process. Most fermions will decay by a weak interaction over time. Such decay also makes radiocarbon dating possible, as carbon-14 decays through the weak interaction to nitrogen-14. It can also create radioluminescence, commonly used in tritium illumination, and in the related field of betavoltaics.[2]

Quarks, which make up composite particles like neutrons and protons, come in six "flavours" – up, down, strange, charm, top and bottom – which give those composite particles their properties. The weak interaction is unique in that it allows for quarks to swap their flavour for another. For example, during beta minus decay, a down quark decays into an up quark, converting a neutron to a proton. Also the weak interaction is the only fundamental interaction that breaks parity-symmetry, and similarly, the only one to break CP-symmetry.

14.1 History

In 1933, Enrico Fermi proposed the first theory of the weak interaction, known as Fermi's interaction. He suggested that beta decay could be explained by a four-fermion interaction, involving a contact force with no range.[3][4]

However, it is better described as a non-contact force field having a finite range, albeit very short. In 1968, Sheldon Glashow, Abdus Salam and Steven Weinberg unified the electromagnetic force and the weak interaction by showing them to be two aspects of a single force, now termed the electro-weak force.

The existence of the W and Z bosons was not directly confirmed until 1983.

14.2 Properties

The weak interaction is unique in a number of respects:

1. It is the only interaction capable of changing the flavor of quarks (i.e., of changing one type of quark into another).

2. It is the only interaction that violates **P** or parity-symmetry. It is also the only one that violates **CP** symmetry.

3. It is propagated by carrier particles (known as gauge bosons) that have significant masses, an unusual feature which is explained in the Standard Model by the Higgs mechanism.

Due to their large mass (approximately 90 GeV/c^{2}[5]) these carrier particles, termed the W and Z bosons, are short-lived: they have a lifetime of under 1×10^{-24} seconds.[6] The weak interaction has a coupling constant (an indicator of interaction strength) of between 10^{-7} and 10^{-6}, compared to the strong interaction's coupling constant of about 1 and the electromagnetic coupling constant of about 10^{-2};[7] consequently the weak interaction is weak in terms of strength.[8] The weak interaction has a very short range (around 10^{-17}–10^{-16} m[8]).[7] At distances around 10^{-18} meters, the weak interaction has a strength of a similar magnitude to the electromagnetic force, but this starts to decrease exponentially with increasing distance. At distances of around 3×10^{-17} m, the weak interaction is 10,000 times weaker than the electromagnetic.[9]

The weak interaction affects all the fermions of the Standard Model, as well as the Higgs boson; neutrinos interact through gravity and the weak interaction only, and neutrinos were the original reason for the name *weak force*.[8] The weak interaction does not produce bound states (nor does it involve binding energy) – something that gravity does on an astronomical scale, that the electromagnetic force does at the atomic level, and that the strong nuclear force does inside nuclei.[10]

Its most noticeable effect is due to its first unique feature: flavor changing. A neutron, for example, is heavier than a proton (its sister nucleon), but it cannot decay into a proton without changing the flavor (type) of one of its two *down*

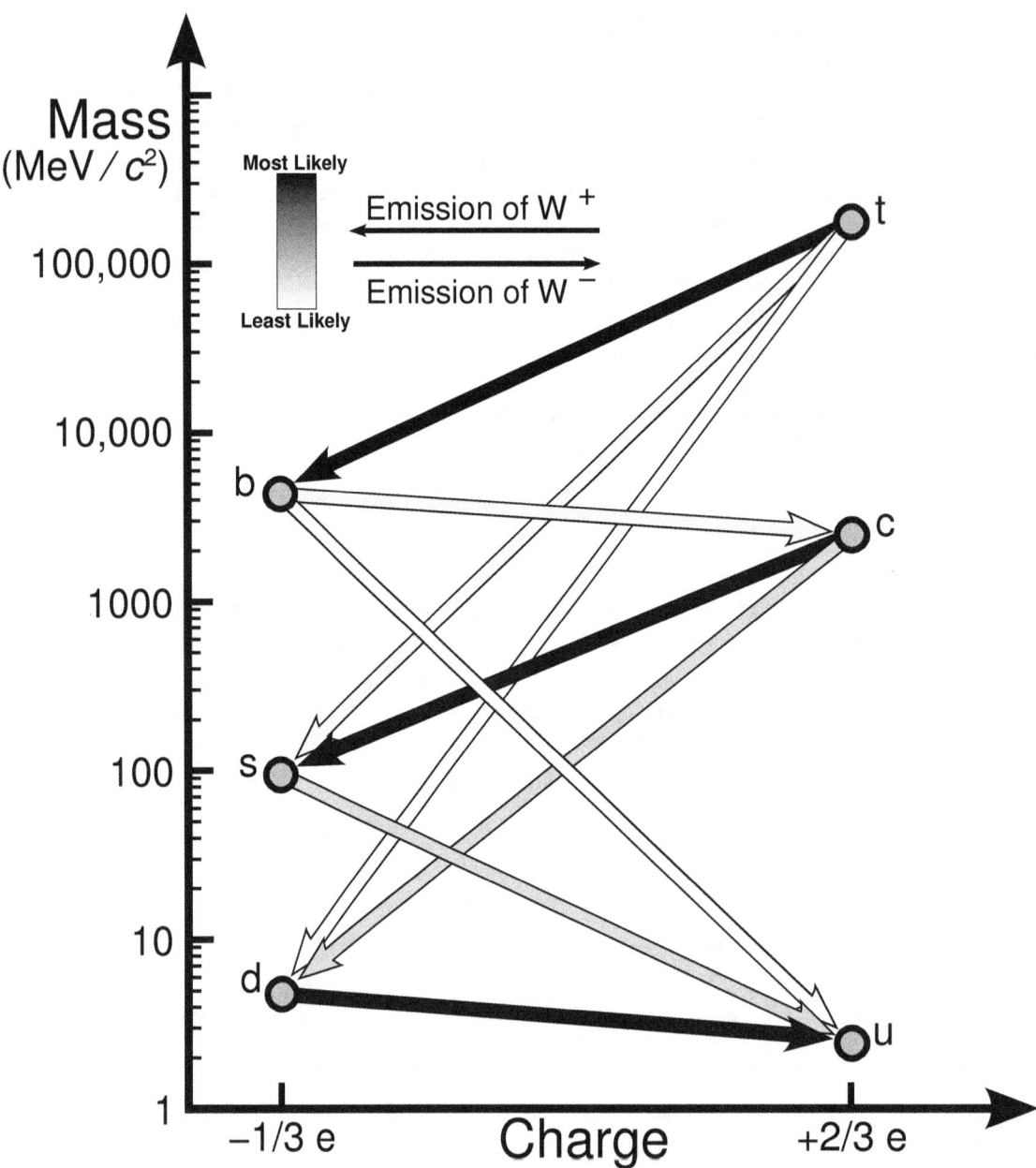

A diagram depicting the various decay routes due to the weak interaction and some indication of their likelihood. The intensity of the lines are given by the CKM parameters.

quarks to *up*. Neither the strong interaction nor electromagnetism permit flavour changing, so this must proceed by **weak decay**; without weak decay, quark properties such as strangeness and charm (associated with the quarks of the same name) would also be conserved across all interactions. All mesons are unstable because of weak decay.[11] In the process known as beta decay, a *down* quark in the neutron can change into an *up* quark by emitting a virtual W− boson which is then converted into an electron and an electron antineutrino.[12] Another example is the electron capture, a common variant of radioactive decay, where a proton (up quark) and an electron within an atom interact, and are changed to a neutron (down quark) and an electron neutrino.

Due to the large mass of a boson, weak decay is much more unlikely than strong or electromagnetic decay, and hence occurs less rapidly. For example, a neutral pion (which decays electromagnetically) has a life of about 10^{-16} seconds, while a charged pion (which decays through the weak interaction) lives about 10^{-8} seconds, a hundred million times longer.[13] In contrast, a free neutron (which also decays through the weak interaction) lives about 15 minutes.[12]

14.2.1 Weak isospin and weak hypercharge

Main article: Weak isospin

All particles have a property called weak isospin (T_3), which serves as a quantum number and governs how that particle interacts in the weak interaction. Weak isospin therefore plays the same role in the weak interaction as electric charge does in electromagnetism, and color charge in the strong interaction. All fermions have a weak isospin value of either $+\frac{1}{2}$ or $-\frac{1}{2}$. For example, the up quark has a T_3 of $+\frac{1}{2}$ and the down quark $-\frac{1}{2}$. A quark never decays through the weak interaction into a quark of the same T_3: quarks with a T_3 of $+\frac{1}{2}$ decay into quarks with a T_3 of $-\frac{1}{2}$ and vice versa.

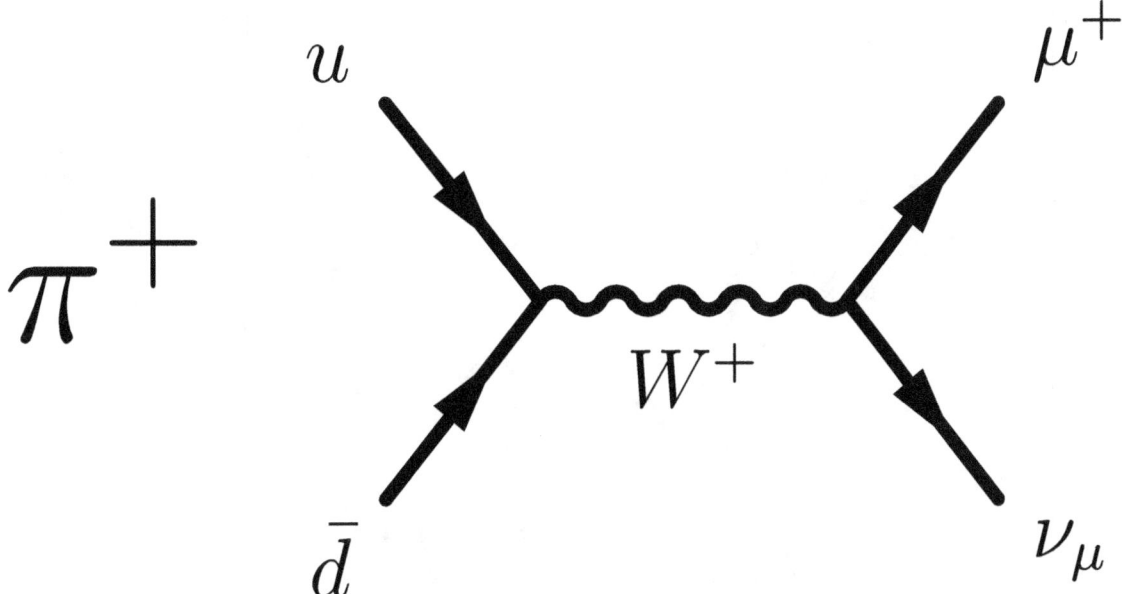

π+ decay through the weak interaction

In any given interaction, weak isospin is conserved: the sum of the weak isospin numbers of the particles entering the interaction equals the sum of the weak isospin numbers of the particles exiting that interaction. For example, a (left-handed) π+, with a weak isospin of 1 normally decays into a ν
μ (+1/2) and a μ+ (as a right-handed antiparticle, +1/2).[13]

Following the development of the electroweak theory, another property, weak hypercharge, was developed. It is dependent on a particle's electrical charge and weak isospin, and is defined as:

$$Y_W = 2(Q - T_3)$$

where YW is the weak hypercharge of a given type of particle, Q is its electrical charge (in elementary charge units) and T_3 is its weak isospin. Whereas some particles have a weak isospin of zero, all particles, except gluons, have non-zero weak hypercharge. Weak hypercharge is the generator of the U(1) component of the electroweak gauge group.

14.3   Interaction types

There are two types of weak interaction (called *vertices*). The first type is called the "charged-current interaction" because it is mediated by particles that carry an electric charge (the W+ or W− bosons), and is responsible for the beta decay phenomenon. The second type is called the "neutral-current interaction" because it is mediated by a neutral particle, the Z boson.

14.3.1   Charged-current interaction

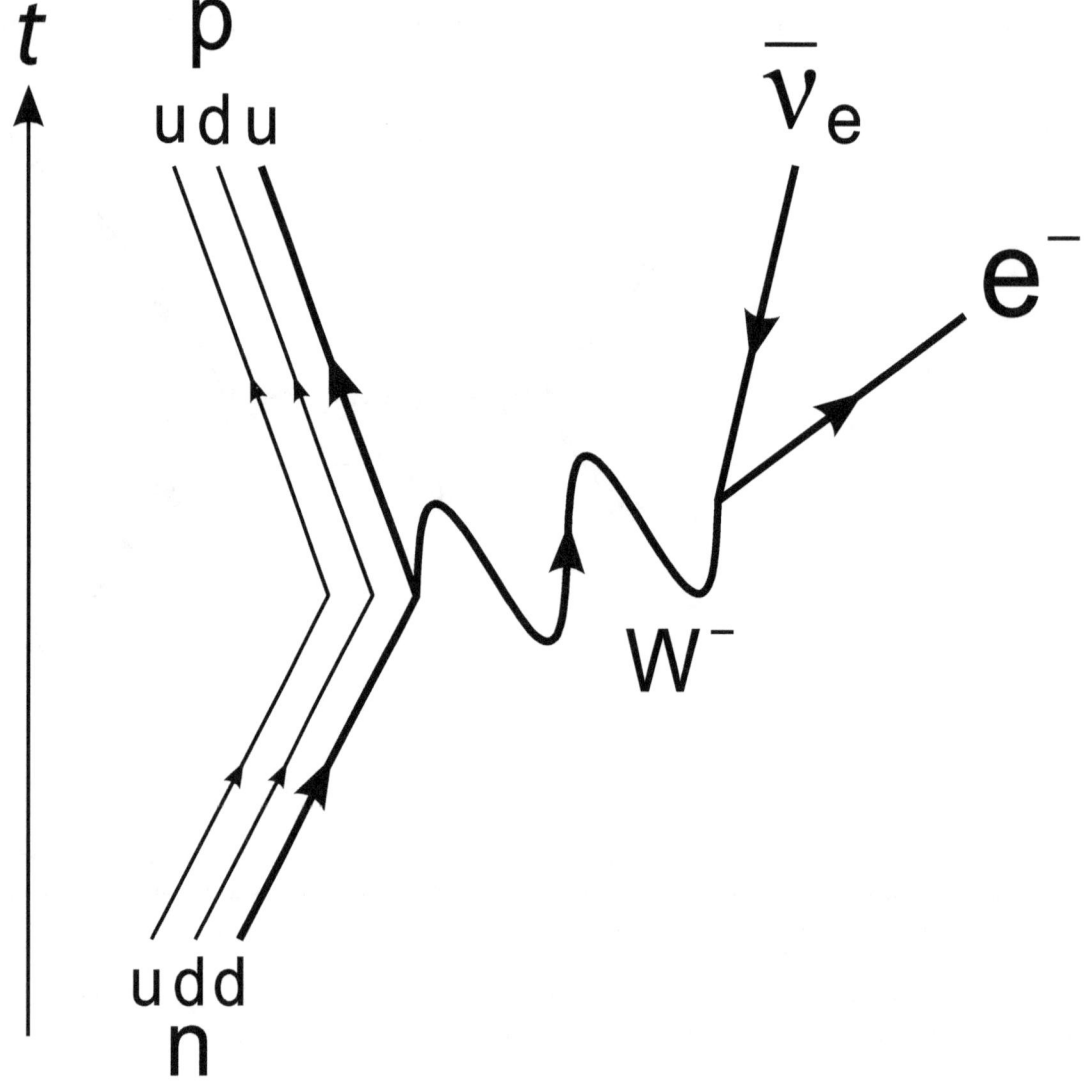

The Feynman diagram for beta-minus decay of a neutron into a proton, electron and electron anti-neutrino, via an intermediate heavy W− boson

In one type of charged current interaction, a charged lepton (such as an electron or a muon, having a charge of −1) can absorb a W+ boson (a particle with a charge of +1) and be thereby converted into a corresponding neutrino (with a charge of 0), where the type ("family") of neutrino (electron, muon or tau) is the same as the type of lepton in the interaction, for example:

$$\mu^- + W^+ \rightarrow \nu_\mu$$

Similarly, a down-type quark (*d* with a charge of $-\frac{1}{3}$) can be converted into an up-type quark (*u*, with a charge of $+\frac{2}{3}$), by emitting a W− boson or by absorbing a W+ boson. More precisely, the down-type quark becomes a quantum superposition of up-type quarks: that is to say, it has a possibility of becoming any one of the three up-type quarks, with the probabilities given in the CKM matrix tables. Conversely, an up-type quark can emit a W+ boson – or absorb a W− boson – and thereby be converted into a down-type quark, for example:

$$d \rightarrow u + W^-$$
$$d + W^+ \rightarrow u$$
$$c \rightarrow s + W^+$$
$$c + W^- \rightarrow s$$

The W boson is unstable so will rapidly decay, with a very short lifetime. For example:

$$W^- \rightarrow e^- + \bar{\nu}_e$$
$$W^+ \rightarrow e^+ + \nu_e$$

Decay of the W boson to other products can happen, with varying probabilities.[15]

In the so-called beta decay of a neutron (see picture, above), a down quark within the neutron emits a virtual W− boson and is thereby converted into an up quark, converting the neutron into a proton. Because of the energy involved in the process (i.e., the mass difference between the down quark and the up quark), the W− boson can only be converted into an electron and an electron-antineutrino.[16] At the quark level, the process can be represented as:

$$d \rightarrow u + e^- + \bar{\nu}_e$$

14.3.2 Neutral-current interaction

In neutral current interactions, a quark or a lepton (e.g., an electron or a muon) emits or absorbs a neutral Z boson. For example:

$$e^- \rightarrow e^- + Z^0$$

Like the W boson, the Z boson also decays rapidly,[15] for example:

$$Z^0 \rightarrow b + \bar{b}$$

14.4 Electroweak theory

Main article: Electroweak interaction

The Standard Model of particle physics describes the electromagnetic interaction and the weak interaction as two different aspects of a single electroweak interaction, the theory of which was developed around 1968 by Sheldon Glashow, Abdus Salam and Steven Weinberg. They were awarded the 1979 Nobel Prize in Physics for their work.[17] The Higgs mechanism provides an explanation for the presence of three massive gauge bosons (the three carriers of the weak interaction) and the massless photon of the electromagnetic interaction.[18]

According to the electroweak theory, at very high energies, the universe has four massless gauge boson fields similar to the photon and a complex scalar Higgs field doublet. However, at low energies, gauge symmetry is spontaneously broken down to the **U**(1) symmetry of electromagnetism (one of the Higgs fields acquires a vacuum expectation value). This symmetry breaking would produce three massless bosons, but they become integrated by three photon-like fields (through the Higgs mechanism) giving them mass. These three fields become the W+, W− and Z bosons of the weak interaction, while the fourth gauge field, which remains massless, is the photon of electromagnetism.[18]

This theory has made a number of predictions, including a prediction of the masses of the Z and W bosons before their discovery. On 4 July 2012, the CMS and the ATLAS experimental teams at the Large Hadron Collider independently announced that they had confirmed the formal discovery of a previously unknown boson of mass between 125–127 GeV/c^2, whose behaviour so far was "consistent with" a Higgs boson, while adding a cautious note that further data and analysis were needed before positively identifying the new boson as being a Higgs boson of some type. By 14 March 2013, the Higgs boson was tentatively confirmed to exist .[19]

14.5 Violation of symmetry

Left- and right-handed particles: p is the particle's momentum and S is its spin. Note the lack of reflective symmetry between the states.

The laws of nature were long thought to remain the same under mirror reflection, the reversal of one spatial axis. The results of an experiment viewed via a mirror were expected to be identical to the results of a mirror-reflected copy of the experimental apparatus. This so-called law of parity conservation was known to be respected by classical gravitation, electromagnetism and the strong interaction; it was assumed to be a universal law.[20] However, in the mid-1950s Chen Ning Yang and Tsung-Dao Lee suggested that the weak interaction might violate this law. Chien Shiung Wu and collaborators in 1957 discovered that the weak interaction violates parity, earning Yang and Lee the 1957 Nobel Prize in Physics.[21]

Although the weak interaction used to be described by Fermi's theory, the discovery of parity violation and renormalization theory suggested that a new approach was needed. In 1957, Robert Marshak and George Sudarshan and, somewhat later, Richard Feynman and Murray Gell-Mann proposed a **V−A** (vector minus axial vector or left-handed) Lagrangian for weak interactions. In this theory, the weak interaction acts only on left-handed particles (and right-handed antiparticles). Since the mirror reflection of a left-handed particle is right-handed, this explains the maximal violation of parity. Interestingly, the **V−A** theory was developed before the discovery of the Z boson, so it did not include the right-handed fields that enter in the neutral current interaction.

However, this theory allowed a compound symmetry **CP** to be conserved. **CP** combines parity **P** (switching left to right) with charge conjugation **C** (switching particles with antiparticles). Physicists were again surprised when in 1964, James

Cronin and Val Fitch provided clear evidence in kaon decays that CP symmetry could be broken too, winning them the 1980 Nobel Prize in Physics.[22] In 1973, Makoto Kobayashi and Toshihide Maskawa showed that CP violation in the weak interaction required more than two generations of particles,[23] effectively predicting the existence of a then unknown third generation. This discovery earned them half of the 2008 Nobel Prize in Physics.[24] Unlike parity violation, CP violation occurs in only a small number of instances, but remains widely held as an answer to the difference between the amount of matter and antimatter in the universe; it thus forms one of Andrei Sakharov's three conditions for baryogenesis.[25]

14.6 See also

- Weakless Universe – the postulate that weak interactions are not anthropically necessary

- Gravity

- Nuclear force

- Electromagnetism

14.7 References

14.7.1 Citations

[1] Griffiths, David (2009). *Introduction to Elementary Particles*. pp. 59–60. ISBN 978-3-527-40601-2.

[2] "The Nobel Prize in Physics 1979: Press Release". *NobelPrize.org*. Nobel Media. Retrieved 22 March 2011.

[3] Fermi, Enrico (1934). "Versuch einer Theorie der β-Strahlen. I". *Zeitschrif t für Physik A* **88** (3–4): 161–177 doi:10.1007/BF01351864.

[4] Wilson, Fred L. (December 1968). "Fermi's Theory of Beta Decay". *American Journal of Physics* **36** (12): 1150–1160. Bibcode:1968AmJPh..36.1150W. doi:10.1119/1.1974382.

[5] W.-M. Yao *et al.* (Particle Data Group) (2006). "Review of Particle Physics: Quarks" (PDF). *Journal of Physics G* **33**: 1–1232. arXiv:astro-ph/0601168. Bibcode:2006JPhG...33....1Y. doi:10.1088/0954-3899/33/1/001.

[6] Peter Watkins (1986). *Story of the W and Z*. Cambridge: Cambridge University Press. p. 70. ISBN 978-0-521-31875-4.

[7] "Coupling Constants for the Fundamental Forces". *HyperPhysics*. Georgia State University. Retrieved 2 March 2011.

[8] J. Christman (2001). "The Weak Interaction" (PDF). *Physnet*. Michigan State University.

[9] "Electroweak". *The Particle Adventure*. Particle Data Group. Retrieved 3 March 2011.

[10] Walter Greiner; Berndt Müller (2009). *Gauge Theory of Weak Interactions*. Springer. p. 2. ISBN 978-3-540-87842-1.

[11] Cottingham & Greenwood (1986, 2001), p.29

[12] Cottingham & Greenwood (1986, 2001), p.28

[13] Cottingham & Greenwood (1986, 2001), p.30

[14] Baez, John C.; Huerta, John (2009). "The Algebra of Grand Unified Theories". *Bull. Am. Math. Soc.* **0904**: 483–552 Bibcode:2009arXiv0904.1556B. doi:10.1090/s0273-0979-10-01294-2. Retrieved 15 October 2013.

[15] K. Nakamura *et al.* (Particle Data Group) (2010). "Gauge and Higgs Bosons" (PDF). *Journal of Physics G* **37**. doi:10.1088/0954-3899/37/7a/075021.

[16] K. Nakamura *et al.* (Particle Data Group) (2010). "n" (PDF). *Journal of Physics G* **37**: 7. doi:10.1088/0954-3899/37/7a/075021.

[17] "The Nobel Prize in Physics 1979". *NobelPrize.org*. Nobel Media. Retrieved 26 February 2011.

[18] C. Amsler *et al.* (Particle Data Group) (2008). "Review of Particle Physics – Higgs Bosons: Theory and Searches" (PDF). *Physics Letters B* **667**: 1–6. Bibcode:2008PhLB..667....1P. doi:10.1016/j.physletb.2008.07.018.

[19] "New results indicate that new particle is a Higgs boson | CERN". Home.web.cern.ch. Retrieved 20 September 2013.

[20] Charles W. Carey (2006). "Lee, Tsung-Dao". *American scientists*. Facts on File Inc. p. 225. ISBN 9781438108070.

[21] "The Nobel Prize in Physics 1957". *NobelPrize.org*. Nobel Media. Retrieved 26 February 2011.

[22] "The Nobel Prize in Physics 1980". *NobelPrize.org*. Nobel Media. Retrieved 26 February 2011.

[23] M. Kobayashi, T. Maskawa (1973). "CP-Violation in the Renormalizable Theory of Weak Interaction". *Progress of Theoretical Physics* **49** (2): 652–657. Bibcode:1973PThPh..49..652K. doi:10.1143/PTP.49.652.

[24] "The Nobel Prize in Physics 1980". *NobelPrize.org*. Nobel Media. Retrieved 17 March 2011.

[25] Paul Langacker (2001) [1989]. "Cp Violation and Cosmology". In Cecilia Jarlskog. *CP violation*. London, River Edge: World Scientific Publishing Co. p. 552. ISBN 9789971505615.

14.7.2   General readers

- R. Oerter (2006). *The Theory of Almost Everything: The Standard Model, the Unsung Triumph of Modern Physics*. Plume. ISBN 978-0-13-236678-6.

- B.A. Schumm (2004). *Deep Down Things: The Breathtaking Beauty of Particle Physics*. Johns Hopkins University Press. ISBN 0-8018-7971-X.

14.7.3   Texts

- D.A. Bromley (2000). *Gauge Theory of Weak Interactions*. Springer. ISBN 3-540-67672-4.

- G.D. Coughlan, J.E. Dodd, B.M. Gripaios (2006). *The Ideas of Particle Physics: An Introduction for Scientists* (3rd ed.). Cambridge University Press. ISBN 978-0-521-67775-2.

- W. N. Cottingham; D. A. Greenwood (2001) [1986]. *An introduction to nuclear physics* (2nd ed.). Cambridge University Press. p. 30. ISBN 978-0-521-65733-4.

- D.J. Griffiths (1987). *Introduction to Elementary Particles*. John Wiley & Sons. ISBN 0-471-60386-4.

- G.L. Kane (1987). *Modern Elementary Particle Physics*. Perseus Books. ISBN 0-201-11749-5.

- D.H. Perkins (2000). *Introduction to High Energy Physics*. Cambridge University Press. ISBN 0-521-62196-8.

Chapter 15

Semi-empirical mass formula

In nuclear physics, the **semi-empirical mass formula** (**SEMF**) (sometimes also called **Weizsäcker's formula**, or the **Bethe–Weizsäcker formula**, or the **Bethe–Weizsäcker mass formula** to distinguish it from the Bethe–Weizsäcker process) is used to approximate the mass and various other properties of an atomic nucleus from its number of protons and neutrons. As the name suggests, it is based partly on theory and partly on empirical measurements. The theory is based on the **liquid drop model** proposed by George Gamow, which can account for most of the terms in the formula and gives rough estimates for the values of the coefficients. It was first formulated in 1935 by German physicist Carl Friedrich von Weizsäcker, and although refinements have been made to the coefficients over the years, the structure of the formula remains the same today.[1][2]

The SEMF gives a good approximation for atomic masses and several other effects, but does not explain the appearance of magic numbers of protons and neutrons, and the extra binding-energy and measure of stability that are associated with these numbers of nucleons.

15.1   The liquid drop model and its analysis

The liquid drop model in nuclear physics treats the nucleus as a drop of incompressible nuclear fluid. It was first proposed by George Gamow and then developed by Niels Bohr and John Archibald Wheeler. The fluid is made of nucleons (protons and neutrons), which are held together by the strong nuclear force. This is a crude model that does not explain all the properties of the nucleus, but does explain the spherical shape of most nuclei. It also helps to predict the nuclear binding energy and to assess how much is available for consumption.

Mathematical analysis of the theory delivers an equation which attempts to predict the binding energy of a nucleus in terms of the numbers of protons and neutrons it contains. This equation has five terms on its right hand side. These correspond to the cohesive binding of all the nucleons by the strong nuclear force, a surface energy term, the electrostatic mutual repulsion of the protons, an asymmetry term (derivable from the protons and neutrons occupying independent quantum momentum states) and a pairing term (partly derivable from the protons and neutrons occupying independent quantum spin states).

If we consider the sum of the following five types of energies, then the picture of a nucleus as a drop of incompressible liquid roughly accounts for the observed variation of binding energy of the nucleus:

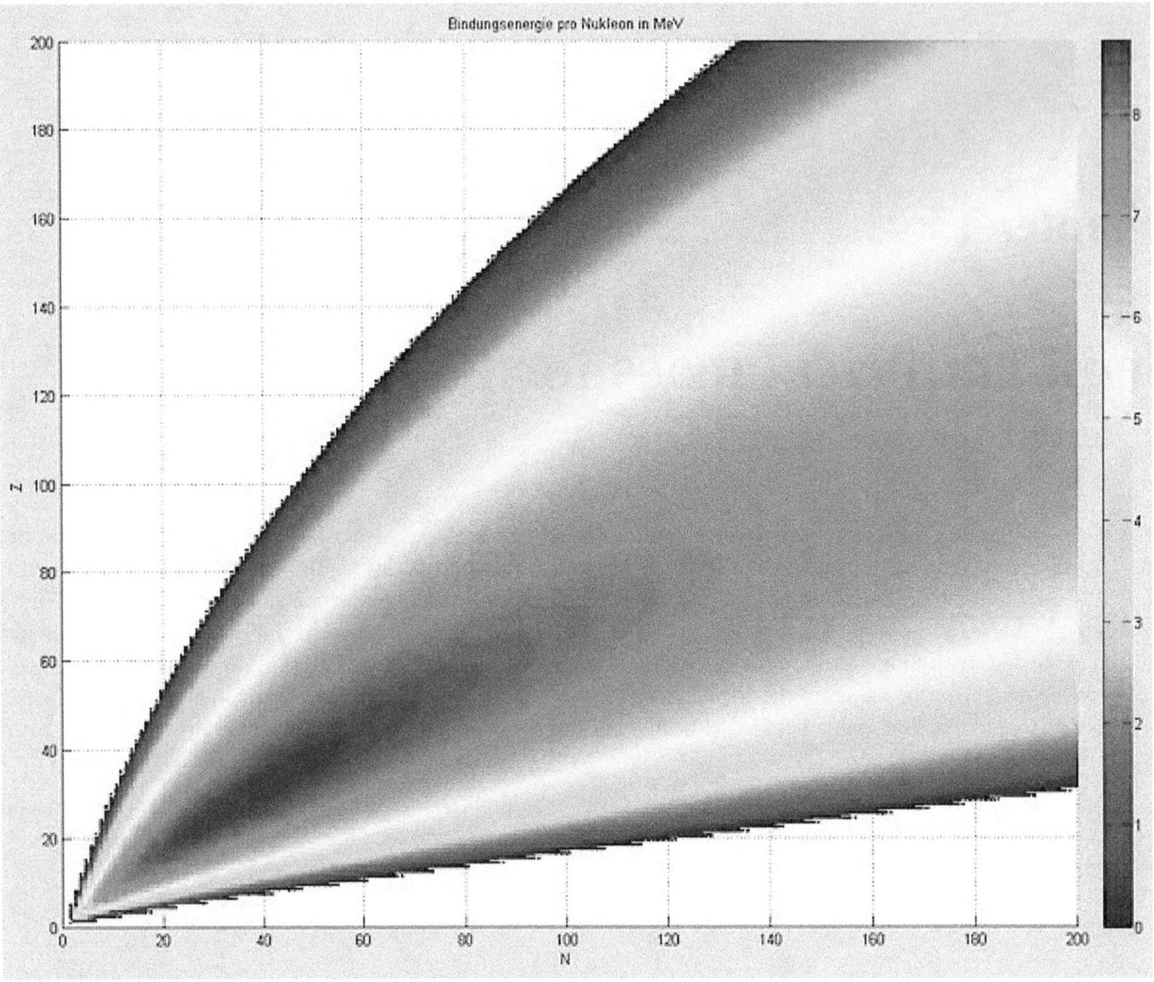

Bindungsenergie pro Nukleon in MeV

A graphical representation of the semi-empirical binding energy formula. The binding energy per nucleon in MeV (highest numbers in dark red, in excess of 8.5 MeV per nucleon) is plotted for various nuclides as a function of Z, the atomic number (on the y-axis), vs. N, the neutron number (on the x-axis). The highest binding energies are seen for Z = 26 (iron).

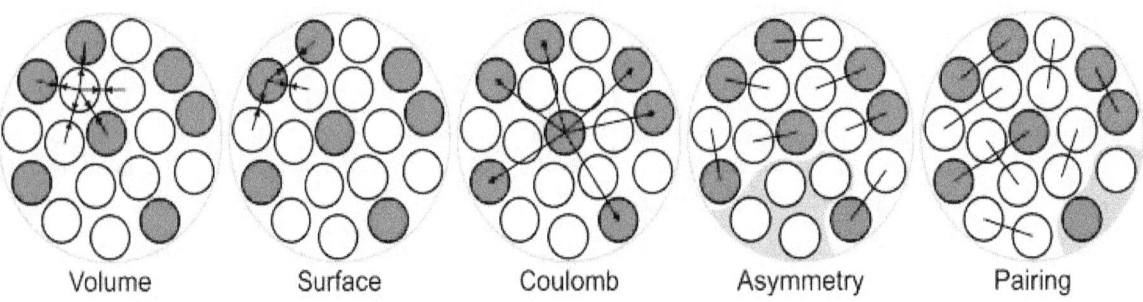

Volume   Surface   Coulomb   Asymmetry   Pairing

Volume energy. When an assembly of nucleons of the same size is packed together into the smallest volume, each interior nucleon has a certain number of other nucleons in contact with it. So, this nuclear energy is proportional to the volume.

Surface energy. A nucleon at the surface of a nucleus interacts with fewer other nucleons than one in the interior of the nucleus and hence its binding energy is less. This surface energy term takes that into account and is therefore negative and is proportional to the surface area.

Coulomb Energy. The electric repulsion between each pair of protons in a nucleus contributes toward decreasing its binding energy.

Asymmetry energy (also called Pauli Energy). An energy associated with the Pauli exclusion principle. Were it not for the Coulomb energy, the most stable form of nuclear matter would have the same number of neutrons as protons, since unequal numbers of neutrons and protons imply filling higher energy levels for one type of particle, while leaving lower energy levels vacant for the other type.

Pairing energy. An energy which is a correction term that arises from the tendency of proton pairs and neutron pairs to occur. An even number of particles is more stable than an odd number.

15.2 The formula

In the following formulae, let A be the total number of nucleons, Z the number of protons, and N the number of neutrons, so that $A=Z+N$.

The mass of an atomic nucleus is given by

$$m = Zm_p + Nm_n - \frac{E_B}{c^2}$$

where m_p and m_n are the rest mass of a proton and a neutron, respectively, and E_B is the binding energy of the nucleus. The semi-empirical mass formula states that the binding energy will take the following form:

$$E_B = a_V A - a_S A^{2/3} - a_C \frac{Z^2}{A^{1/3}} - a_A \frac{(A-2Z)^2}{A} \pm \delta(A, Z) \text{ [3]}$$

Each of the terms in this formula has a theoretical basis, as will be explained below. The coefficients a_V, a_S, a_C, a_A and a coefficient that appears in the formula for $\delta(A, Z)$ are determined empirically.

15.3 Terms

15.3.1 Volume term

The term $a_V A$ is known as the *volume term*. The volume of the nucleus is proportional to A, so this term is proportional to the volume, hence the name.

The basis for this term is the strong nuclear force. The strong force affects both protons and neutrons, and as expected, this term is independent of Z. Because the number of pairs that can be taken from A particles is $\frac{A(A-1)}{2}$, one might expect a term proportional to A^2. However, the strong force has a very limited range, and a given nucleon may only interact strongly with its nearest neighbors and next nearest neighbors. Therefore, the number of pairs of particles that actually interact is roughly proportional to A, giving the volume term its form.

The coefficient a_V is smaller than the binding energy of the nucleons to their neighbours E_b, which is of order of 40 MeV. This is because the larger the number of nucleons in the nucleus, the larger their kinetic energy is, due to the Pauli exclusion principle. If one treats the nucleus as a Fermi ball of A nucleons, with equal numbers of protons and neutrons, then the total kinetic energy is $\frac{3}{5}A\epsilon_F$, with ϵ_F the Fermi energy which is estimated as 28 MeV. Thus the expected value of a_V in this model is $E_b - \frac{3}{5}\epsilon_F \sim 17$ MeV, not far from the measured value.

15.3.2 Surface term

The term $a_S A^{2/3}$ is known as the *surface term*. This term, also based on the strong force, is a correction to the volume term.

The volume term suggests that each nucleon interacts with a constant number of nucleons, independent of A. While this is very nearly true for nucleons deep within the nucleus, those nucleons on the surface of the nucleus have fewer nearest neighbors, justifying this correction. This can also be thought of as a surface tension term, and indeed a similar mechanism creates surface tension in liquids.

If the volume of the nucleus is proportional to A, then the radius should be proportional to $A^{1/3}$ and the surface area to $A^{2/3}$. This explains why the surface term is proportional to $A^{2/3}$. It can also be deduced that a_S should have a similar order of magnitude as a_V.

15.3.3 Coulomb term

The term $a_C \frac{Z(Z-1)}{A^{1/3}}$ or $a_C \frac{Z^2}{A^{1/3}}$ is known as the *Coulomb* or *electrostatic term*.

The basis for this term is the electrostatic repulsion between protons. To a very rough approximation, the nucleus can be considered a sphere of uniform charge density. The potential energy of such a charge distribution can be shown to be

$$E = \frac{3}{5} \left(\frac{1}{4\pi\epsilon_0} \right) \frac{Q^2}{R}$$

where Q is the total charge and R is the radius of the sphere. Identifying Q with Ze, and noting as above that the radius is proportional to $A^{1/3}$, we get close to the form of the Coulomb term. However, because electrostatic repulsion will only exist for more than one proton, Z^2 becomes $Z(Z-1)$. The value of a_C can be approximately calculated using the equation above:

Empirical nuclear radius:

$$R \approx r_0 A^{\frac{1}{3}}.$$

Quantum charge integers:

$$Q = Ze$$

$$Z^2 \approx Z(Z-1).$$

Integration by substitution:

$$E = \frac{3}{5} \left(\frac{1}{4\pi\epsilon_0} \right) \frac{Q^2}{R} = \frac{3}{5} \left(\frac{1}{4\pi\epsilon_0} \right) \frac{(Ze)^2}{(r_0 A^{\frac{1}{3}})} = \frac{3e^2 Z^2}{20\pi\epsilon_0 r_0 A^{\frac{1}{3}}} \approx \frac{3e^2 Z(Z-1)}{20\pi\epsilon_0 r_0 A^{\frac{1}{3}}} = a_C \frac{Z(Z-1)}{A^{1/3}}$$

Potential energy of charge distribution:

$$E = \frac{3e^2 Z(Z-1)}{20\pi\epsilon_0 r_0 A^{\frac{1}{3}}}$$

Electrostatic Coulomb constant:

$$a_C = \frac{3e^2}{20\pi\epsilon_0 r_0}$$

The value of a_C using the fine structure constant:

$$a_C = \frac{3}{5} \left(\frac{\hbar c \alpha}{r_0} \right) = \frac{3}{5} \left(\frac{R_P}{r_0} \right) \alpha m_p c^2$$

where α is the fine structure constant and $r_0 A^{1/3}$ is the radius of a nucleus, giving r_0 to be approximately 1.25 femtometers. R_P is the proton Compton radius and m_p the proton mass. This gives a_C an approximate theoretical value of 0.691 MeV, not far from the measured value.

$$a_C = 0.691 \text{ MeV}$$

15.3.4   Asymmetry term

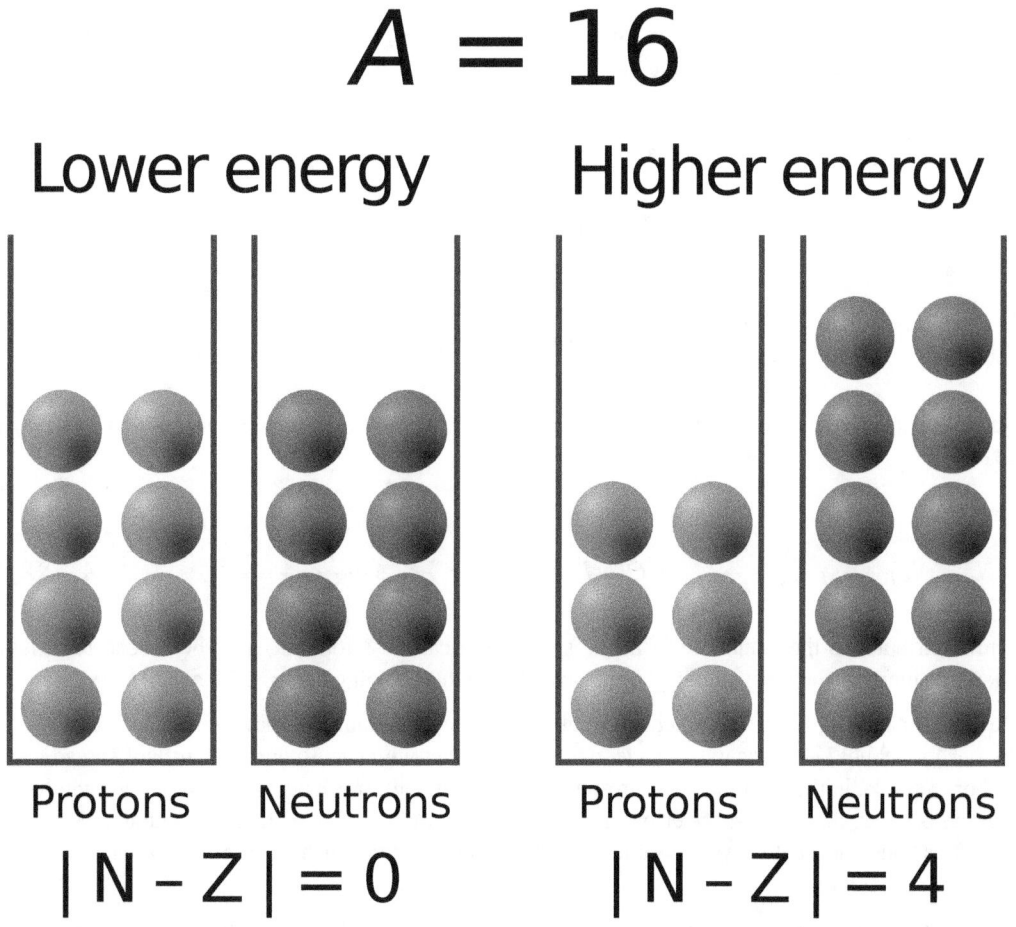

Illustration of basis for asymmetric term

The term $a_A \frac{(A-2Z)^2}{A}$ or $4a_A \frac{((A/2)-Z)^2}{A}$ is known as the *asymmetry term*. Note that as $A = N + Z$, the parenthesized expression can be rewritten as $(N - Z)$. The form $(A - 2Z)$ is used to keep the dependence on A explicit, as will be important for a number of uses of the formula.

The theoretical justification for this term is more complex. The Pauli exclusion principle states that no two fermions can occupy exactly the same quantum state in an atom. At a given energy level, there are only finitely many quantum states available for particles. What this means in the nucleus is that as more particles are "added", these particles must occupy higher energy levels, increasing the total energy of the nucleus (and decreasing the binding energy). Note that this effect is not based on any of the fundamental forces (gravitational, electromagnetic, etc.), only the Pauli exclusion principle.

Protons and neutrons, being distinct types of particles, occupy different quantum states. One can think of two different "pools" of states, one for protons and one for neutrons. Now, for example, if there are significantly more neutrons than protons in a nucleus, some of the neutrons will be higher in energy than the available states in the proton pool. If we could move some particles from the neutron pool to the proton pool, in other words change some neutrons into protons, we would significantly decrease the energy. The imbalance between the number of protons and neutrons causes the energy to be higher than it needs to be, *for a given number of nucleons*. This is the basis for the asymmetry term.

The actual form of the asymmetry term can again be derived by modelling the nucleus as a Fermi ball of protons and neutrons. Its total kinetic energy is

$$E_k = \frac{3}{5}(N_p \epsilon_{Fp} + N_n \epsilon_{Fn})$$

where N_p, N_n are the numbers of protons and neutrons and ϵ_{Fp}, ϵ_{Fn} are their Fermi energies. Since the latter are proportional to $N_p^{2/3}$ and $N_n^{2/3}$, respectively, one gets

$$E_k = C(N_p^{5/3} + N_n^{5/3}) \text{ for some constant } C.$$

The leading expansion in the difference $N_n - N_p$ is then

$$E_k = \frac{C}{2^{2/3}}\left((N_p + N_n)^{5/3} + \frac{5}{9}\frac{(N_n - N_p)^2}{(N_p + N_n)^{1/3}}\right) + O((N_n - N_p)^2).$$

At the zeroth order expansion the kinetic energy is just the Fermi energy $\epsilon_F \equiv \epsilon_{Fp} = \epsilon_{Fn}$ multiplied by $\frac{3}{5}(N_p + N_n)^{2/3}$. Thus we get

$$E_k = \frac{3}{5}\epsilon_F(N_p + N_n) + \frac{1}{3}\epsilon_F\frac{(N_n - N_p)^2}{(N_p + N_n)} + O((N_n - N_p)^4) = \frac{3}{5}\epsilon_F A + \frac{1}{3}\epsilon_F\frac{(A - 2Z)^2}{A} + O((A - 2Z)^4).$$

The first term contributes to the volume term in the semi-empirical mass formula, and the second term is minus the asymmetry term (remember the kinetic energy contributes to the total binding energy with a *negative* sign).

ϵ_F is 38 MeV, so calculating a_A from the equation above, we get only half the measured value. The discrepancy is explained by our model not being accurate: nucleons in fact interact with each other, and are not spread evenly across the nucleus. For example, in the shell model, a proton and a neutron with overlapping wavefunctions will have a greater strong interaction between them and stronger binding energy. This makes it energetically favourable (i.e. having lower energy) for protons and neutrons to have the same quantum numbers (other than isospin), and thus increase the energy cost of asymmetry between them.

One can also understand the asymmetry term intuitively, as follows. It should be dependent on the absolute difference $|N - Z|$, and the form $(A - 2Z)^2$ is simple and differentiable, which is important for certain applications of the formula. In addition, small differences between Z and N do not have a high energy cost. The A in the denominator reflects the fact that a given difference $|N - Z|$ is less significant for larger values of A.

15.3.5 Pairing term

The term $\delta(A, Z)$ is known as the *pairing term* (possibly also known as the pairwise interaction). This term captures the effect of spin-coupling. It is given by:[4]

$$\delta(A, Z) = \begin{cases} +\delta_0 & Z, N \text{ even } (A \text{ even}) \\ 0 & A \text{ odd} \\ -\delta_0 & Z, N \text{ odd } (A \text{ even}) \end{cases}$$

where

$$\delta_0 = \frac{a_P}{A^{1/2}}.$$

Due to Pauli exclusion principle the nucleus would have a lower energy if the number of protons with spin up were equal to the number of protons with spin down. This is also true for neutrons. Only if both Z and N are even can both protons and neutrons have equal numbers of spin up and spin down particles. This is a similar effect to the asymmetry term.

The factor $A^{-1/2}$ is not easily explained theoretically. The Fermi ball calculation we have used above, based on the liquid drop model but neglecting interactions, will give an A^{-1} dependence, as in the asymmetry term. This means that the actual effect for large nuclei will be larger than expected by that model. This should be explained by the interactions between nucleons; For example, in the shell model, two protons with the same quantum numbers (other than spin) will have completely overlapping wavefunctions and will thus have greater strong interaction between them and stronger binding energy. This makes it energetically favourable (i.e. having lower energy) for protons to pair in pairs of opposite spin. The same is true for neutrons.

15.4 Calculating the coefficients

The coefficients are calculated by fitting to experimentally measured masses of nuclei. Their values can vary depending on how they are fitted to the data. Several examples are as shown below, with units of megaelectronvolts.

The semi-empirical mass formula provides a good fit to heavier nuclei, and a poor fit to very light nuclei, especially ^{4}He. This is because the formula does not consider the internal shell structure of the nucleus. For light nuclei, it is usually better to use a model that takes this structure into account.

15.5 Examples for consequences of the formula

By maximizing $B(A,Z)$ with respect to Z, we find the best neutrons to protons ratio N/Z for a given atomic weight A.[6] We get

$$N/Z \approx 1 + \frac{a_C}{2a_A} A^{2/3}.$$

This is roughly 1 for light nuclei, but for heavy nuclei the ratio grows in good agreement with nature.

By substituting the above value of Z back into B one obtains the binding energy as a function of the atomic weight, $B(A)$. Maximizing $B(A)/A$ with respect to A gives the nucleus which is most strongly bound, i.e. most stable. The value we get is $A = 63$ (copper), close to the measured values of $A = 62$ (nickel) and $A = 58$ (iron).

15.6 Notes

[1] von Weizsäcker, C. F. (1935). "Zur Theorie der Kernmassen". *Zeitschrift für Physik* (in German) **96** (7–8): 431–458. Bibcode:1935ZPhy...96..431W. doi:10.1007/BF01337700.

[2] Bailey, D. "Semi-empirical Nuclear Mass Formula". *PHY357: Strings & Binding Energy*. University of Toronto. Retrieved 2011-03-31.

[3] Oregon State University. "Nuclear Masses and Binding Energy Lesson 3" (PDF). Archived from the original (PDF) on 30 September 2015. Retrieved 30 September 2015.

[4] Krane, K. (1988). *Introductory Nuclear Physics*. John Wiley & Sons. p. 68. ISBN 0-471-85914-1.

[5] Wapstra, A. H. (1958). "Atomic Masses of Nuclides". *External Properties of Atomic Nuclei*. Springer. pp. 1–37. doi:10.1007/978-3-642-45901-6_1. ISBN 978-3-642-45902-3.

[6] Rohlf, J. W. (1994). *Modern Physics from α to Z^0*. John Wiley & Sons. ISBN 978-0471572701.

15.7   References

- Freedman, R.; Young, H. (2004). *Sears and Zemanskey's University Physics with Modern Physics* (11th ed.). pp. 1633–1634. ISBN 0-8053-8768-4.

- Liverhant, S. E. (1960). *Elementary Introduction to Nuclear Reactor Physics*. John Wiley & Sons. pp. 58–62. LCCN 60011725.

- Choppin, G.; Liljenzin, J.-O.; Rydberg, J. (2002). "Nuclear Mass and Stability" (PDF). *Radiochemistry and Nuclear Chemistry* (3rd ed.). Butterworth-Heinemann. pp. 41–57. ISBN 978-0-7506-7463-8.

15.8   External links

- Nuclear liquid drop model

- The semi-empirical mass formula

- Liquid drop model in the hyperphysics online reference at Georgia State University.

- Liquid drop model with parameter fit from *First Observations of Excited States in the Neutron Deficient Nuclei* $^{160,161}W$ *and* ^{159}Ta, Alex Keenan, PhD thesis, University of Liverpool, 1999 (HTML version).

15.9 Text and image sources, contributors, and licenses

15.9.1 Text

- **Nuclear binding energy** *Source:* https://en.wikipedia.org/wiki/Nuclear_binding_energy?oldid=684035729 *Contributors:* Darkwind, Glenn, Rhombus, Dratman, Miranche, Eric Kvaalen, Woohookitty, Tabletop, Bgwhite, Wavelength, Jimp, Limulus, SCZenz, Dhollm, Moe Epsilon, SamuelRiv, Boivie, CWenger, FyzixFighter, Eigenlambda, Gilliam, Sbharris, Ascentury, Noleander, Shorespirit, Headbomb, Nick Number, Morngnstar, Philg88, R'n'B, Trusilver, DAID, Fylwind, Dirkbb, SoopahMan, Kbrose, Flyer22 Reborn, Dtvjho, Hamiltondaniel, Dolphin51, Bschaeffer~enwiki, Panatomic, Niceguyedc, Addbot, Mortense, LarryJeff, KitemanSA, AnomieBOT, Materialscientist, Citation bot, Arthur-Bot, LilHelpa, Aa77zz, Lithopsian, NoRad, A. di M., FrescoBot, Steve Quinn, Martinvl, Achim1999, Artem Korzhimanov, Double sharp, Cmdrrnvr, K6ka, Hhhippo, Wayne Slam, Petrb, ClueBot NG, Muon, WikiPuppies, Reify-tech, Iste Praetor, Titodutta, Calabe1992, DBigXray, Larrew, Vanischenu, Khazar2, Webclient101, Joeinwiki, Spyglasses, Tjraptis20, Mahusha, Yamikaze Akatsuka, DSCrowned, Hrishikesh M Kulkarni, TMA-1701 and Anonymous: 79

- **Atomic nucleus** *Source:* https://en.wikipedia.org/wiki/Atomic_nucleus?oldid=685784371 *Contributors:* Kpjas, Andre Engels, XJaM, Merphant, Graft, Stevertigo, Patrick, JohnOwens, Michael Hardy, Tim Starling, Nixdorf, Bcrowell, Mcarling, Looxix~enwiki, Ellywa, Ahoerstemeier, Александъ, Andres, Smack, Rednblu, The Anomebot, Topbanana, Cvaneg, Palefire, Gentgeen, Robbot, Sander123, Arkuat, Merovingian, Meelar, Wikibot, GarnetRChaney, Anthony, Giftlite, Graeme Bartlett, Christopher Parham, Awolf002, Mikez, Fastfission, Xerxes314, Bensaccount, Yath, Antandrus, Beland, OverlordQ, Icairns, Joyous!, JohnArmagh, Deglr6328, Trevor MacInnis, Mike Rosoft, Chris Howard, Perey, Guanabot, Igorivanov~enwiki, Vsmith, Gianluigi, Paul August, El C, Koenige, Madhu p, Shanes, Bookofjude, Mickeymousechen~enwiki, Foobaz, Obradovic Goran, Nsaa, Jumbuck, Disneyfreak96, Alansohn, Gintautasm, Neonumbers, Riana, Kfitzgib, Bart133, Velella, EvenT, H2g2bob, DV8 2XL, Mattbrundage, Kay Dekker, Flying fish, Stemonitis, Firsfron, AndriyK, Mandarax, Graham87, FreplySpang, Martinevos~enwiki, Rjwilmsi, Syndicate, Strait, Tangotango, Ttwaring, Lcolson, Nivix, OSt~enwiki, Maustrauser, Fresheneesz, Srleffler, Chobot, DVdm, Ahpook, Gwernol, Roboto de Ajvol, YurikBot, RobotE, Bambaiah, JWB, RussBot, Sillybilly, SpuriousQ, Hellbus, Okedem, Gaius Cornelius, CambridgeBayWeather, Wiki alf, FFLaguna, Dbfirs, Bota47, Supspirit, Tachyon01, Jess Riedel, Zzuuzz, Tsunaminoai, CWenger, Anclation~enwiki, Curpsbot-unicodify, Kungfuadam, RG2, AssistantX, SmackBot, FocalPoint, Incnis Mrsi, Bomac, Wogsland, Jrockley, Edgar181, Gilliam, Betacommand, Schmiteye, Chris the speller, Keegan, SchfiftyThree, DHN-bot~enwiki, Sbharris, Klacquement, Blake-, Itchjones, Dreadstar, DMacks, Daniel.Cardenas, Mion, Sadi Carnot, Bdushaw, Pilotguy, Kukini, Clicketyclack, Serein (renamed because of SUL), Kuru, Olin, Zarniwoot, NongBot~enwiki, Ekrub-ntyh, Funnybunny, MTSbot~enwiki, Cbuckley, Caiaffa, Dan Gluck, Kelvinaom, Joseph Solis in Australia, Lottamiata, Tubezone, Tawkerbot2, Flubeca, Sxim, Ale jrb, Scohoust, Shernren, Rowellcf, Engelmann15~enwiki, Kanags, MC10, Gogo Dodo, My Flatley, Christian75, DumbBOT, Bieeanda, Thijs!bot, Headbomb, Marek69, Tellyaddict, Wildthing61476, CTZMSC3, AntiVandalBot, Widefox, Quintote, TimVickers, Ilovescience, Gdo01, Gmarsden, JAnDbot, D99figge, Leuko, MER-C, Gfsheppard, .anacondabot, Bongwarrior, VoABot II, Rajb245, JamesBWatson, CalamusFortis, Mother.earth, Dirac66, Kopovoi, Dravick, Vssun, JaGa, Philg88, Hbent, Goodynotion, Akhil999in, MartinBot, Schmloof, Tiger-Smith, Roastytoast, J.delanoy, Classicalclarinet, Trusilver, Fnordius, Maurice Carbonaro, WarthogDemon, Lol nubs, Fylwind, Eshywiki, Tygrrr, TraceyR, Dylan bossart, VolkovBot, Indubitably, Stefan Kruithof, The Original Wildbear, Rei-bot, Anonymous Dissident, Piperh, Anna Lincoln, Shonenknifefan1, Venny85, Synthebot, Antixt, Enviroboy, PGWG, EmxBot, NEIL4737, SieBot, Sonicology, Scarian, Gerakibot, Caltas, Tiptoety, Onesspite, BenoniBot~enwiki, Afernand74, Cyfal, ClueBot, The Thing That Should Not Be, Jan1nad, Jekatz, Drmies, Boing! said Zebedee, Lainy8, CounterVandalismBot, Excirial, Alexbot, LordFoppington, SpikeToronto, Rhododendrites, Brews ohare, PhySusie, Tinymonty, Kakofonous, Versus22, InternetMeme, XLinkBot, Rangel lucy, Mr beeg lol, Addbot, Peyton.gaumer, Praseprase, Some jerk on the Internet, Leszek Jańczuk, Cst17, LinkFA-Bot, 5 albert square, CuteHappyBrute, Numbo3-bot, Tide rolls, Jan eissfeldt, Luckas-bot, Yobot, WikiDan61, 2D, TaBOT-zerem, Anypodetos, IW.HG, محبوب عالم, Podlif, Synchronism, AnomieBOT, Kingpin13, Flewis, Bluerasberry, Materialscientist, The High Fin Sperm Whale, Citation bot, Carlsotr, Quebec99, Xqbot, SouthH, Sionus, Capricorn42, Jeffrey Mall, RibotBOT, BSTemple, Shrikeangel, A. di M., Ironboy11, Kobewetnaps, Citation bot 1, Rylee118, Pinethicket, Tinton5, Games 101 wiki, RedBot, Minivip, White Shadows, FoxBot, Lionslayer, TheBFG, Dinamik-bot, Reaper Eternal, Specs112, Onel5969, RjwilmsiBot, Killaoftoast, DASHBot, EmausBot, Ornithikos, Mariov0288, The Pineapple, Cedar T., Thecheesykid, Hhhippo, JSquish, StringTheory11, Arbnos, Wayne Slam, Ocaasi, Kim cupcake, Sunshine4921, Mjbmrbot, Petrb, ClueBot NG, Magic Wizard, MelbourneStar, Fukushimayoshiho, Schunck, IOPhysics, Widr, Reify-tech, Helpful Pixie Bot, Downtowntrollin, Bibcode Bot, Lowercase sigmabot, AvocatoBot, Flying hippo705, Jordanf7, Gimp 11, Zedshort, Uopchem2511, Jburk711, DarafshBot, Akbask, JYBot, BrightStarSky, Dexbot, TwoTwoHello, Lugia2453, Bulba2036, Marekich, SassyLilNugget, Wlad2000, Mark viking, Siddhantsingh123, DennouNeko, DavidLeighEllis, Sladeb, Spyglasses, Y-S.Ko, KasparBot, Aless Val M and Anonymous: 442

- **Nucleon** *Source:* https://en.wikipedia.org/wiki/Nucleon?oldid=681131517 *Contributors:* Kpjas, Andre Engels, Peterlin~enwiki, Pichai Asokan, Twilsonb, Fruge~enwiki, Ellywa, Evercat, Jusjih, BenRG, Gromlakh, Donarreiskoffer, Robbot, Waelder, Merovingian, Ojigiri~enwiki, Kagredon, Herbee, Xerxes314, Karol Langner, Icairns, D6, Rich Farmbrough, Igorivanov~enwiki, Vsmith, El C, Whosyourjudas, La goutte de pluie, Jumbuck, Msh210, Keenan Pepper, Kwikwag, Kusma, Gene Nygaard, Linas, Palica, Marudubshinki, Emerson7, Kbdank71, JIP, Miq, Rjwilmsi, Strait, Ttwaring, FlaBot, Eubot, OSt~enwiki, Srleffler, Wrightbus, Guliolopez, YurikBot, Dirigible, Bambaiah, Chuck Carroll, Salsb, Welsh, Długosz, Daniel Mietchen, CPColin, Bota47, Dna-webmaster, Katieh5584, GrinBot~enwiki, Sbyrnes321, SmackBot, Melchoir, SashatoBot, MTSbot~enwiki, Peyre, Heartofgoldfish, WeggeBot, Headbomb, John254, Wiki fanatic, Aadal, KrakatoaKatie, JAnDbot, Gcm, BenB4, DanPMK, Nyttend, Mother.earth, TheEgyptian, Leyo, Mike.lifeguard, TraceyR, VolkovBot, FantasticAsh, JhsBot, Don4of4, Antixt, AlleborgoBot, SieBot, Citizen, OKBot, Anchor Link Bot, Li4kata, Florentyna, Djr32, Keithbowden, SoxBot, Qwfp, SkyLined, Addbot, Hakan Kayı, Chamal N, Lightbot, Verazzano, Luckas-bot, Yobot, Tannkrem, The High Fin Sperm Whale, Citation bot, Bci2, ArthurBot, Xqbot, Melmann, DJWolfy, 嘉嘉, FrescoBot, LucienBOT, Citation bot 1, Merongb10, Thinking of England, Trappist the monk, PNG, Jonkerz, Begoon, RjwilmsiBot, EmausBot, WikitanvirBot, Dcirovic, StringTheory11, Ethaniel, Quondum, RolteVolte, Efiiamagus, ClueBot NG, Gilderien, Satellizer, Bibcode Bot, 4Jays1034, ChrisGualtieri, TwoTwoHello, Marekich, Trompedo, Rayhartung, Monkbot, Internucleon, KasparBot, Snackbag and Anonymous: 69

- **Nuclear force** *Source:* https://en.wikipedia.org/wiki/Nuclear_force?oldid=686682891 *Contributors:* Agtx, Jakohn, Owain, GreatWhiteNortherner, Xerxes314, D6, Hidaspal, ESkog, Rbj, Nk, Riana, Kocio, Kusma, Forteblast, Linas, Tabletop, Scroipt, RE, Yamamoto Ichiro, Eubot, Nihiltres, Chobot, Cshay, WriterHound, YurikBot, Wavelength, Borgx, Mushin, Bambaiah, Hairy Dude, Phmer, Jimp, Mconst, SCZenz,

Jrockley, Dauto, Chris the speller, Jjalexand, Complexica, Sbharris, Colonies Chris, Richard001, TTE, Spiritia, Titus III, FrozenMan, Newone, Happy-melon, Conrad.Irwin, Rowellcf, Chrisahn, Cydebot, Danny Bierek, Mtpaley, Wannabe Runny, Irigi, Headbomb, Niduzzi, KP Botany, Tlabshier, Hanzoro5, Dougher, Steelpillow, JAnDbot, Supertheman, Roleplayer, Magioladitis, Kopovoi, Vssun, Hoverfish, Khalid Mahmood, Pan Dan, Robin S, Vortimer, Sujaybhu, Natsirtguy, Peter Chastain, Cpiral, Tygrrr, Treisijs, Alpvax, VolkovBot, JohnBlackburne, TXiKiBoT, Marskuzz, Muro de Aguas, Qxz, Sintaku, SieBot, Gerakibot, Escape Artist Swyer, Proton666, ObfuscatePenguin, ClueBot, GorillaWarfare, Jackey0105, DragonBot, Jefflayman, Nownownow, Cenarium, Razorflame, Zahnrad, Silvercromagnon, InternetMeme, Rreagan007, WikHead, Drogs630, Addbot, Guoguo12, Omega Squad, ThisIsMyWikipediaName, Seratna, CarsracBot, Purple Emu, CosmiCarl, AgadaUrbanit, Tide rolls, Lightbot, Luckas-bot, Timeroot, Donthedev, Rifter0x0000, Umnum, AnomieBOT, VanishedUser sdu9aya9fasdsopa, Orange Knight of Passion, Piano non troppo, Citation bot, Obersachsebot, Xqbot, DSisyphBot, Barelistido, Almabot, GrouchoBot, RibotBOT, SassoBot, Mnmngb, CES1596, Gummer85, Citation bot 1, Boulaur, RedBot, Jauhienij, Surf5270, ElPeste, Slon02, EmausBot, John of Reading, Mnkyman, JSquish, Cogiati, Bamyers99, Rexprimoris, Donner60, ChuispastonBot, ClueBot NG, Jj1236, Helpful Pixie Bot, Bolatbek, ElphiBot, J.wong.wiki, Glevum, Zedtwitz, Zedshort, Nishantkumar19, Kisokj, YFdyh-bot, Andyhowlett, Reatlas, CsDix, EvergreenFir, Aurelianjh, Jwratner1, Diggerh, Kshitizarora2993, Tetra quark, KasparBot and Anonymous: 171

- **Mass–energy equivalence** *Source:* https://en.wikipedia.org/wiki/Mass%E2%80%93energy_equivalence?oldid=686867731 *Contributors:* Tarquin, Stevertigo, Edward, Ubiquity, Patrick, Michael Hardy, SebastianHelm, Ahoerstemeier, Darkwind, Julesd, Charles Matthews, Stone, Kbk, Andrewman327, Evgeni Sergeev, Doradus, Tpbradbury, Dragons flight, McKay, AnonMoos, Eugene van der Pijll, BenRG, Twang, Robbot, Owain, ZimZalaBim, Gandalf61, Postdlf, Btljs, Tobias Bergemann, Enochlau, Jimpaz, Giftlite, Muzzle, C2357, Kpalion, Jackol, ConradPino, Antandrus, OverlordQ, Thorwald, Mike Rosoft, Discospinster, Rich Farmbrough, FT2, Pjacobi, Vsmith, Ponder, SpookyMulder, Chadlupkes, Bender235, ESkog, ZeroOne, JustinWick, Ben Webber, El C, Carlon, Lycurgus, Haxwell, Causa sui, Bobo192, Longhair, Savvo, Jojit fb, Kjkolb, Officiallyover, Yalbik, Landroni, Alansohn, Gary, Anthony Appleyard, Tek022, Riana, Ashley Pomeroy, Scarecroe, Mysdaao, Stillnotelf, Bart133, Melaen, Clubmarx, Danhash, Count Iblis, RainbowOfLight, Dirac1933, Mikeo, H2g2bob, Bsadowski1, GabrielF, Gene Nygaard, Ron Ritzman, Zntrip, Mindmatrix, StradivariusTV, Kzollman, Robert K S, ^demon, WadeSimMiser, Qwertyman~enwiki, GregorB, Zzyzx11, Mandarax, Jclemens, Rjwilmsi, Koavf, Jake Wartenberg, Vary, Bob A, SMC, Bfigura, Yamamoto Ichiro, FayssalF, Wikiliki, Eubot, RobertG, Gurch, Fresheneesz, Chobot, DVdm, Bgwhite, McGinnis, Vyroglyph, Wavelength, Hairy Dude, Huw Powell, Sarranduin, Zafiroblue05, Ericorbit, Bhny, Gaius Cornelius, CambridgeBayWeather, Bovineone, Thane, NawlinWiki, Arichnad, JoeBruno, RazorICE, Dureo, CecilWard, RUL3R, Tony1, Dbfirs, BOT-Superzerocool, Woscafrench, Poochy, WAS 4.250, Enormousdude, 21655, 2over0, Zzuuzz, Dspradau, Staxringold, RG2, NeilN, Finell, Sardanaphalus, SmackBot, Unschool, Ashill, InverseHypercube, Hydrogen Iodide, McGeddon, Shoy, Frasor, ASarnat, Canthusus, Gilliam, Skizzik, GwydionM, Chris the speller, Bluebot, JCSantos, Rakela, Oli Filth, Miquonranger03, Silly rabbit, Sbharris, Colonies Chris, Darth Panda, Derekt75, Can't sleep, clown will eat me, Timothy Clemans, Onorem, Rrburke, Addshore, Stevenmitchell, Jmnbatista, Wen D House, Cybercobra, Nakon, Andrew c, DMacks, Acdx, Aftertheend, Ohconfucius, Angela26, SashatoBot, Lambiam, ArglebargleIV, Kuru, Mgiganteus1, Zarniwoot, Aleenf1, Ben Moore, 16@r, Smith609, Dr Greg, Slakr, MarcAurel, Dicklyon, Waggers, Spiel496, Cbuckley, Dan Gluck, Iridescent, Joseph Solis in Australia, R~enwiki, Blehfu, Courcelles, Achoo5000, JForget, Sakurambo, CmdrObot, Ninetyone, Editorius, Green caterpillar, Gdbiederman, Cydebot, Kanags, MC10, Subwoofer, Gogo Dodo, Yuzz, Tkynerd, Edgerck, Capedia, Christian75, DumbBOT, DarkLink, JSal, Malleus Fatuorum, Epbr123, Barticus88, Biruitorul, Pajz, Headbomb, Marek69, Davidlawrence, John254, NorwegianBlue, James086, X201, Davidhorman, Thljcl, D.H, Klausness, Ellid021, Mentifisto, Majorly, Seaphoto, Orionus, Elmoosecapitan, Smittycity42, Edokter, TimVickers, Joe Schmedley, Naveen Sankar, Farosdaughter, Spencer, Spartaz, Gökhan, Res2216firestar, DOSGuy, JAnDbot, Aheyfromhome, MER-C, CosineKitty, Txomin, Thenub314, Andonic, Hut 8.5, Cvkline, Casmith 789, Magioladitis, Puellanivis, Pedro, Bongwarrior, VoABot II, Sekfetenmet, Sikory, Rimibchatterjee, Jatkins, Twsx, DAGwyn, Tristan Horn, Zanibas, Fabrictramp, Catgut, Indon, Crunchy Numbers, JJ Harrison, 28421u2232nfenfcenc, DerHexer, JNF Tveit, An Sealgair, G.A.S, MartinBot, Ariel., Lelandrb, Sm8900, Keith D, R'n'B, Dgcaste, CommonsDelinker, Onixz100, Jaredroussel, J.delanoy, Pharaoh of the Wizards, GoatGuy, C. Trifle, Maurice Carbonaro, AngleWyrm, DD2K, Lantonov, Ajmint, Dispenser, Nsigniacorp, Uranium grenade, Greater mind, NewEnglandYankee, Rominandreu, Zojj, DavidCBryant, Remember the dot, JohnOdhner, Barraki, WillPF, Scott Illini, JavierMC, Useight, Halmstad, SoCalSuperEagle, Funandtrvl, X!, VolkovBot, Jeff G., AlnoktaBOT, TXiKiBoT, Oshwah, Drhtl, NPrice, Mieszko the first, Sintaku, Graham Wellington, Martin451, Solo1234, Jackfork, UnitedStatesian, Madhero88, Enigmaman, Jacobandrew2012, Antixt, Enviroboy, Insanity Incarnate, Dufo, Why Not A Duck, HiDrNick, Logan, Tvinh, Vegardo, Xgllo, SieBot, Ivan Štambuk, Madman, Timb66, Euryalus, Ziolkovsky, VVVBot, Arpose, Triwbe, Jason Patton, RatnimSnave, AvengedSevenfold00, Keilana, Happysailor, Likebox, Flyer22 Reborn, Qst, Le Pied-bot~enwiki, Aly89, Oxymoron83, Robertfreemanfund, Harry~enwiki, Beast of traal, GaryColemanFan, Steven Crossin, Lightmouse, Wackedout, Techman224, OKBot, Onopearls, Coldcreation, Sapoty, Susan118, Ascidian, Dolphin51, Denisarona, Bschaeffer~enwiki, Loren.wilton, Elassint, ClueBot, Orangedolphin, Chalmersss, Binksternet, Jackollie, The Thing That Should Not Be, Marek zielinski, NunchuckJack, VQuakr, Mild Bill Hiccup, MathGeek123, Polyamorph, Chwilliamson, Boing! said Zebedee, Yamakiri, Unitfreak, Adamslattery54, EricTN, Estevoaei, Blanchardb, NakedEye71, Agge1000, Dazzafar, Oxnard27, Hi777, Paulcmnt, Excirial, -Midorihana-, Jusdafax, Leonard^Bloom, Sgroupace, Brews ohare, NuclearWarfare, LongLiveRock72, Animalality, Lumpy27, Cartledge555, Noosentaal, Dekisugi, GluonBall, Mikaey, Melkijad, La Pianista, Fernandinho1000, Thingg, Zeekyb00gydoog wrongwhom7, Versus22, LieAfterLie, Djk3, AC+79 3888, Vanished user uih38riiw4hjlsd, Ashish16328, Bgeelhoed, BendersGame, XLinkBot, Hotcrocodile, Jovianeye, Rror, Dthomsen8, Avoided, WikHead, Thewho65, Alexius08, Kodster, Tayste, Xp54321, Cxz111, JPINFV, Willking1979, DOI bot, Tcncv, Captain Ref Desk, Friginator, Valejo10005, DougsTech, Ymath, Ronhjones, CanadianLinuxUser, Jaeger123, NjardarBot, TomTyldesley, OliverTwisted, Yujie1, Download, Brett37, LaaknorBot, The yoster1, Epik phale, Glane23, Favonian, Playerace5, Jaf24, LinkFA-Bot, Jasper Deng, 5 albert square, Amoskowitz, Agathor222, Squandermania, Botbotkins, Slushieeater, Dayewalker, Bigzteve, Tide rolls, Lightbot, Cesiumfrog, Teles, Gail, Stevek 85, CountryBot, StarLight, Angrysockhop, Luckas-bot, Yobot, Fraggle81, JakeH07, Paepaok, Spysdudeqazwsx, THEN WHO WAS PHONE?, Gunnar Hendrich, Myktk, Prometheusindisguise, Backslash Forwardslash, AnomieBOT, VanishedUser sdu9aya9fasdsopa, FatAndSassy4, Joule36e5, Jim1138, IRP, Piano non troppo, A09fa2, Kingpin13, ConsciousUniverse, Fredd2374, Visiting Guest, Materialscientist, Are you ready for IPv6?, Citation bot, Tonytony9, ArthurBot, Monkeybutts93, Pyrodude431, Gravityforce, Rittigai, Xqbot, Aman2007007, Ertebatbama, Drilnoth, Acebulf, Dandelion Jane, Pvkeller, DSisyphBot, Nitrxgen, Jeffwang, Runaway9995, Laa Careon, Gap9551, Srich32977, Coretheapple, Domjm, Från-KlarhetTillKlarhet, UNCLE ROCKA, Pmlineditor, Loosah, Omnipaedista, Anhydrobiosis, Elbigger1, RibotBOT, Amaury, Ace111, Paul maul123, Aaron35510, IShadowed, Mikesoc28, DanielDisastrous, N419BH, Shadowjams, A. di M., Databytecorp, CES1596, Hegaldi, FrescoBot, Paine Ellsworth, Ryryrules100, Dogposter, Maxamilliona, Yaser soleimani, Alex-c-johnson, Rymmen, Gfjohnsn, Evalowyn, PirateSmackK, Masked Turk, Citation bot 1, Aditya narain srivastava, Dterp, Pinethicket, I dream of horses, 10metreh, Jonesey95, A412, Tom.Reding, Mekeretrig, Triplestop, Random editor, RedBot, Artem Korzhimanov, Rausch, Reconsider the static, IVAN3MAN, Meier99, Odenjr, Nishant1997nishu, Vrenator, Anti-Nationalist, Mr.98, Reaper Eternal, Xobekil, Linguisticgeek, Tbhotch, Lolmanz, DARTH SIDIOUS 2, Whisky drinker, Onel5969, Harryking177, Mineadwaly, RjwilmsiBot, Skipstar7, Rajettan, WildBot, Wintonian, Slon02, Spamking93, EmausBot, Williamthomasandrews, Immunize, Gfoley4, Mikey12348, Britannic124, RA0808, Twisindia, Lja514, Havoc606 Wakka-Pakka, Blink'em, Youncej, Zeusiscute, Bjwill13, Wikipelli, Sepguilherme, JSquish, ZéroBot, Cogiati, Daonguyen95, 1howardsr1, SomDood, A930913, Quondum, WinstonsDomain, SporkBot, ThatBird, Wayne Slam, Tercerista, The Talking Toaster, Maschen, Donner60, Carmichael, Dmyasish, RockMagnetist, TYelliot, Llightex, Scooter12345, MFDMICRO, Xanchester, ClueBot NG, Tpain1776, BriiGarcia, Kenny90655, Caute AF, Mkoconnor, SusikMkr, Carbon editor, Frietjes, Cntras, Nikhileditor, Widr, Antiqueight, Eimeardoneapoo, NO FUSE HERE, Helpful Pixie Bot, Martin Berka, Smoothieking7, Bibcode Bot, BG19bot, Albert012101, Xtfcr7, Hallows AG, Who.was.phone, Mark Arsten, Zachaysan, Ivor Ludlam, F=q(E+v^B), Eelkeher, 🔲🔲🔲 🔲🔲🔲🔲🔲, Harizotoh9, Tremere2, Lee Kyle Jay, Shawn Worthington Laser Plasma, SirTobiasII, Zeegeorge, Mitch H. Waylee, RudolfRed, BattyBot, N lasters, Timothy Gu, NatalieAvigailL, Smartdonkey, Khazar2, EuroCarGT, Dexbot, Physicsmaster 1 3, Jayster294, Josepht404, Thebaconhawk, The Nuke, Numbermaniac, Lugia2453, Mihir John, Cobalt174, Sui docuit, Adwait.a.raste, Reatlas, Thebaconhawk69, WorldWideJuan, Liugaila, CsDix, Curtis P.... Heimberg, Eyesnore, PhantomTech, Zelliej, Dr DonZi, NeapleBerlina, Nigellwh, The Herald, Surfscoter, Ginsuloft, 8i347g8gl, DavRosen, Physikerwelt, Punit chaudhary, Frinthruit, Tssbender, Monkbot, Johnnyideal, Anuvarshanw, Gianluca Di Fiore, Lomtucas, Akifumii, CoolOppo, Asdklf;, TranquilHope, JMP EAX, Narky Blert, Samangivian, Arisht Aveiro, Beckzilla178, Infogamer12345, Videogamefreak43rv, Knaveknight, Karlswag, Brobroswagens, Alleballeeeeeeeeeee, Corsairio, Tetra quark, Isambard Kingdom, Samfart20, Swaglord908199920088, SWAGlordLOLZ, Supdiop, Praneeth Sarvade, Pac6mon9, Zarrus.rasaili94, Kdkddkkdkekksksjdj, Adeptussoratis stormlord1, SayanChakraborty1234567890, Undolie, Brekkestewart, Amrit kushwaha, ARUL ASHRI, Masterredstoner, Alexanrdio, Aman Rajak, Sumit Dhariwal, XXOMGitsBarryXx and Anonymous: 1037

- **Nuclear fusion** *Source:* https://en.wikipedia.org/wiki/Nuclear_fusion?oldid=686726342 *Contributors:* AxelBoldt, Magnus Manske, Chenyu, Trelvis, Mav, Bryan Derksen, The Anome, AstroNomer~enwiki, Taw, Malcolm Farmer, Verloren, Andre Engels, Ted Longstaffe, Jkominek, Youssefsan, XJaM, Peterlin~enwiki, Ben-Zin~enwiki, Maury Markowitz, Heron, Tobin Richard, Stevertigo, Patrick, JohnOwens, Tim Starling, Ixfd64, Looxix~enwiki, Ellywa, Ahoerstemeier, William M. Connolley, Angela, Andrewa, Aarchiba, Glenn, Kaihsu, Jedidan747, Ghewgill, Pizza Puzzle, Mulad, Rob.derosa, Stismail, Pladask, Furrykef, Rei, Omegatron, Bevo, Pstudier, Finlay McWalter, Jni, Gentgeen, Robbot, Altenmann, Naddy, Securiger, Danhuby, Sverdrup, Justanyone, Rursus, Litefantastic, Bkell, Hadal, Pifactorial, Diberri, Cyrius, Matt Gies, Giftlite, DocWatson42, Inter, Wolfkeeper, Art Carlson, Fastfission, MadmanNova, Wwoods, Everyking, Wikibob, Bobblewik, Mooquackwooftweetmeow, Neilc, OldakQuill, ChicXulub, Utcursch, Slowking Man, Pcarbonn, Star controller, Karol Langner, AlexanderWinston, Rdsmith4, Anythingyouwant, Bosmon, Zfr, Urhixidur, Tsemii, Irpen, Bbpen, JohnArmagh, Deglr6328, Grunt, Mike Rosoft, D6, DanielCD, Imaglang, Discospinster, Solitude, Rich Farmbrough, FT2, Hippojazz, Vsmith, Jpk, ArnoldReinhold, Slipstream, Mani1, Bender235, Rubicon, Kaisershatner, Neko-chan, MBisanz, Ben Webber, El C, Huntster, Worldtraveller, Art LaPella, RoyBoy, Euyyn, Alxndr, Noren, Bobo192, Longhair, Smalljim, Jag123, Zwilson, La goutte de pluie, Nk, Ben@liddicott.com, Rje, BW52, Tos~enwiki, Nsaa, Alansohn, Gerweck, Mo0, Free Bear, Eric Kvaalen, Arthena, Keenan Pepper, Hipocrite, ABCD, Riana, Yamla, MarkGallagher, InShaneee, Wdfarmer, SMesser, Velella, BRW, KingTT, Wtshymanski, Amnesiac, Cal 1234, RainbowOfLight, Sfacets, A.Kurtz, DV8 2XL, Gene Nygaard, Alai, Blaxthos, Ultramarine, Nuno Tavares, Firsfron, CryoCone, Mindmatrix, Wdyoung, Fingers-of-Pyrex, Borb, Benbest, MGTom, Vorn, Atomicarchive, Dysepsion, ObsidianOrder, Mandarax, Graham87, KyuuA4, Martinevos~enwiki, Rjwilmsi, Syndicate, Panoptical, JedRothwell, Vary, Jmcc150, SMC, Oblivious, ElKevbo, Erkcan, Williamborg, Oo64eva, Yamamoto Ichiro, Lcolson, Nihiltres, Nivix, Itinerant1, RexNL, Gurch, Karelj, Valermos, Jrtayloriv, Wingsandsword, EronMain, Smithbrenon, Danielfong, Chobot, Scoops, Sharkface217, DVdm, Mhking, Dstrozzi, Elfguy, Roboto de Ajvol, Hairy Dude, Xhyljen, Charles Gaudette, Midgley, Phantomsteve, Sillybilly, Postglock, Bhny, Splash, Gaius Cornelius, CambridgeBayWeather, Pseudomonas, Hubert Wan, Thane, Brian Sisco, David R. Ingham, NawlinWiki, Wiki alf, ErkDemon, Nad, LiamE, Ragesoss, Sangwine, Matticus78, Bobak, Moe Epsilon, Eltwarg, Mlouns, Xiroth, Brat32, Black Falcon, Enormousdude, 2over0, Zzuuzz, Nemu, ColinMcMillen, Dspradau, Petri Krohn, CWenger, Ilmari Karonen, Serendipodous, Nekura, Splendidtorch, Tom Morris, That Guy, From That Show!, SpLoT, SmackBot, Ashill, Saravask, KnowledgeOfSelf, K-UNIT, Ufundo, Tonyr68uk, David.Mestel, Elminster Aumar, Blue520, Davewild, Chumtoad, Jrockley, Delldot, Dr.Science, Cessator, Rjanson, Canthusus, RonaldHayden, Geoff B, Onebravemonkey, Edgar181, TimTim, IstvanWolf, Gaff, Yamaguchi🔲🔲, Cool3, Gilliam, Hmains, Skizzik, Chris the speller, Quinsareth, Miquonranger03, SchfiftyThree, Moshe Constantine Hassan Al-Silverburg, Complexica, Wykis, DHN-bot~enwiki, Croquant, Sbharris, Colonies Chris, Rogermw, Can't sleep, clown will eat me, Mike J., Wikipedia brown, Kcordina, Edivorce, Charlieb63, Puddle~enwiki, Zirconscot, Fuhghettaboutit, Makemi, Nakon, Jklin, Sadi Carnot, Zimmy2000, Srikeit, John, Dungeonmaster, EDUCA33E, Gobonobo, JoshuaZ, Kirk Grabowski, Bezenek, CaptainVindaloo, Ckatz, NNemec, Special-T, Munita Prasad, SQGibbon, Dicklyon, Otac0n, Stickboy42, ILovePlankton, Iridescent, Luzu, JoeBot, UncleDouggie, ErWenn, Courcelles, Tawkerbot2, Ouishoebean, Lahiru k, Deathcrap, Delphwhite, Jackzhp, Rockcutter88, Fite ez then, FlyingToaster, Avillia, RockMaster, Karenjc, Malamockq, Cydebot, Vanished user vjhsduheuiui4t5hjri, Gogo Dodo, Travelbird, JFreeman, A Softer Answer, ANTIcarrot, Pascal.Tesson, Roberta F., Nsaum75, Kozuch, Mtpaley, Casliber, EvocativeIntrigue, Epbr123, Jadahl, Markus Pössel, T.C.Thornberry, Pepperbeast, Headbomb, Neil916, Second Quantization, Tellyaddict, E. Ripley, Obuolys~enwiki, Ozzah, FreeKresge, Sikkema, MichaelMaggs, WhaleyTim, Uruiamme, Noclevername, Pie Man 360, FireHorse, AntiVandalBot, Gioto, Luna Santin, Why My Fleece?, Seaphoto, Orionus, Yongrenjie, Opelio, Stepan Roucka, Morngnstar, Samuel Erau, Res2216firestar, JAnDbot, SeanTater, CosineKitty, The Transhumanist, Andonic, Igodard, Mauricio Maluff, Magioladitis, Bongwarrior, VoABot II, Alta-Snowbird, JamesBWatson, Think outside the box, Inmate20, Kaiserkarl13, Kevinmon, ClaudeSB, Indon, Thechangster, Beagel, Schumi555, Kopovoi, Laur2ro, Glen, DerHexer, AtomicZebra, Ztobor, DancingPenguin, ClubOranje, MartinBot, STBot, Theron110, Babur8, Mschel, Kostisl, AlexiusHoratius, Brothejr, Cyrus Andiron, J.delanoy, Pharaoh of the Wizards, Uncle Dick, Maurice Carbonaro, WindAndConfusion, Bluesquareapple, Rod57, Skinny McGee, Nextai13, Tokamac, Austin512, Skier Dude, Gurchzilla, Giacona, Aqm2241, NewEnglandYankee, Wesino, Murderbydeath222, Jorfer, DAID, Tanaats, Blckavnger, Pandawelch, KylieTastic, Juliancolton, Cometstyles, Rachel McPhearson, DorganBot, R. A. C., Yoyomin, Elenseel, Itmakesmehappy, Close2reality, Useight, Fusion Power, Nottrue, CardinalDan, Sheliak, Funandtrvl, Remi0o, ACSE, Objectivist, VolkovBot, Johnfos, SergeyKurdakov, QuackGuru, Philip Trueman, Drunkenmonkey, TXiKiBoT, Destroyer 65, The Original Wildbear, Tr-the-maniac, Vipinhari, StevenBKrivit, Andrius.v, Qxz, Finestblade33, Loboguy, Piperh, Anna Lincoln, Lradrama, Mihaip, Tpre007, BotKung, SpecMode, Makelifecheap, Venny85, Larklight, Coolbromley, Crested Penguin, Enviroboy, Monty845, Sue Rangell, Cderoose, Chuck Sirloin, NHRHS2010, EmxBot, Jwilson14, Cryonic07, Maddiemoo39, EJF, SieBot, WereSpielChequers, PanagosTheOther, Gerakibot, Dawn Bard, Polio18, Gravitan, Bentogoa, Quest for Truth, Flyer22 Reborn, Csblack, Jpr2x, JLKrause, Antonio Lopez, Faradayplank, R J Sutherland, Thisnamestaken, Afernand74, Searmemcmxciii, Mg1967cup20011, Maelgwnbot, Gregie156, Anchor Link Bot, Hamiltondaniel, Anyeverybody, Magma828, Jamous77, Susan118, Ascidian, Neo., Denisarona, Jordan 1972, Francvs, Martarius, Elassint, ClueBot, Binksternet, Fasettle, Panoptik, The Thing That Should Not Be, VsBot, JohnAspinall, Wwheaton, Eiland, Regibox, Blanchardb, Rotational, Maxtitan, Rex360,

TOO9, Djr32, Excirial, Jusdafax, Brews ohare, HoudiniMan, Setoor g, Mjj4, PhySusie, Ember of Light, Snacks, Razorflame, ChrisHodges-UK, Thehelpfulone, Thingg, Jonverve, Ranjithsutari, Terminator484, Rhinocerous Ranger, Akaszynski, Bobbobls, SoxBot III, HumphreyW, Tylerdfreeze11, Crowsnest, Alastair Carnegie, DumZiBoT, Pitt, BarretB, XLinkBot, Spitfire, Rror, Ost316, Little Mountain 5, Avoided, SlimX, Mifter, PL290, Alexius08, Vianello, SkyLined, Shoemaker's Holiday, Nuclear fusion man, Gameboy lbl, Addbot, Xp54321, Uruk2008, Jo-jhutton, Tcncv, SunDragon34, Blethering Scot, Shanee753, Infobloat, CanadianLinuxUser, JakeDodd, Cst17, SoSaysChappy, Williaml123, PranksterTurtle, Viewgray3, Mundo0987, LinkFA-Bot, Jasper Deng, JustinFLeighton, Quamarquazi, ProfessorToomin, AgadaUrbanit, Offs-Blink, Tide rolls, Smeagol 17, Krano, Loupeter, Caroliano, Frehley, Ben Ben, Luckas-bot, Yobot, Fraggle81, Grebaldar, Mmxx, KamikazeBot, MessiahBenDavid, Eric-Wester, AnomieBOT, DemocraticLuntz, SamuraiBot, MuhalaC, GrimFang4, 1exec1, Daniel.dalegowski, Jim1138, Doomcookie222, JackieBot, AdjustShift, Kingpin13, Ulric1313, Ubergeekguy, Materialscientist, The High Fin Sperm Whale, Citation bot, OllieFury, E2eamon, Neurolysis, LilHelpa, Marshallsumter, Xqbot, Sketchmoose, Tripodian, StaanJacobsen, Engineering Guy, JimVC3, Moonphase95, Capricorn42, Drilnoth, Nickkid5, DSisyphBot, Blindgrapefruit2, GrouchoBot, Dsjeisl, Abce2, Riotrocket8676, Franco3450, Linkman21, Doulos Christos, Sheeson, TanSuey, FrescoBot, Johnb96, Ryryrules100, Cookyes, Frankw101, Rkr1991, Foxbull, Arpadko-rossy, Killface55, Pinethicket, I dream of horses, Tanweer Morshed, Anden21, A412, MJ94, Calmer Waters, Michalsmid, Serols, Île flot-tante, Casimir9999, Jrobbinz123, Xeworlebi, Saayiit, Horst-schlaemma, Comet Tuttle, Destineyyyy, Miracle Pen, Mr.98, TheGrimReaper NS, Davish Krail, Gold Five, Suffusion of Yellow, Sampathsris, Mramz88, Bobby122, Jakedakac, Djgbradley, Bento00, Bhawani Gautam, Ryancherry4, Joed269, DASHBot, EmausBot, Immunize, DoranisGallant, IncognitoErgoSum, Bonvoyage123, Kulmeetster, PantsPhantom, Jakedog730, Wikipelli, K6ka, Hhhippo, AvicBot, JSquish, ZéroBot, Daonguyen95, (o^^0r, Fæ, H3llBot, Olliesherlock101, Christina Silver-man, Wayne Slam, Olhp, Rcsprinter123, UmJumFlum, L Kensington, Donner60, ChuispastonBot, Foolonthehill135, Y.r.agrawal, DASH-BotAV, Whoop whoop pull up, AMD, Warharmer, Jharrell2010, ClueBot NG, Mechanical digger, Gareth Griffith-Jones, Syphallitic mon-key, Gilderien, Misshamid, Snotbot, Blackbear12, NuclearEnergy, Helpful Pixie Bot, Thaw.htet.pat, HMSSolent, Gob Lofa, Bibcode Bot, Therevsyn, BG19bot, Mr. Nuke, Bobo360, Jramgo, Juro2351, MusikAnimal, Nsinger13, Ahj97, StuGeiger, Mark Arsten, Op47, Altaïr, The-masterwriter, Chuckstvns, RMA1129, Blaspie55, Zedshort, Vz25, Refrencecard, Shawn Worthington Laser Plasma, Soccerluva872, Qasaur, Hello162626, Thom801, Drclaptop, Ddude1969, Pffeifer, Promptjump, Rub117, Cyberbot II, The Illusive Man, Tarafauss, ChrisGualtieri, Lesvesla, Angelagibson11, SD5bot, Aschuess, Sapce Cowboy, Mahesh gandikota, Lsclear, Dexbot, Cyro43, Erjablow, Teleohapsis, An-cdefg, PeerRevision, Lugia2453, Agrrules, Reatlas, Joeinwiki, Akihabarabankinya, Kalyanpadmandar, Aj7s6, Hellogj, Teraminato, Samhg, Momnkey1997, Drewvillines, Morg00, Brett6781, WikiHelper2134, Hevatroid, LieutenantLatvia, HAKANYASARKAYA, Bill theuser88, Crow, Lette Sgo, JaconaFrere, Epic Failure, Sethcp, Wyn.junior, Clubclubclub, Tysonf3, Heathflugruger, PaulZapata, Meinneger, Dilipku-mardk, CharlieFerry, Aguy77, PedroGodoyP, DewDewey, Borgieporgie, Oiyarbepsy, Dgasparri, Strongjam, Wijowa, Xelevationzx, Udont-knowmynamerandom, Hater lov hotdogs, Cratter the matter, Chickennuggets886, KasparBot, Yeddyeggel, Vul Vokun Ah and Anonymous: 1158

- **Nuclear power** *Source:* https://en.wikipedia.org/wiki/Nuclear_power?oldid=687225380 *Contributors:* Trelvis, Robert Merkel, Malcolm Farmer, Xaonon, Deb, Ray Van De Walker, Atlan, Tedernst, Edward, D, Bewildebeast, Fred Bauder, Wwwwolf, Ixfd64, Shoaler, Gbleem, Paul A, Aho-erstemeier, Mac, Andrewa, Aarchiba, Julesd, Bogdangiusca, Netsnipe, Susurrus, Andres, Smack, Ec5618, Stone, Andrewman327, Whisper-ToMe, Katana0182, IceKarma, DJ Clayworth, Tpbradbury, Dragons flight, Furrykef, LMB, SEWilco, Omegatron, Geraki, Raul654, Pstudier, David.Monniaux, Pollinator, Jni, Twang, Nufy8, Moncrief, Modulatum, Mayooranathan, Academic Challenger, Rhombus, Aleron235, Sunray, Sindri~enwiki, Dina, Alan Liefting, Giftlite, DocWatson42, Cokoli, Massysett, Tom harrison, Fastfission, Peruvianllama, Wwoods, Average Earthman, Everyking, Curps, Michael Devore, Sietse, Tweenk, AdamJacobMuller, Steve802, Wmahan, Stevietheman, Gadfium, Utcursch, Andycjp, Knutux, Slowking Man, Yath, Antandrus, Beland, Kusunose, Paddyez, HistoryBA, Rdsmith4, DragonflySixtyseven, Tatarize, Kevin B12, Gscshoyru, Asbestos, Fintor, McCart42, Grm wnr, Deglr6328, Adashiel, Canterbury Tail, Freakofnurture, Spiffy sperry, CALR, Dis-cospinster, William Pietri, Rich Farmbrough, NeuronExMachina, Andros 1337, Inkypaws, Wrp103, Vsmith, Zen-master, Flatline, Mjpieters, Berkut, Michael Zimmermann, Kenb215, Bender235, Thnr, Kbh3rd, Kaisershatner, Zscout370, El C, RoyBoy, Femto, Kaveh, Dalf, Jpgor-don, Adambro, Bobo192, Nigelj, Flxmghvgvk, Cohesion, Geocachernemesis~enwiki, Kjkolb, Nk, Tritium6, Nickolay Stelmashenko, Mpulier, C-4, QuantumEleven, Sanmartin, Jumbuck, Disneyfreak96, Storm Rider, Stephen G. Brown, Bob rulz, Wereldburger758, Alansohn, Qwe, Mo0, Interiot, Atlant, Rd232, Jeltz, Andrewpmk, Ronline, Cjthellama, AjAldous, Lectonar, Lightdarkness, Phocks, Malo, Katefan0, Rwend-land, Super-Magician, Colin Kimbrell, Dabbler, Wtshymanski, Uffish, Runtime, Tony Sidaway, Amorymeltzer, Sciurinæ, BlastOButter42, Mmsarfraz, DV8 2XL, Ent, Alai, Drbreznjev, Recury, Vadim Makarov, Bookandcoffee, Dwiki, Forteblast, Ultramarine, RyanGerbil10, Dismas, RPIRED, Tariqabjotu, Mahanga, Ron Ritzman, Stephen, Lkinkade, Roland2~enwiki, Simetrical, Distantbody, Mel Etitis, Linas, Wdyoung, CyrilleDunant, Thivierr, Rukkyg, Erikpatt, Mazca, Zealander, ^demon, Peter Beard, MGTom, MONGO, Exxolon, Miss Made-line, Lawe, Kgrr, Wikiklrsc, Damicatz, Sengkang, Lizard1959, Eras-mus, Karmosin, Male1979, JohnC, Wayward, Hgd4th, Silverwood, PeregrineAY, Mandarax, Rnt20, BD2412, Galwhaa, Ligar~enwiki, Kbdank71, Vanderdecken, Sjakkalle, Rjwilmsi, Scandum, Joe Decker, Hitssquad, TitaniumDreads, Breenius, Jake Wartenberg, Joffan, Bob A, Tangotango, Sdornan, SMC, Aronomy, Durin, Bhadani, Ttwaring, JamesEG, DirkvdM, Yamamoto Ichiro, Firebug, Taskinen, Lcolson, Wobble, FayssalF, RainR, Titoxd, Sgkay, Fivemack, Ian Pitchford, Mirror Vax, SchuminWeb, Ground Zero, Old Moonraker, Doc glasgow, AndrewStuckey, Harmil, Nivix, Benplaut, RexNL, Kolbasz, Common Man, UnlimitedAccess, Imnotminkus, Chobot, Theo. Pardilla, Benjamin Gatti, Stephen Compall, Simesa, Gwernol, Protarion, Wiserd911, The Rambling Man, Wavelength, Angus Lepper, Retaggio, Sceptre, Phmer, RussBot, Arado, Junky, Sillybilly, Anonymous editor, TheDoober, ChristianEdwardGruber, GLaDOS, Netscott, SpuriousQ, Rada, Skydot, Lar, Stephenb, Alex Ramon, Gaius Cornelius, Ihope127, Kyoro-suke, Tungsten, Wimt, Kerry Raymond, NawlinWiki, Anomie, Wiki alf, Bachrach44, Uberisaac, Damnfuct, NickBush24, Joel7687, Dureo, Tokachu, Adamrush, Nick, Xdenizen, Dmoss, Boadrummer, Davemck, KatzMotel, Tony1, Aaron Schulz, Mieciu K, PrimeCupEevee, Tasty-Cakes, DeadEyeArrow, WCX, Nick123, Wknight94, Xabian40409, Boivie, WAS 4.250, FF2010, Poppy, Zzuuzz, Vertigre, Bayerischermann, Theda, Closedmouth, Drogue, KGasso, Nemu, Terryc, Wsiegmund, MaNeMeBasat, JoanneB, Jor70, Smurrayinchester, Peter, ArielGold, Stuhacking, Bluezy, Kungfuadam, Revengeofthynerd, RG2, Mardus, ModernGeek, That Guy, From That Show!, Luk, Sardanaphalus, Waulf-gang, SmackBot, Amcbride, Burtonpe, Ratarsed, Ashenai, Prodego, KnowledgeOfSelf, TestPilot, Marc Lacoste, McGeddon, Col 10022, Prthealien, Ariedartin, Pgk, Goldfishbutt, Od Mishehu, Joelnish, Aliensvortex, Jfurr1981, BlackCow, KVDP, Delldot, StefanoC, Edgar181, Man with two legs, Xaosflux, Chef Ketone, Aksi great, Gilliam, Ohnoitsjamie, Hmains, Betacommand, Jushi, Oscarthecat, Carbon-16, Pow-erco, Benoman, SlimJim, Agateller, MK8, Lordkazan, WikiFlier, SchfiftyThree, CSWarren, Effer, Ctbolt, Colonies Chris, Dual Freq, Fset-tle, Danielnez1, Rogermw, J00tel, Can't sleep, clown will eat me, KG6YKN, Penrithguy, Shalom Yechiel, Frap, ScottishPinko, Onorem, Terry Oldberg, GreatBigCircles, Matthew, TheKMan, Rrburke, Run!, Britmax, Aces lead, Rsm99833, Edivorce, Renegade Lisp, Kyle sb, Mrdempsey, Khoikhoi, Pepsidrinka, Theanphibian, Wen D House, Emre D., Smooth O, VegaDark, Richard001, RandomP, Tompsci, Eran of Arcadia, Nrcprm2026, Dcamp314, Enr-v, Lcarscad, Virtualsim, DMacks, Drc79, KeithB, Andeggs, Daniel.Cardenas, Mion, Pilotguy, 图图图, Glacier109, Tburke261, Nishkid64, Dono, AThing, Gloriamarie, DA3N, NotMuchToSay, Srikeit, Sophia, Kuru, John, KenFehling, Euchiasmus, Zaphraud, Diemunkiesdie, Writtenonsand, Vgy7ujm, Gizzakk, Loodog, Disavian, JohnI, Soumyasch, Lazylaces, Sir Nicholas de Mimsy-Porpington, KathyR@aol.com, Linnell, This user has left wikipedia, JoshuaZ, CaptainVindaloo, NYCJosh, Cielomobile, PseudoSudo, Cronos2546, Camilo Sanchez, CyrilB, Chuck Simmons, Gjp23, Stupid Corn, Special-T, Sombrera, AdultSwim, Ryulong, Zapvet, Beck162, Fromeout11, TimTL, Kvng, Hu12, JYi, OnBeyondZebrax, Iridescent, Zootsuits, Michaelbusch, Tamino, LeyteWolfer, Myrtone86, Shoeofdeath, Newone, Lord E, Daxmaryrussel, CP\M, Bruinfan12, Pjbflynn, Tawkerbot2, Fugger89, Banancanard, Ldukes, Dlohcierekim, Daniel5127, Chelydra, Samnuva, GerryWolff, The Haunted Angel, Kotepho, FatalError, SkyWalker, Rondack, Dia^, JForget, Twipie, Ishanv, DangerousPanda, Urutapu, CmdrObot, Wafulz, Pmyteh, Scohoust, John Riemann Soong, Delose, Styler 13, TaranMoltu, Davidjfarrell, GHe, Kylu, Green caterpillar, Mak Thorpe, OMGsplosion, Zosimus, Zinjixmaggir, Location, MarkusQ, HonztheBusDriver, Myasuda, M0nkmaster, Fl295, Varinawalker, Fable of flame, Mierlo, Diegom809, Masterchiefjedimaster, Abeg92, Tntnnbltn, Gogo Dodo, HPaul, Bridgecross, JFreeman, Denghu, Corpx, ST47, Eft160, Chasingsol, Lugnuts, Nick2253, Rracecarr, Weavrmom, BillySharps, Dancter, DumbBOT, Praetor jon, Kozuch, Lewisskinner, Afinebalance, JodyB, Realrahul, Dudeirock34, The machine512, Gordonmcdowell, Thickycat, Epbr123, Barticus88, Cimbalom, Eggsyntax, Aravindashwin, Alfredo22, Andyjsmith, Mrcaseyj, J.Ring, Gralo, Hugo.arg, Headbomb, Marek69, Bagnewauckland, Cool Blue, Ttussok, Marcushatesgirls, Eljamoquio, Kaaveh Ahangar~enwiki, Muaddeeb, Abdel Hameed Nawar, SvenAERTS, Escarbot, Dalliance, Dzubint, Mentifisto, Hmrox, Thadius856, Cyclonenim, AntiVandalBot, Polypmaster, Luna Santin, Opelio, Peter50, Rabqa1, CJSquire, Czj, Fru1tbat, Mrshaba, Jbaranao, Sweart1, Dakatzpajamas, MontanNito, Mack2, TexMurphy, LibLord, Darklilac, Poolman09, Kennard2, Richiez, Gökhan, Ioeth, JAnDbot, Niagara, MER-C, BlindEagle, The Transhumanist, Robidy, Mcorazao, IanOsgood, Mark Rizo, Miti gta, Paxuscalta, Hut 8.5, AlmostReadytoFly, Kaonslau~enwiki, SiobhanHansa, Meeples, Magioladitis, Pedro, Bongwarrior, VoABot II, Canyonwren, Trnj2000, Estonofunciona~enwiki, Richrobison, Cobrachen, Nblanton, Blackicehorizon, ClaudeSB, Recurring dreams, Jmartinsson, Zephyr2k~enwiki, CraZyBob, Bubba hotep, Cardc, Engineman, Indon, Animum, KumfyKittyKlub, Nposs, Dinohunter, XMog, Gldavies, Beagel, Curlydave200, Spellmaster, Glen, DerHexer, Khalid Mahmood, Zoidstar, Textorus, Johnbrownsbody, WLU, Lord Pheasant, Connor Behan, Geboy, Gjd001, Akhil0095, FisherQueen, Sire22, Pauly04, E.vondarkmoor, MartinBot, Rukaribe, Gandydancer, Mattjs, Pupster21, Rettetast, Sm8900, Lmakin, KonaScout, Mschel, Beantownbomber33, R'n'B, Giachen, CommonsDelinker, Lightbulbs~enwiki, Patar knight, Squidraider, Smokizzy, Lilac Soul, JeremyGordon, Stonecoldclassics, J.delanoy, Nev1, Brastein, EscapingLife, AndrewBolt, Valvicus, Hans Dunkelberg, Leaflet, Cymbalta, NerdyNSK, SubwayEater, Rod57, Dispenser, Enuja, Ragnaroks, Scweiner, Aonrotar, Afrojim, Skier Dude, Jayden54, Milkyface, Beyonder1, Zenpher, Plasticup, Belovedfreak, SJP, Aannddrreeww, Jae-24, Supe95, Non Curat Lex, Scoterican, Haljackey, SirJibby, Day 1, WJBscribe, Jamesontai, Wthoang, Epistemenical, Guyzero, Xaxx, Lookingood, Sworded, Xpanzion, Inwind, Useight, Woohoo11, DASonnenfeld, Squids and Chips, Specter01010, Levydav, ThePointblank, Shaunus4, Idioma-bot, Funandtrvl, Spellcast, AndrewTJ31, ACSE, Chromancer, Lights, Maniaphobic, Deor, Shiggity, VolkovBot, Johnfos, ICE77, Blubba78, Alexandria, Alt.pured, Bovineboy2008, DavidMIA, Barneca, Slashe50, Marekzp, PBAJ, TXiKiBoT, Oshwah, Eve Hall, Gwib, Oconnor663, Pwnage8, Nucengineer, Ann Stouter, Nsougia2, Bookworm007, Ask123, Macslacker, Taimaster, Mkolva, Steven J. Anderson, Sintaku, To Catch a Thief, Dlae, PDFbot, Skateboardingrulz, Henryisamazingyeh, Smeghead2007, GProcter, Cremepuff222, Oogly boogles, Corona2666, Mishlai, Wingedsubmariner, Joshers13, Mattfin, Plazak, Vaubin, Cantiorix, Marijuanarchy, Altermike, Freebiegrabber, Nayarsk, Blaynew, Temporaluser, Alfrodull, Thisismyrofl, Kmapeterson, Spitfire8520, HiDrNick, NPguy, JonathanSSMann, Symane, NHRHS2010, Signsolid, Jiele, HybridBoy, Gilawson, Red, Steven Weston, D. Recorder, SieBot, Primal400, Pizzachicken, Scarian, Weeliljimmy, Sparrowman980, Fanra, Gerakibot, Da Joe, Caltas, Universe=atom, Wwiilleeyybb, LookingYourBest, Grundle2600, Sch0013sch0013, Tareksamy, Aillema, AdmiralCustard, Joegolland, Quest for Truth, Radon210, DiscoStuMan, Wooo100, Raulcleary, Cheahiamjohn, Wombatcat, Fredgina, Oxymoron83, Crazy isreali, AngelOfSadness, Mhollinshead, Quicksilver1444, Sacre14, Lightmouse, RW Marloe, SH84, WacoJacko, AWeishaupt, Mhollinshead2k7, Hyperhullo06, N96, LonelyMarble, Phippsyuk, NameThatWorks, Shedseven, Iknowyourider, Jacob.jose, Mygerardromance, Bseay, Anyeverybody, Mikedsd, HexB, Jwardy00, BassFace50, Ghetsmith, Miyokan, Hannanan, Hello9393, HubcapD, MBK004, Acdoyle2000, ClueBot, Mac9092, SummerWithMorons, Binksternet, Garthhh, Panoptik, Ctiefel, Nailedtooth, The Thing That Should Not Be, Rjd0060, Arwad, Tipdrill, VsBot, Wwheaton, Farolif, Mr sean meers, Mild Bill Hiccup, Axe27, Eiland, Watti Renew, DanielDeibler, Polyamorph, Wikisteff, Lwnf360, CounterVandalismBot, Kattania, Niceguyedc, Boardin8, Wikifast1991, Gyrcompass, Renven3, Trivialist, Puchiko, Bobob167, Hossein756, Anne Prouse, Alexbot, Nizzzle, Eeekster, Muenda, CraigWenner, Arjayay, Coolio 226, Dn9ahx, Kaecyy, Iohannes Animosus, Smartypants2612, Pat010, Darthozzan, Nukeless, Mustufailed, Pluto 239~enwiki, Togokill, Dilanfor3, Flower Priest, Versus22, MelonBot, Mirkin man, Qwfp, SoxBot III, Lastchancepowerdrive, Gotmeonmyknees, Nobodyknowsyouwhenyouredownandout, Slayerteez, DumZiBoT, Escientist, Jamesscottbrown, BarretB, XLinkBot, Ladsgroup, FellGleaming, LeaW, Qgil-WMF, Sb616~enwiki, JinJian, Red1001802, Eledille, Niftyneb, The Rationalist, CàlcuIntegral, Emilyisaac, HexaChord, JoeJoe11~enwiki, Engkamalzack, Kajabla, Addbot, Speer320, Krawndawg, Linepm, C6541, Manuel Trujillo Berges, Awes141, Chum2, Tcncv, M.nelson, GeneralAtrocity, Older and ... well older, Ronhjones, Artivist, Leszek Jańczuk, WFPM, Conningcris, Download, CarsracBot, BepBot, Jreconomy, In'nOutfan90210, Granitethighs, LinkFA-Bot, Tobyrox9, Tassedethe, Delphi234, I Wake Up Screaming, ScAvenger, Gail, TeH nOmInAtOr, Greyhood, Zimmy, ChNPP, Legobot, Publicly Visible, Yobot, Twexcom, Enviro1, Shrikrishnabhardwaj, THEN WHO WAS PHONE?, Vrinan, Eric-Wester, Magog the Ogre, Bility, Juliancolton Alternative, AnomieBOT, Ichwan Palongengi, Tavrian, Jeni, Galoubet, Ipatrol, Ufim, Materialscientist, OllieFury, DynamoDegsy, LilHelpa, Apjohns54, Xqbot, Susiemorgan, Ali944rana, Sionus, Cureden, Capricorn42, Drilnoth, SomebodyOnInternet, Nicholas Frost, Tom132m, Mike3050, James Wenham, Totemruby, StealthCopyEditor, RibotBOT, WallabieJoey, Rb88guy, Deanolympics, Fionaclee, Ascottlane88, Zfrag, Shadowjams, 图图图图图图图, Astatine-210, Marioo1182, A. di M., Sesu Prime, Margoward, PercussionPirate, Fotaun, Hzhzhzhz, Jack B108, GliderMaven, Pkur0wnz1, Edward130603, FrescoBot, Illustria, Tobby72, Horseychick4life, Lampunvaihtaja, Hacky360, Tranletuhan, YOKOTA Kuniteru, KerryO77, Compoundinterestisboring, Michael93555, Gibby is gibby, Samuelsidler, Iamwerid, Spanish32, Blubbaloo, Citation bot 1, Kopiersperre, Boulaur, Hard Sin, Andynct, RedBot, Brian Everlasting, Giosue' Campi, Brett R. Stone, Plasticspork, Jauhienij, Cnwilliams, Orenburg1, Ttuku, Mr.98, TheCaroline!, RjwilmsiBot, Louiselives, Jackehammond, AlexRexR, John of Reading, Sparks1911, Trofobi, Clark42, Alex3yoyo, Boundarylayer, Dewritech, Giornorosso, GoingBatty, Pavlo Chemist, Rndm85, Challisrussia, RIS cody, Stripar, Druzhnik, Ckeiderling, China Dialogue Net, H3llBot, UrbanNerd, AlexH555, AManWithNoPlan, Mehrdad1900s, Aschwole, Rcsprinter123, Sebastian barnes, Sachinvenga, Noodleki, Teapeat, Ivolocy, Miradre, Rememberway, ClueBot NG, Oyauguru, Deano8216, Physics is all gnomes, OpenInfoForAll, Violetbonmua, Rezabot, Costesseyboy, Redrit, CopperSquare, Cfrahm, Paulzubrinich, Anon5791, Helpful Pixie Bot, Intheeventofstructuralfailure, HMSSolent, Wbm1058, Bibcode Bot, Sokavik, BG19bot, NewsAndEventsGuy, Provessor, Guy vandegrift, Lisamccabe, CityOfSilver, Andol, MaxHerz, Compfreak7, Blaspie55, ConradMayhew, Mikepabell, BattyBot, Factsearch, Acadēmica Orientālis, Cyberbot II, ChrisGualtieri, Shokioto22, Electricmuffin11, Khazar2, Aschuess, EagerToddler39, Dexbot, NewebNL, Mogism,

TwoTwoHello, Stratoprutser, Reatlas, Rfassbind, Desterpot, Limnalid, Mfb, Marikanessa, CarnivorousBunny, Monkbot, Trackteur, Pedro-Madrid1976, Nimrainayat6290, Insertcleverphrasehere, Guegreen, IamM1rv, KasparBot, ILikeTau and Anonymous: 1358

- **Weak interaction** *Source:* https://en.wikipedia.org/wiki/Weak_interaction?oldid=679571094 *Contributors:* AxelBoldt, Chenyu, Sodium, Bryan Derksen, Tarquin, AstroNomer~enwiki, Andre Engels, XJaM, Heron, JohnOwens, Gdarin, Delirium, Andrewa, Andres, Emperorbma, Timwi, Fibonacci, Phys, Phil Boswell, Lowellian, Mayooranathan, Tobias Bergemann, Giftlite, Sj, Herbee, Xerxes314, Jcobb, Mckaysalisbury, Munkee, Toby Woodwark, Bbbl67, Icairns, AmarChandra, Lumidek, Jørgen Friis Bak, Discospinster, ArnoldReinhold, Roybb95~enwiki, Gianluigi, Joanjoc~enwiki, Shanes, AJP, AtomicDragon, Danski14, Alansohn, Arthena, Axl, SidneySM, Hwefhasvs, DV8 2XL, Nightstallion, Kazvorpal, Linas, StradivariusTV, Benbest, Bbatsell, Palica, Tevatron~enwiki, Graham87, BD2412, Ketiltrout, Rjwilmsi, Strait, Erkcan, The wub, FlaBot, Naraht, Itinerant1, Srleffler, Chobot, Krishnavedala, YurikBot, Borgx, Bambaiah, Hairy Dude, Jimp, Sillybilly, Conscious, Epolk, JabberWok, Gaius Cornelius, Shaddack, SCZenz, Irishguy, Shimei, Willtron, RG2, Phr en, That Guy, From That Show!, Luk, SmackBot, David Kernow, Tom Lougheed, WookieInHeat, Dauto, Chris the speller, Philosopher, Moshe Constantine Hassan Al-Silverburg, Complexica, DHNbot~enwiki, Zirconscot, BIL, Wen D House, "alyosha", Maxwahrhaftig, Akriasas, Vina-iwbot~enwiki, Bdushaw, TTE, SashatoBot, Fontenello, Herr apa, Condem, Tony Fox, MottyGlix, JRSpriggs, Heartofgoldfish, Calmargulis, Green caterpillar, Joelholdsworth, Cydebot, Michael C Price, Mtpaley, Thijs!bot, ChKa, Kichwa Tembo, Headbomb, Hcobb, Icep, Escarbot, AntiVandalBot, Jimeree, Steelpillow, JAnDbot, Magioladitis, Swpb, باسم, Wormcast, DAGwyn, Giggy, Khalid Mahmood, Gah4, Tarotcards, 2help, Lighted Match, DorganBot, Halmstad, Idiomabot, VolkovBot, Jcuadros, Hilarious Bookbinder, TXiKiBoT, Rei-bot, CaptinJohn, Awl, Shenanegins, BotKung, Wingedsubmariner, Antixt, Xxxlilbritxxx, Ptrslv72, Monty845, AlleborgoBot, SieBot, Paolo.dL, Skyentist, Ptr123, ClueBot, Bondchic007, SuperHamster, Erudecorp, Rotational, Jackey0105, Alexbot, Cenarium, Zomno, Zahnrad, He6kd, TimothyRias, InternetMeme, Timo Metzemakers, Stephen Poppitt, Addbot, Some jerk on the Internet, Markdman, ChenzwBot, Ehrenkater, Tide rolls, Luckas-bot, Yobot, Les boys, Kilom691, THEN WHO WAS PHONE?, Rifter0x0000, Duping Man, Dickdock, Magog the Ogre, AnomieBOT, Materialscientist, Citation bot, Quebec99, Kreigiron, Xqbot, Drilnoth, BurntSynapse, GrouchoBot, Omnipaedista, RibotBOT, Workanode, Jaz1305, Mnmngb, Dave3457, FrescoBot, Charles.walker, LucienBOT, Ionutzmovie, Grandiose, Pinethicket, Boulaur, Rameshngbot, RedBot, 23790AD, Tea with toast, Jauhienij, FoxBot, Earthandmoon, RjwilmsiBot, Itamarhason, Newty23125, EmausBot, WikitanvirBot, GA bot, GoingBatty, Splibubay, StringTheory11, Braswiki, Git2010, Wayne Slam, Jsayre64, Maschen, ChuispastonBot, ClueBot NG, VinculumMan, Physics is all gnomes, Fjpyanez, Mouse20080706, Helpful Pixie Bot, Geo7777, Bibcode Bot, Junaid2754, Bolatbek, Phbarnacle, Neutral current, Glevum, Idenshi, Marioedesouza, Dexbot, Spray787, Reatlas, CsDix, Jamesmcmahon0, Ihatedirac2k13, Kharkiv07, Jwratner1, YimmyYohnson, Monkbot, BalderdashVonDrivel, ASCarretero, Malerisch, Lachlan Newland, Tetra quark, KasparBot and Anonymous: 155

- **Semi-empirical mass formula** *Source:* https://en.wikipedia.org/wiki/Semi-empirical_mass_formula?oldid=683400907 *Contributors:* AxelBoldt, Charles Matthews, Dcoetzee, Giftlite, CyborgTosser, Rich Farmbrough, Vsmith, Xezbeth, Longhair, ABCD, Burn, Hdeasy, H2g2bob, Gene Nygaard, Christopher Thomas, Mike Peel, Hashproduct, Zapateria, Philten, YurikBot, Jimp, Krea, BirgitteSB, Nate1481, Sliggy, SmackBot, Dauto, Yurigerhard, Sbharris, Radagast83, Pwjb, Postscript07, NotoriousTF, Inquisitus, Dan Gluck, UberScienceNerd, Barticus88, TDF, Headbomb, Lovibond, Morngnstar, Dirac66, Mollwollfumble, Jim.henderson, Pamputt, YonaBot, Droog Andrey, ArdClose, Biggerj1, Niel.Bowerman, Tangoludwig, Addbot, Yobot, Quasar1826, AnomieBOT, ^musaz, Citation bot, Capricorn42, Dafirenze, Maggyero, Shashwat986, Calmer Waters, Pbrower2a, Hullernuc, McSaks, Marsupilami (DE), GoingBatty, BredoteauU2, Jlemans89, Marechal Ney, Reify-tech, MerlIwBot, AdventurousSquirrel, KKloepfer, BattyBot, Mogism, Aryaindia, Marekich, Joeinwiki, Bramlap92, Femmtom and Anonymous: 68

15.9.2 Images

- **File:2005_Energy_Policy_Act.jpg** *Source:* https://upload.wikimedia.org/wikipedia/commons/5/52/2005_Energy_Policy_Act.jpg *License:* Public domain *Contributors:* http://georgewbush-whitehouse.archives.gov/news/releases/2005/08/images/20050808-6_f1g3456-515h.html *Original artist:* Eric Draper

- **File:2011-05-10_18-57-46_Switzerland_-_Wil.jpg** *Source:* https://upload.wikimedia.org/wikipedia/commons/1/1f/2011-05-10_18-57-46_Switzerland_-_Wil.jpg *License:* CC BY-SA 3.0 *Contributors:* Own work: Hansueli Krapf (User Simisa (talk · contribs)) *Original artist:* Hansueli Krapf

- **File:ANTIAKW.jpg** *Source:* https://upload.wikimedia.org/wikipedia/commons/4/40/ANTIAKW.jpg *License:* CC BY-SA 2.0 de *Contributors:* Own work *Original artist:* Hans Weingartz (Leonce49 at de.wikipedia)

- **File:Ambox_current_red.svg** *Source:* https://upload.wikimedia.org/wikipedia/commons/9/98/Ambox_current_red.svg *License:* CC0 *Contributors:* self-made, inspired by Gnome globe current event.svg, using Information icon3.svg and Earth clip art.svg *Original artist:* Vipersnake151, penubag, Tkgd2007 (clock)

- **File:Ambox_important.svg** *Source:* https://upload.wikimedia.org/wikipedia/commons/b/b4/Ambox_important.svg *License:* Public domain *Contributors:* Own work, based off of Image:Ambox scales.svg *Original artist:* Dsmurat (talk · contribs)

- **File:Annual_electricity_net_generation_from_renewable_energy_in_the_world.svg** *Source:* https://upload.wikimedia.org/wikipedi f/fa/Annual_electricity_net_generation_from_renewable_energy_in_the_world.svg *License:* CC0 *Contributors:* EIA *Original artist:* Lery007

- **File:Annual_electricity_net_generation_in_the_world.svg** *Source:* https://upload.wikimedia.org/wikipedia/commons/d/df/Annual_elec net_generation_in_the_world.svg *License:* CC0 *Contributors:* EIA *Original artist:* Lery007

- **File:Atom-Moratorium.svg** *Source:* https://upload.wikimedia.org/wikipedia/commons/5/58/Atom-Moratorium.svg *License:* CC BY-SA 2.5 *Contributors:* Kernkraftwerke in Deutschland.svg *Original artist:* Kernkraftwerke in Deutschland.svg: **Lencer**

- **File:Beta-minus_Decay.svg** *Source:* https://upload.wikimedia.org/wikipedia/commons/a/aa/Beta-minus_Decay.svg *License:* Public domain *Contributors:* This vector image was created with Inkscape. *Original artist:* Inductiveload

- **File:Beta_Negative_Decay.svg** *Source:* https://upload.wikimedia.org/wikipedia/commons/8/89/Beta_Negative_Decay.svg *License:* Public domain *Contributors:* This vector image was created with Inkscape. *Original artist:* Joel Holdsworth (Joelholdsworth)

- **File:Betheweizsaecker.jpg** *Source:* https://upload.wikimedia.org/wikipedia/commons/c/c0/Betheweizsaecker.jpg *License:* Public domain *Contributors:* Own work *Original artist:* Aluminium

- **File:Flag_of_Europe.svg** *Source:* https://upload.wikimedia.org/wikipedia/commons/b/b7/Flag_of_Europe.svg *License:* Public domain *Contributors:*

- File based on the specification given at [1]. *Original artist:* User:Verdy p, User:-xfi-, User:Paddu, User:Nightstallion, User:Funakoshi, User: Jeltz, User:Dbenbenn, User:Zscout370

- **File:Flag_of_France.svg** *Source:* https://upload.wikimedia.org/wikipedia/en/c/c3/Flag_of_France.svg *License:* PD *Contributors:* ? *Original artist:* ?

- **File:Flag_of_Germany.svg** *Source:* https://upload.wikimedia.org/wikipedia/en/b/ba/Flag_of_Germany.svg *License:* PD *Contributors:* ? *Original artist:* ?

- **File:Flag_of_India.svg** *Source:* https://upload.wikimedia.org/wikipedia/en/4/41/Flag_of_India.svg *License:* Public domain *Contributors:* ? *Original artist:* ?

- **File:Flag_of_Iran.svg** *Source:* https://upload.wikimedia.org/wikipedia/commons/c/ca/Flag_of_Iran.svg *License:* Public domain *Contributors:* URL http://www.isiri.org/portal/files/std/1.htm and an English translation / interpretation at URL http://flagspot.net/flags/ir'.html *Original artist:* Various

- **File:Flag_of_Italy.svg** *Source:* https://upload.wikimedia.org/wikipedia/en/0/03/Flag_of_Italy.svg *License:* PD *Contributors:* ? *Original artist:* ?

- **File:Flag_of_Japan.svg** *Source:* https://upload.wikimedia.org/wikipedia/en/9/9e/Flag_of_Japan.svg *License:* PD *Contributors:* ? *Original artist:* ?

- **File:Flag_of_Kazakhstan.svg** *Source:* https://upload.wikimedia.org/wikipedia/commons/d/d3/Flag_of_Kazakhstan.svg *License:* Public domain *Contributors:* own code, construction sheet *Original artist:* -xfi-

- **File:Flag_of_Mexico.svg** *Source:* https://upload.wikimedia.org/wikipedia/commons/f/fc/Flag_of_Mexico.svg *License:* Public domain *Contributors:* This vector image was created with Inkscape. *Original artist:* **Alex Covarrubias**, 9 April 2006

- **File:Flag_of_Pakistan.svg** *Source:* https://upload.wikimedia.org/wikipedia/commons/3/32/Flag_of_Pakistan.svg *License:* Public domain *Contributors:* The drawing and the colors were based from flagspot.net. *Original artist:* User:Zscout370

- **File:Flag_of_Portugal.svg** *Source:* https://upload.wikimedia.org/wikipedia/commons/5/5c/Flag_of_Portugal.svg *License:* Public domain *Contributors:* http://jorgesampaio.arquivo.presidencia.pt/pt/republica/simbolos/bandeiras/index.html#imgs *Original artist:* Columbano Bordalo Pinheiro (1910; generic design); Vítor Luís Rodrigues; António Martins-Tuválkin (2004; this specific vector set: see sources)

- **File:Flag_of_Russia.svg** *Source:* https://upload.wikimedia.org/wikipedia/en/f/f3/Flag_of_Russia.svg *License:* PD *Contributors:* ? *Original artist:* ?

- **File:Flag_of_South_Korea.svg** *Source:* https://upload.wikimedia.org/wikipedia/commons/0/09/Flag_of_South_Korea.svg *License:* Public domain *Contributors:* Ordinance Act of the Law concerning the National Flag of the Republic of Korea, Construction and color guidelines (Russian/English) ← This site is not exist now.(2012.06.05) *Original artist:* Various

- **File:Flag_of_Spain.svg** *Source:* https://upload.wikimedia.org/wikipedia/en/9/9a/Flag_of_Spain.svg *License:* PD *Contributors:* ? *Original artist:* ?

- **File:Flag_of_Sweden.svg** *Source:* https://upload.wikimedia.org/wikipedia/en/4/4c/Flag_of_Sweden.svg *License:* PD *Contributors:* ? *Original artist:* ?

- **File:Flag_of_Switzerland.svg** *Source:* https://upload.wikimedia.org/wikipedia/commons/f/f3/Flag_of_Switzerland.svg *License:* Public domain *Contributors:* PDF Colors Construction sheet *Original artist:* User:Marc Mongenet

Credits:

- **File:Flag_of_Ukraine.svg** *Source:* https://upload.wikimedia.org/wikipedia/commons/4/49/Flag_of_Ukraine.svg *License:* Public domain *Contributors:* ДСТУ 4512:2006 - Державний прапор України. Загальні технічні умови

  SVG: 2010

  Original artist: України

- **File:Flag_of_the_Czech_Republic.svg** *Source:* https://upload.wikimedia.org/wikipedia/commons/c/cb/Flag_of_the_Czech_Republic.svg *License:* Public domain *Contributors:*

  - -xfi-'s file
  - -xfi-'s code
  - Zirland's codes of colors

  Original artist:
  (of code): SVG version by cs:-xfi-.

- **File:Flag_of_the_Netherlands.svg** *Source:* https://upload.wikimedia.org/wikipedia/commons/2/20/Flag_of_the_Netherlands.svg *License:* Public domain *Contributors:* Own work *Original artist:* Zscout370

- **File:Flag_of_the_People'{}s_Republic_of_China.svg** *Source:* https://upload.wikimedia.org/wikipedia/commons/f/fa/Flag_of_the_People%27s_Republic_of_China.svg *License:* Public domain *Contributors:* Own work, http://www.protocol.gov.hk/flags/eng/n_flag/design.html *Original artist:* Drawn by User:SKopp, redrawn by User:Denelson83 and User:Zscout370

15.9.3 Content license

www.ingramcontent.com/pod-product-compliance
Lightning Source LLC
Chambersburg PA
CBHW080653190526
45169CB00006B/2099